Recent Advances in Condensed Matter Physics

Recent Advances in Condensed Matter Physics

Edited by
Jaron Finley

MURPHY & MOORE
www.murphy-moorepublishing.com

Published by Murphy & Moore Publishing,
1 Rockefeller Plaza,
New York City, NY 10020, USA

ISBN: 978-1-63987-475-0

Cataloging-in-Publication Data

Recent advances in condensed matter physics / edited by Jaron Finley.
   p. cm.
Includes bibliographical references and index.
ISBN 978-1-63987-475-0
1. Condensed matter. 2. Matter. 3. Solid state physics. 4. Physics. I. Finley, Jaron.
QC173.454 .A38 2022
530.41--dc23

For information on all Murphy & Moore Publications
visit our website at www.murphy-moorepublishing.com

 MURPHY & MOORE

Contents

Permissions

List of Contributors

Index

Preface

Condensed matter physics is the study of microscopic and macroscopic physical properties of materials. It deals with the solid-state of substances by applying concepts from quantum mechanics, electromagnetism, statistical mechanics, and looks after the theoretical concepts of materials science. It also applies the principles of physics such as quantum physics laws, and statistical mechanics. Condensed matter physics has a wide variety of branches such as crystallography, metallurgy, elasticity, and magnetism. It finds applications in jet turbines, modern tennis rackets, MRI tomography, space exploration, homeland security, and various medical procedures. This book attempts to understand the multiple branches that fall under the discipline of condensed matter physics and how such concepts have practical applications Different approaches, evaluations, methodologies and advanced studies in this field have been included herein. This book will provide comprehensive knowledge to the readers.

The researches compiled throughout the book are authentic and of high quality, combining several disciplines and from very diverse regions from around the world. Drawing on the contributions of many researchers from diverse countries, the book's objective is to provide the readers with the latest achievements in the area of research. This book will surely be a source of knowledge to all interested and researching the field.

In the end, I would like to express my deep sense of gratitude to all the authors for meeting the set deadlines in completing and submitting their research chapters. I would also like to thank the publisher for the support offered to us throughout the course of the book. Finally, I extend my sincere thanks to my family for being a constant source of inspiration and encouragement.

Editor

Zero-Temperature Equation of State of a Two-Dimensional Bosonic Quantum Fluid with Finite-Range Interaction

Andrea Tononi![ORCID]

Dipartimento di Fisica e Astronomia "Galileo Galilei" and CNISM, Università di Padova, Via Marzolo 8, I-35131 Padova, Italy; andrea.tononi@phd.unipd.it

Abstract: We derive the two-dimensional equation of state for a bosonic system of ultracold atoms interacting with a finite-range effective interaction. Within a functional integration approach, we employ a hydrodynamic parameterization of the bosonic field to calculate the superfluid equations of motion and the zero-temperature pressure. The ultraviolet divergences, naturally arising from the finite-range interaction, are regularized with an improved dimensional regularization technique.

Keywords: Bose–Einstein condensation; ultracold atoms; finite-range; equation of state; two-dimensional

1. Introduction

A fluid perspective on the study of bosonic gases made of ultracold atoms may have originated in the pioneering work of Landau [1], who correctly described the superfluid behavior of He-4 as observed in his experiments [2]. From those years, several technical advances allowed a precise experimental control of the atomic gases, culminating in the achievement of Bose–Einstein condensation in 1995 [3–5]. Despite the rich phenomenology deriving from the variety of different trapping potentials [6], the theoretical and experimental study of uniform gases gives a fundamental insight into the intrinsic properties of the condensates. Particularly interesting is the case of two spatial dimensions, in which quantum and thermal fluctuations play a fundamental role [7,8], justifying the necessity of a beyond mean field theory. Historical results for a two-dimensional Bose gas were obtained by Schick, who calculated the thermodynamics of a gas of hard-spheres [9], improved by Popov derivation of the equation of state for a weakly-interacting superfluid [10]. Recent works provide an extension of the Popov approach [11], while the nonuniversal corrections to the equation of state in $D = 2$, arising from a finite-range interaction between the atoms, have been studied [12,13]. Thank to the tunability of interparticle interactions, it is possible to investigate the static and dynamical properties of homogeneous quantum fluids in D-spatial dimensions in regimes where finite-range corrections are relevant [14–16]. In this work, we provide an alternative derivation of the zero-temperature equation of state by adopting an explicit superfluid parametrization of the bosonic field. In particular, we develop an improved dimensional regularization technique to regularize the zero-temperature pressure of a bosonic quantum fluid, whose particles interact with a finite-range interaction.

2. Zero-Temperature Equation of State of a Two-Dimensional Bosonic Quantum Fluid

2.1. Superfluid Parametrization of the Bosonic Field

We introduce the Euclidean Lagrangian $\mathcal{L}$ of a uniform quantum fluid of bosonic particles with mass m, described by the complex field $\psi(\vec{r}, t)$, namely [17]

$$\mathcal{L} = \bar{\psi}(\vec{r}, \tau) \left(\hbar \partial_\tau - \frac{\hbar^2 \nabla^2}{2m} - \mu \right) \psi(\vec{r}, \tau) + \frac{1}{2} \int d^D r' \, |\psi(\vec{r}', \tau)|^2 \, V(|\vec{r} - \vec{r}'|) \, |\psi(\vec{r}, \tau)|^2, \qquad (1)$$

where $\hbar$ is Planck's constant, μ is the chemical potential, and we suppose that the particles interact with the isotropic two-body potential $V(|\vec{r} - \vec{r}'|)$. The imaginary time τ is introduced for uniformity with a functional integration approach, but real time t can be recovered at any moment by performing the Wick rotation $\tau \to it$. According to the least action principle, the Euler–Lagrange equations of the system are obtained as the functional derivative of the action $S[\bar{\psi}, \psi]$, which reads

$$S[\bar{\psi}, \psi] = \int_0^{\beta \hbar} d\tau \int_{L^D} d^D r \, \mathcal{L} \qquad (2)$$

where L^D is the volume in D dimensions containing the particles and $\beta = 1/(k_B T)$, with T the absolute temperature, and k_B the Boltzmann constant. Until the end of this paper, for reasons connected to the dimensional regularization of the final results, we will not explicitly fix the spatial dimension to $D = 2$. The minimization of the action of Equation (2) gives the Gross–Pitaevski equation for the complex field $\psi(\vec{r}, \tau)$, which constitutes the macroscopic wavefunction of the condensate [18]. In this work, however, we adopt a superfluid perspective through the following phase-amplitude parametrization of the bosonic field [19]

$$\psi(\vec{r}, \tau) = \sqrt{\rho(\vec{r}, \tau)} \, e^{i\theta(\vec{r}, \tau)}, \qquad (3)$$

where $\rho(\vec{r}, \tau) = |\psi(\vec{r}, \tau)|^2$ is a real field describing the system local density and $\theta(\vec{r}, \tau)$ is the phase field, which must be included because of the complex nature of the order parameter $\psi(\vec{r}, \tau)$. This field transformation allows us to introduce the superfluid velocity $\vec{v}_s$, which is proportional to the gradient of the phase, namely

$$\vec{v}_s = \frac{\hbar}{m} \vec{\nabla} \theta. \qquad (4)$$

We emphasize that the phase field $\theta(\vec{r}, \tau)$ is defined in the compact interval $[0, 2\pi]$ and is therefore periodic of 2π: this fact constitutes the origin of many topological phenomena in condensed-matter physics. Indeed, here we focus on two-dimensional systems, where the singularities of the phase field—the vortices—are responsible for the Berezinskii–Kosterlitz–Thouless (BKT) transition [20,21]. However, in the following, we will study only the zero-temperature properties, for which the vortex–antivortex phenomenology does not play a fundamental role and can be neglected. We will then assume that the domain of definition of the phase field $\theta(\vec{r}, \tau)$ can be extended to $\mathbb{R}$ and that its spatial and time derivatives are well defined everywhere. In this case, the superfluid flow is irrotational, thus it has zero vorticity

$$\vec{\nabla} \times \vec{v}_s = 0. \qquad (5)$$

We now substitute the parametrization of Equation (3) in the Lagrangian (1), obtaining

$$\mathcal{L} = -\mu \rho + i\hbar \rho \partial_\tau \theta + \frac{\hbar^2}{8m\rho} (\nabla \rho)^2 + \frac{\hbar^2 \rho}{2m} (\nabla \theta)^2 + \frac{1}{2} \int d^D r' \, \rho(\vec{r}', \tau) \, V(|\vec{r} - \vec{r}'|) \, \rho(\vec{r}, \tau), \qquad (6)$$

where we omit for simplicity the dependence of the fields on their coordinates $(\vec{r}, \tau)$. The minimization of the action (2), which now becomes a functional of ρ and θ: $S = S[\rho, \theta]$, leads to the Euler–Lagrange equations for these fields. Recovering the real time $t \to -i\tau$, we get the hydrodynamic equations

$$\frac{\partial \rho}{\partial t} + \vec{\nabla} \cdot (\rho \vec{v}_s) = 0 \tag{7}$$

and

$$m \frac{\partial \vec{v}_s}{\partial t} + \vec{\nabla} \left(\frac{1}{2} m v_s^2 - \mu + \int d^D r' \, \rho(\vec{r}', \tau) \, V(|\vec{r} - \vec{r}'|) - \frac{\hbar^2}{2m\rho^{1/2}} \nabla^2 (\rho^{1/2}) \right) = 0, \tag{8}$$

which are the continuity equation and the equation of motion of a superfluid with velocity $\vec{v}_s$ and number density ρ. Notice that, inside the parenthesis, Equation (8) contains the quantum pressure term: if the density ρ of the fluid is slowly varying and this contribution can be neglected, Equation (8) reduces to the familiar Euler equation of motion for an irrotational fluid without viscosity. For a self-consistent derivation of these equations from the Gross–Pitaevski equation we refer the reader to Reference [22].

2.2. Zero-Temperature Equation of State

We now adopt the superfluid parametrization of Equation (3) to derive the zero-temperature equation of state of the quantum fluid, namely a relation at $T = 0$ between the pressure P and the chemical potential μ. For the finite-range interaction, an explicit implementation of this relation will be given in the next section.

In the grand canonical ensemble, we calculate the pressure of the bosonic fluid as $P = -\Omega/L^D$, where Ω is the grand potential

$$\Omega = -\frac{1}{\beta} \ln(\mathcal{Z}) \tag{9}$$

and $\mathcal{Z}$ is the grand canonical partition function, which, within a functional integration perspective, can be calculated as

$$\mathcal{Z} = \int \mathcal{D}[\rho, \theta] \, e^{-\frac{S[\rho, \theta]}{\hbar}}. \tag{10}$$

To perform the explicit functional integration of the Lagrangian (6), we rewrite the local density field $\rho(\vec{r}, \tau)$ as

$$\rho(\vec{r}, \tau) = \rho_0 + \delta\rho(\vec{r}, \tau), \tag{11}$$

where ρ_0 is the condensate density of the system in the broken-symmetry phase and $\delta\rho(\vec{r}, \tau)$ is a real field describing the density fluctuations.

We substitute the field transformation (11) in the Lagrangian of Equation (6), obtaining

$$\mathcal{L} = - \mu\rho_0 - \mu \, \delta\rho + i\hbar\rho_0\partial_\tau\theta + \frac{\hbar^2}{8m\rho_0}(\nabla\delta\rho)^2 + \frac{\hbar^2\rho_0}{2m}(\nabla\theta)^2 +$$
$$\frac{1}{2} \int d^D r' \, V(|\vec{r} - \vec{r}'|)(\rho_0^2 + \delta\rho(\vec{r}, \tau) + \delta\rho(\vec{r}', \tau) + \rho(\vec{r}, \tau)\delta\rho(\vec{r}', \tau)), \tag{12}$$

where we keep only terms up to the second order in the fluctuation fields $\delta\rho(\vec{r}, \tau)$ and $\theta(\vec{r}, \tau)$, thus making a Gaussian (one-loop) approximation.

Considering the Lagrangian of Equation (12) inside the action S, which now becomes the functional $S = S[\delta\rho, \theta]$, it is particularly convenient to express it in terms of the Fourier series of the fluctuation fields, namely

$$
\begin{aligned}
\delta\rho(\vec{r}, \tau) &= \frac{1}{\sqrt{L^D}} \sum_{\vec{k}\,\omega_n} e^{i\vec{k}\cdot\vec{r}} e^{-i\omega_n\tau}\, \delta\rho(\vec{k}, \omega_n) \\
\theta(\vec{r}, \tau) &= \frac{1}{\sqrt{L^D}} \sum_{\vec{k}\,\omega_n} e^{i\vec{k}\cdot\vec{r}} e^{-i\omega_n\tau}\, \theta(\vec{k}, \omega_n) \\
\delta\rho(\vec{k}, \omega_n) &= \frac{1}{\beta\hbar\sqrt{L^D}} \int_0^{\beta\hbar} d\tau \int_{L^D} d^D r\, e^{-i\vec{k}\cdot\vec{r}} e^{i\omega_n\tau}\, \delta\rho(\vec{r}, \tau) \\
\theta(\vec{k}, \omega_n) &= \frac{1}{\beta\hbar\sqrt{L^D}} \int_0^{\beta\hbar} d\tau \int_{L^D} d^D r\, e^{-i\vec{k}\cdot\vec{r}} e^{i\omega_n\tau}\, \theta(\vec{r}, \tau),
\end{aligned}
\tag{13}
$$

where $\omega_n = 2\pi n/(\beta\hbar)$ are the bosonic Matsubara frequencies. Notice that, since we are supposing that the phase field $\theta(\vec{r}, \tau)$ is defined on $\mathbb{R}$, its Fourier components are non-numerable and can assume continuous values, thus they can be treated like ordinary functional integral variables. The action in the Fourier space is obtained by simply substituting these expressions in S and using the definition of the $D + 1$-dimensional delta function. Moreover, we also substitute the Fourier series $\tilde{V}(k)$ of the real space interaction potential and we define with g_0 the zero-range interaction strength $g_0 = \tilde{V}(k = 0)$. In this way, the action can be rewritten as the sum of two contributions

$$
S = S_0 + S_g.
\tag{14}
$$

The first is the action of the homogeneous system S_0, namely

$$
S_0 = \beta\hbar L^D\left(-\mu\rho_0 + \frac{1}{2}g_0\rho_0^2\right),
\tag{15}
$$

which does not depend on the functional integration variables: using Equations (9) and (10), one can employ S_0 to calculate Ω_0, the mean field contribution to the grand potential

$$
\Omega_0 = \left(-\mu\rho_0 + \frac{1}{2}g_0\rho_0^2\right)L^D.
\tag{16}
$$

The second contribution to the action S is the Gaussian action S_g, which is given by

$$
S_g = \beta\hbar \sum_{\vec{k}\,\omega_n}\left[\frac{\hbar^2 k^2 \rho_0}{2m}\theta(k)\theta(-k) + \left(\frac{\hbar^2 k^2}{8m\rho_0} + \frac{\tilde{V}(k)}{2}\right)\delta\rho(k)\delta\rho(-k) + \hbar\omega_n\theta(k)\delta\rho(-k)\right],
\tag{17}
$$

where, for simplicity of notation, we define $\delta\rho(\pm k) = \delta\rho(\pm\vec{k}, \pm\omega_n)$ and $\theta(\pm k) = \theta(\pm\vec{k}, \pm\omega_n)$. Since S_g is quadratic in the fluctuation fields $\delta\rho(k)$ and $\theta(k)$, one can rewrite it in the following matricial form:

$$
S_g = \frac{\hbar}{2} \sum_{\vec{k}\,\omega_n} \begin{pmatrix} \theta(k) & \theta(-k) & \delta\rho(k) & \delta\rho(-k) \end{pmatrix} \mathbf{M}(k) \begin{pmatrix} \theta(k) \\ \theta(-k) \\ \delta\rho(k) \\ \delta\rho(-k) \end{pmatrix},
\tag{18}
$$

where $\mathbf{M}(k)$, the inverse of the propagator, is the 4×4 matrix

$$\mathbf{M}(k) = \beta \begin{pmatrix} 0 & \frac{\hbar^2 k^2 \rho_0}{m} & 0 & \hbar\omega_n \\ \frac{\hbar^2 k^2 \rho_0}{m} & 0 & -\hbar\omega_n & 0 \\ 0 & -\hbar\omega_n & 0 & \frac{\hbar^2 k^2}{4m\rho_0} + \tilde{V}(k) \\ \hbar\omega_n & 0 & \frac{\hbar^2 k^2}{4m\rho_0} + \tilde{V}(k) & 0 \end{pmatrix}. \tag{19}$$

The functional integral of the real fluctuation fields $\theta(k)$ and $\delta\rho(k)$ can be performed explicitly [23], obtaining the corresponding Gaussian grand canonical partition function $\mathcal{Z}_g$ as

$$\mathcal{Z}_g = \prod_{\substack{\vec{k}\,\omega_n \\ k_z > 0}} [\det \mathbf{M}(k)]^{-1/2}, \tag{20}$$

which, considering the definition of the grand potential of Equation (9), leads to the Gaussian contribution to the grand potential

$$\Omega_g = \frac{1}{2\beta} \sum_{\vec{k}\,\omega_n} \ln[\beta^2(\hbar^2\omega_n^2 + E_k^2)]. \tag{21}$$

Here, we find the gapless excitation spectrum E_k of the quantum fluid in the form

$$E_k = \sqrt{\frac{\hbar^2 k^2}{2m}\left(\frac{\hbar^2 k^2}{2m} + 2\rho_0 \tilde{V}(k)\right)}, \tag{22}$$

where, within a perturbative approach, ρ_0 is determined by the saddle point condition $\partial\Omega_0/\partial\rho_0 = 0$, which leads to

$$\rho_0 = \frac{\mu}{g_0} \tag{23}$$

and whose substitution in the excitation spectrum gives E_k^B, the renowned Bogoliubov spectrum [24]

$$E_k^B = \sqrt{\frac{\hbar^2 k^2}{2m}\left(\frac{\hbar^2 k^2}{2m} + 2\mu\frac{\tilde{V}(k)}{g_0}\right)}. \tag{24}$$

The sum over the Matsubara frequencies ω_n in the Gaussian grand potential of Equation (21) can be performed according to the prescriptions described in the Appendix A, obtaining the grand potential as the sum of three contributions

$$\Omega = \Omega_0 + \Omega_g^{(0)} + \Omega_g^{(T)}, \tag{25}$$

where $\Omega_0 = -L^D\mu^2/(2g_0)$ due to Equations (16) and (23), and

$$\Omega_g^{(0)} = \frac{1}{2}\sum_{\vec{k}} E_k^B \tag{26}$$

is the zero-temperature Gaussian grand potential encoding quantum fluctuations, while

$$\Omega_g^{(T)} = \frac{1}{\beta}\sum_{\vec{k}} \ln(1 - e^{-\beta E_k^B}) \tag{27}$$

is the finite-temperature Gaussian grand potential, encoding thermal fluctuations. Finally, we explicitly write the zero-temperature equation of state, namely, we calculate the pressure as the opposite of the grand potential of Equation (25) at $T = 0$:

$$P(\mu, T = 0) = \frac{\mu^2}{2g_0} - \frac{1}{2L^D} \sum_{\vec{k}} E_k^B. \tag{28}$$

In the thermodynamic limit of $L \to \infty$, the sum over $\vec{k}$ can be rewritten as a D-dimensional integral in momentum space $(2\pi/L)^D \sum_{\vec{k}} = \int d^D k$, and, substituting again the Bogoliubov spectrum (24), the equation of state becomes

$$P(\mu, T = 0) = \frac{\mu^2}{2g_0} - \frac{1}{2} \int \frac{d^D k}{(2\pi)^D} \sqrt{\frac{\hbar^2 k^2}{2m} \left(\frac{\hbar^2 k^2}{2m} + 2\mu \frac{\tilde{V}(k)}{g_0} \right)}, \tag{29}$$

where the integral can be calculated after the explicit choice of $\tilde{V}(k)$.

2.3. Explicit Implementation for Finite-Range Interaction

We now provide an explicit implementation of the zero-temperature equation of state (29) for a bosonic quantum fluid of particles interacting with the finite-range effective interaction

$$\tilde{V}(k) = g_0 + g_2 k^2, \tag{30}$$

where $g_0 = \tilde{V}(k = 0)$ is the usual zero-range interaction coupling, and

$$g_2 = \frac{1}{2} \int d^2 r \, r^2 \, V(|\vec{r}|) \tag{31}$$

is the first nonzero correction in the gradient expansion of an isotropic interaction potential $V(|\vec{r}|)$. At zero temperature, we expect the finite-range corrections to the equation of state to be detectable, but small with respect to the zero-range result of Reference [25]. By using scattering theory in two spatial dimensions, these couplings can be linked with the s-wave scattering length a_s and the characteristic range R of the real interatomic two-body interaction [12,26,27]

$$g_0 = \frac{4\pi\hbar^2}{m |\ln(na_s^2)|}, \qquad g_2 = \frac{\pi\hbar^2 R^2}{m |\ln(na_s^2)|}, \tag{32}$$

where n is the number density of the system in $D = 2$.

The equation of state (29) becomes, with the finite-range interaction of Equation (30)

$$P(\mu, T = 0) = \frac{\mu^2}{2g_0} + P_g^{(0)}, \tag{33}$$

where we define the zero temperature Gaussian pressure $P_g^{(0)}$ as

$$P_g^{(0)} = -\frac{1}{2} \int \frac{d^D k}{(2\pi)^D} \sqrt{\frac{\hbar^2 k^2}{2m} \left(\frac{\hbar^2 k^2}{2m} \lambda + 2\mu \right)}, \tag{34}$$

with

$$\lambda = 1 + \frac{4m}{\hbar^2} \frac{\mu}{g_0} g_2. \tag{35}$$

Since the integrand function depends only on the modulus of the momentum $|\vec{k}|$, we rewrite the integral in $P_g^{(0)}$ using D-dimensional spherical coordinates, namely

$$P_g^{(0)} = -\frac{S_D}{2(2\pi)^D} \int_0^{+\infty} dk\, k^{D-1} \sqrt{\frac{\hbar^2 k^2}{2m}\left(\frac{\hbar^2 k^2}{2m}\lambda + 2\mu\right)}, \tag{36}$$

where $S_D = 2\pi^{D/2}/\Gamma[D/2]$ is the solid angle in D-dimensions and $\Gamma[D/2]$ is the Euler Gamma function. In order to integrate this equation, we introduce the adimensional variable $t = \hbar^2 k^2 \lambda/(4m\mu)$, obtaining

$$P_g^{(0)} = -\frac{\mu}{\lambda^{1/2}\Gamma[D/2]}\left(\frac{m\mu}{\pi\hbar^2\lambda}\right)^{D/2} \int_0^{+\infty} dt\, t^{\frac{D-1}{2}}(1+t)^{1/2}. \tag{37}$$

As a consequence of the substitution of the real interatomic potential with an effective interaction, the zero-temperature Gaussian pressure $P_g^{(0)}$ is ultraviolet divergent. In our framework, an efficient way to regularize $P_g^{(0)}$ is constituted by the technique of dimensional regularization [28]. The basic idea of this approach is to rewrite a diverging integral in terms of the Euler beta and gamma functions, whose integral representation for $x, y, z > 0$ is given by

$$B(x,y) = \int_0^{+\infty} dt\, \frac{t^{x-1}}{(1+t)^{x+y}}, \tag{38}$$

$$\Gamma(z) = \int_0^{+\infty} dt\, t^{z-1}\, e^{-z}. \tag{39}$$

Thanks to the properties $B(x,y) = \Gamma(x)\Gamma(y)/\Gamma(x+y)$ and $\Gamma[z+1] = z\,\Gamma[z]$, one can extend the domain of definition of the gamma and beta functions by analytic continuation of their arguments x, y, z also to negative values, which usually appear in many physical problems. However, despite this, the dimensional regularization procedure can be successfully used to regularize many ultraviolet diverging integrals, in our peculiar two-dimensional case the procedure described above would lead to a result containing the gamma function evaluated for negative integer values, which is again a diverging quantity. To avoid this residual divergence, we extend the dimensions of the system to the complex value $\mathcal{D} = D - \varepsilon$, and we formally perform the integration of Equation (37). We obtain

$$P_g^{(0)} = \frac{\kappa^\varepsilon}{2}\left(\frac{\mu}{\pi\lambda}\right)^{(\mathcal{D}+1)/2}\left(\frac{m}{\hbar^2}\right)^{\mathcal{D}/2} \frac{\Gamma[(D-\varepsilon+1)/2]\,\Gamma[(\varepsilon-D-2)/2]}{\Gamma[(D-\varepsilon)/2]}, \tag{40}$$

in which the wavevector κ is introduced for dimensional reasons. Notice how in $D = 2$ and for $\varepsilon = 0$ the Gaussian pressure is still divergent. To regularize it, we rely on the following small-ε expansion of the gamma function [29]

$$\Gamma(-n+\varepsilon) = \frac{(-1)^n}{n!}\left[\frac{1}{\varepsilon} + \Psi(n+1) + \frac{\varepsilon}{2}\left(\frac{\pi^2}{3} + \Psi(n+1)^2 - \Psi'(n+1)\right) + o(\varepsilon^2)\right], \tag{41}$$

where $\Psi(n+1)$ is the digamma function and $\Psi'(n+1)$ is its derivative. Moreover, we express the exponentiation of a generic coefficient x^ε for $\varepsilon \to 0$ as

$$x^\varepsilon = \exp(\varepsilon\ln(x)) \sim_{\varepsilon\to 0} 1 + \varepsilon\ln(x) + o(\varepsilon^2). \tag{42}$$

With this recipe, the Gaussian pressure $P_g^{(0)}$ in $D = 2$ gives

$$P_g^{(0)} = \frac{m\mu^2}{2\pi^{3/2}\hbar^2\lambda^{3/2}} \left[\frac{\pi^{1/2}}{2}\frac{1}{\varepsilon} + \frac{\pi^{1/2}}{8}(\ln(16) - 2\gamma - 1) + \frac{\pi^{1/2}}{4}\ln\left(\frac{\pi\lambda\hbar^2\kappa^2}{m\mu}\right) + o(\varepsilon) \right], \quad (43)$$

where $\gamma \approx 0.55722$ is the Euler–Mascheroni constant. Finally, we delete the $o(\varepsilon^{-1})$ divergence in the square bracket [30] and we rewrite the zero-temperature equation of state $P(\mu, T = 0)$ of Equation (29) as

$$P(\mu, T = 0) = \frac{m\mu^2}{8\pi\hbar^2\lambda^{3/2}} \left[\ln\left(\frac{\epsilon_0}{\mu}\lambda\right) - \frac{1}{2} \right], \quad (44)$$

where we define the energy cutoff ϵ_0 as

$$\epsilon_0 = \frac{4\pi\hbar^2\kappa^2}{m\exp(\gamma - \frac{4\pi\hbar^2\lambda^{3/2}}{mg_0})}. \quad (45)$$

The equation of state (44) improves the one derived for bosons with a zero-range interaction [10] by Popov, whose result can be reproduced by setting $\lambda = 1$, i.e., $g_2 = 0$. We emphasize that, with a precise tuning of the interparticle interaction (see Reference [12] for a detailed discussion), the finite-range corrections derived within our Gaussian approximation become larger than the zero-range beyond-Gaussian ones obtained by Mora and Castin [31]. For weakly-interacting bosons with $na_s^2 \ll 1$, where a_s is the two-dimensional s-wave scattering length, we expect that the nonuniversal corrections of Equation (44) arise for $R \geq a_s$, where R is the characteristic range of the interaction. In this intermediate regime, the neglection of higher order terms in the gradient expansion of Equation (30) is justified but, at the same time, the finite-range contributions are of comparable size to the zero-range ones.

3. Conclusions

In this work, we derive the two-dimensional zero-temperature equation of state for a bosonic quantum fluid with a generic isotropic interaction. The superfluid perspective is emphasized by performing the Gaussian functional integration within a phase-amplitude parametrization of the complex order parameter. For a system with zero-range interaction, we reproduce the classical result by Popov. Nonetheless, we apply a novel dimensional regularization recipe to reproduce the nonuniversal corrections for a finite-range interaction potential. We expect, with a fine-tuning of the experimental interaction parameters, the finite-range correction to produce sizable corrections to the thermodynamics of the weakly-interacting superfluid. Our derivation of the zero-temperature equation of state is also valid for other interparticle interactions. In particular, the previous results can be extended for a quasi-two-dimensional system of dipolar bosons whose polarization direction is perpendicular to the plane of confinement. For a generic orientation, however, it is necessary to consider the dependence of the interaction on the in-plane angle between the particles and to include it consistently in the dimensional regularization procedure.

Funding: This research received no external funding.

Acknowledgments: The author thanks Luca Salasnich and Alberto Cappellaro for useful discussions and suggestions.

Conflicts of Interest: The author declares no conflict of interest.

Appendix A

We illustrate here the procedure to calculate the summation over the bosonic Matsubara frequencies ω_n, which are defined as

$$\omega_n = \frac{2\pi n}{\beta \hbar}, \tag{A1}$$

where $n \in \mathbb{Z}$ are integer numbers. The most common sum that one has to perform is in the form

$$I[\xi_{\vec{k}}] = \frac{1}{2\beta} \sum_{n=-\infty}^{+\infty} \ln[\beta^2(\hbar^2 \omega_n^2 + \xi_{\vec{k}}^2)]. \tag{A2}$$

Using the properties of the logarithm and considering that the summation involves all $n \in \mathbb{Z}$ integers, both positive and negative, $I[\xi_{\vec{k}}]$ can also be rewritten in the useful form

$$I[\xi_{\vec{k}}] = \frac{1}{\beta} \sum_{n=-\infty}^{+\infty} \ln[\beta(-i\hbar\omega_n + \xi_{\vec{k}})]. \tag{A3}$$

Taking the derivative of $I[\xi_{\vec{k}}]$ with respect to $\xi_{\vec{k}}$ in Equation (A2), we get

$$\frac{\partial I[\xi_{\vec{k}}]}{\partial \xi_{\vec{k}}} = \frac{1}{\beta} \sum_{n=-\infty}^{+\infty} \frac{\xi_{\vec{k}}}{\hbar^2 \omega_n^2 + \xi_{\vec{k}}^2}. \tag{A4}$$

In the zero temperature limit, the difference

$$\Delta\omega = \omega_n - \omega_{n-1} = \frac{2\pi}{\beta\hbar} \xrightarrow[\beta \gg 1]{} d\omega \tag{A5}$$

becomes infinitesimal and we can substitute the sum over n with an integral over ω, obtaining

$$\frac{\partial I[\xi_{\vec{k}}]}{\partial \xi_{\vec{k}}} = \frac{1}{\beta} \int_{-\infty}^{+\infty} d\omega \, \frac{\beta\hbar}{2\pi} \frac{\xi_{\vec{k}}}{\hbar^2\omega^2 + \xi_{\vec{k}}^2} = \frac{1}{2}, \tag{A6}$$

which is the zero-temperature contribution to $I[\xi_{\vec{k}}]$. If the temperature is relatively low, but non-zero, we cannot substitute the sum in Equation (A4) with an integral, but we can rewrite it as

$$\frac{\partial I[\xi_{\vec{k}}]}{\partial \xi_{\vec{k}}} = \frac{\beta\xi_{\vec{k}}}{(2\pi)^2} \sum_{n=-\infty}^{+\infty} \frac{1}{n^2 + \left(\frac{\beta\xi_{\vec{k}}}{2\pi}\right)^2} \tag{A7}$$

and, using the identity

$$\sum_{n=0}^{+\infty} \frac{1}{n^2 + a^2} = \frac{1 + \pi a \, \coth(\pi a)}{2a^2}, \tag{A8}$$

we obtain

$$\frac{\partial I[\xi_{\vec{k}}]}{\partial \xi_{\vec{k}}} = \frac{1}{2} \coth\left(\frac{\beta\xi_{\vec{k}}}{2}\right) = \frac{1}{2} + \frac{1}{e^{\beta\xi_{\vec{k}}} - 1}. \tag{A9}$$

We integrate this equation on $\xi_{\vec{k}}$, and setting the arbitrary constant resulting from the indefinite integral to zero (it is not dependent on physical parameters), we finally obtain the result of the summation over the Matsubara frequencies

$$I[\xi_{\vec{k}}] = \frac{\xi_{\vec{k}}}{2} + \frac{1}{\beta} \ln(1 - e^{-\beta\xi_{\vec{k}}}), \tag{A10}$$

which is used in this article to obtain Equation (21).

References

1. Landau, L.D. Theory of the Superfluidity of Helium II. *Phys. Rev.* **1941**, *60*, 356–358. [CrossRef]
2. Kapitza, P. Viscosity of Liquid Helium below the λ-Point. *Nature* **1938**, *141*, 74. [CrossRef]
3. Anderson, M.H.; Ensher, J.R.; Matthews, M.R.; Wieman, C.E.; Cornell, E.A. Observation of Bose-Einstein Condensation in a Dilute Atomic Vapor. *Science* **1995**, *269*, 198–201. [CrossRef] [PubMed]
4. Davis, K.B.; Mewes, M.-O.; Andrews, M.R.; van Druten, N.J.; Durfee, D.S.; Kurn, D.M.; Ketterle, W. Bose-Einstein Condensation in a Gas of Sodium Atoms. *Phys. Rev. Lett.* **1995**, *75*, 3969–3973. [CrossRef] [PubMed]
5. Bradley, C.C.; Sackett, C.A.; Tollett, J.J.; Hulet, R.G. Evidence of Bose-Einstein Condensation in an Atomic Gas with Attractive Interactions. *Phys. Rev. Lett.* **1995**, *75*, 1687–1690. [CrossRef] [PubMed]
6. Dalfovo, F.; Giorgini, S.; Pitaevskii, L.P.; Stringari, S. Theory of Bose-Einstein condensation in trapped gases. *Rev. Mod. Phys.* **1999**, *71*, 463–512. [CrossRef]
7. Mermin, N.D.; Wagner, H. Absence of Ferromagnetism or Antiferromagnetism in One- or Two-Dimensional Isotropic Heisenberg Models. *Phys. Rev. Lett.* **1966**, *17*, 1133–1136. [CrossRef]
8. Hohenberg, P.C. Existence of long-range order in one and two dimensions. *Phys. Rev.* **1967**, *158*, 383–386. [CrossRef]
9. Schick, M. Two-dimensional system of hard-core bosons. *Phys. Rev. A* **1971**, *3*, 1067–1073. [CrossRef]
10. Popov, V.N. On the theory of the superfluidity of two- and one-dimensional bose systems. *Theor. Math. Phys.* **1972**, *11*, 565–573. [CrossRef]
11. Pastukhov, V. Ground-State Properties of a Dilute Two-Dimensional Bose Gas. *J. Low Temp. Phys.* **2018**, 1–12. [CrossRef]
12. Salasnich, L. Nonuniversal Equation of State of the Two-Dimensional Bose Gas. *Phys. Rev. Lett.* **2017**, *118*, 130402. [CrossRef] [PubMed]
13. Beane, S.R. Effective-range corrections to the ground-state energy of the weakly-interacting Bose gas in two dimensions. *Eur. Phys. J. D* **2018**, *72*, 55. [CrossRef]
14. Braaten, E.; Hammer, H.-W.; Hermans, S. Nonuniversal effects in the homogeneous Bose gas. *Phys. Rev. A* **2001**, *63*, 063609. [CrossRef]
15. Cappellaro, A.; Salasnich, L. Thermal field theory of bosonic gases with finite-range effective interaction. *Phys. Rev. A* **2017**, *95*, 033627. [CrossRef]
16. Cappellaro, A.; Salasnich, L. Finite-range corrections to the thermodynamics of the one-dimensional Bose gas. *Phys. Rev. A* **2017**, *96*, 063610. [CrossRef]
17. Nagaosa, N. *Quantum Field Theory in Condensed Matter Physics*; Springer: Berlin, Germany, 1999.
18. Leggett, A.J. *Quantum Liquids*; Oxford University Press: Oxford, UK, 2006.
19. Salasnich, L. Hydrodynamics of Bose and Fermi superfluids at zero temperature: The superfluid nonlinear Schrödinger equation. *Laser Phys.* **2009**, *19*, 642–646. [CrossRef]
20. Berezinskii, V.L. Destruction of long-range order in one-dimensional and two-dimensional system possessing a continous symmetry group—II. Quantum systems. *Sov. Phys.-JETP* **1971**, *34*, 1144–1156.
21. Kosterlitz, L.M.; Thouless, D.J. Ordering, metastability and phase transitions in two-dimensional systems. *J. Phys. C* **1973**, *6*, 1181–1203. [CrossRef]
22. Coste, C. Nonlinear Schrödinger equation and superfluid hydrodynamics. *Eur. Phys. J. B* **1998**, *1*, 245–253. [CrossRef]
23. Altland, A.; Simons, B.D. *Condensed Matter Field Theory*; Cambridge University Press: Cambridge, UK, 2006.
24. Bogoliubov, N.N. On the theory of superfluidity. *J. Phys.* **1947**, *11*, 23.
25. Salasnich, L.; Toigo, F. Zero-point energy of ultracold atoms. *Phys. Rep.* **2016**, *640*, 1–29. [CrossRef]
26. Tononi, A.; Cappellaro, A.; Salasnich, L. Condensation and superfluidity of dilute Bose gases with finite-range interaction. *New J. Phys.* **2018**, *20*, 125007. [CrossRef]
27. Astrakharchik, G.E.; Boronat, J.; Casulleras, J.; Kurbakov, I.L.; Lozovik, Yu.E. Equation of state of a weakly interacting two-dimensional Bose gas studied at zero temperature by means of quantum Monte Carlo methods. *Phys. Rev. A* **2009**, *79*, 051602. [CrossRef]

28. 't Hooft, G.; Veltman, M. Regularization and renormalization of gauge fields. *Nucl. Phys. B* **1972**, *44*, 189–213. [CrossRef]

29. Kleinert, H.; Schulte-Frohlinde, V. *Critical Properties of ϕ^4 Theories*; World Scientific: Singapore, 2001.

30. Zeidler, E. *Quantum Field Theory II: Quantum Electrodynamics*; Springer: Berlin, Germany, 2009.

31. Mora, C.; Castin, Y. Ground State Energy of the Two-Dimensional Weakly Interacting Bose Gas: First Correction Beyond Bogoliubov Theory. *Phys. Rev. Lett.* **2009**, *102*, 180404. [CrossRef] [PubMed]

An Equation of State for Metals at High Temperature and Pressure in Compressed and Expanded Volume Regions

S. V. G. Menon [1,*,†] ⓘ **and Bishnupriya Nayak** [2]

[1]   Shiv Enclave, 304, 31-B-Wing, Tilak Nagar, Mumbai 400089, India

[2]   High Pressure and Synchroton Radiation Physics Division, Bhabha Atomic Research Centre, Mumbai 400085, India

*   Correspondence: menon.svg98@gmail.com; Tel.: +91-887-939-4488

†   Retired from Bhabha Atomic Research Centre, Mumbai 400085, India.

Abstract: A simple equation of state model for metals at high temperature and pressure is described. The model consists of zero-temperature isotherm, thermal ionic components, and thermal electronic components, and is applicable in compressed as well as expanded volume regions. The three components of the model, together with appropriate correction terms, are described in detail using Cu as a prototype example. Shock wave Hugoniot, critical point parameters, liquid–vapor phase diagram, isobaric expansion, etc., are evaluated and compared with experimental data for Cu. The semianalytical model is expected to be useful to prepare extended tables for use in hydrodynamics calculations in high-energy-density physics.

Keywords: metals; equation of state; high pressure physics; shock waves; liquid–vapor phase; critical point parameters; soft-sphere model

---

1. Introduction

Equation of State (EOS) of materials is an inevitable ingredient in several fields of solid state science like geophysics, hydrodynamic applications for the analysis of inertial confinement fusion systems, stellar structures, nuclear weapons, etc. Other applications include fast reactor accident analysis and study of weapon effects in various media. Euler equations of hydrodynamics, which expresses conservation laws of mass, momentum, and energy, are routinely used to describe the dynamical behavior of materials [1]. However, these equations describe the space-time evolution of four thermodynamic variables—viz., mass density (or specific volume), material velocity, specific internal energy, and pressure. The system of these equations is then closed with the addition of equation of state (EOS), which provides pressure when specific internal energy and density are given. The Mie–Grüneisen EOS [1] with an empirical specification for Grüneisen parameter is the most commonly used EOS of this type. There is also Tillotson's EOS [2] which has a larger range of validity. However, a more complete EOS is specified by providing pressure and specific internal energy as functions of density and temperature. This temperature corresponds to thermodynamic equilibrium in the material, and can be eliminated from the expression for pressure in favor of specific internal energy to obtain the above mentioned relation between pressure, specific internal, energy, and density. An EOS of similar class, which treats pressure as independent variable, was proposed by Rice and Walsh [3] to model water. Here, specific volume is expressed in terms of enthalpy using the enthalpy–parameter which depends on pressure. This class of EOS, generally called enthalpy-based EOS, has been developed to model shock compression of porous materials [4], including explicit accounting of electronic effects [5].

In the following sections, we describe a semi-analytical EOS model of the general type, where volume and temperature are independent variables. The different components of the EOS model, including correction terms, are discussed in detail. Experimental data for Cu on Shock-Hugoniot, critical point parameters, liquid–vapor phase diagram, and isobaric expansion are used to test the model. Good agreement obtained shows that the model can be employed to prepare extended EOS tables for use in hydrodynamics calculations.

2. Three Component EOS Model

In general, EOS models consist of three components [6], which describe (i) the zero-temperature (or cold) isotherm, (ii) thermal ionic effects, and (iii) thermal electronic effects. Pressure and specific internal energy are then expressed as functions of volume (V) and temperature (T):

$$P(V,T) = P_c(V) + P_{ti}(V,T) + P_{te}(V,T),\tag{1}$$

$$E(V,T) = E_c(V) + E_{ti}(V,T) + E_{te}(V,T),\tag{2}$$

where the terms denote, respectively, the three components mentioned above. The subscripts c, ti, and te stand for the terms 'cold', 'thermal-ion', and 'thermal-electron', respectively. There are interaction effects between ionic and electronic motion, however, these effects contribute only a few percent to pressure and energy and so may be neglected in high-pressure physics applications [6]. Models for the different terms in Equations (1) and (2) of varying degrees of sophistication are currently available in the literature [7].

First, principle methods using Density-functional theory (DFT) for E_c and quasi-harmonic approximation (QHA), based on density-functional perturbation theory, for lattice vibration contributions to E_{ti} are now quite common [8]. Such methods have proven to be extremely useful in thermodynamic studies of compounds of interest in Earth sciences [9]. Further, electronic density of states determined from DFT calculations yield accurate estimates of E_{te} for lower ranges of temperature [10]. Availability of efficient computer implementation [11] of QHA, which uses DFT-generated data on E_c and volume-dependent vibration frequencies, make QHA the method of choice for detailed studies of thermodynamic properties of materials in the solid phase. However, for developing global EOS models, which deal with very high temperatures ($\sim$keV) and pressures ($\sim$ tens of megabar), it is necessary to take only the relevant information from DFT computations and supplement it with other models [10] to incorporate effects of melting, extreme pressure and thermal ionization, expanded volume states, etc. The general approach is to use empirical fits to the cold isotherm [12], Debye–Grüneisen model for lattice thermal motion [13] and Thomas–Fermi model for thermal electron excitation [14]. Such extended models are essential for some of the hydrodynamic applications mentioned in the beginning.

The global EOS model we describe below, which is applicable even at very high temperatures and pressures in the compressed as well as expanded volume states, indeed uses different parameters obtained from DFT analyses, particularly when accurate experimental data on these are unavailable.

3. Zero-Temperature Isotherm

The zero-temperature isotherm is a manifestation of the Fermi-pressure developed in degenerate electron systems, and is a quantum effect just like zero-point vibration energy. This contributes significantly to the total pressure in compressed solids, and becomes the dominant contribution at extreme compression. A variety of approximate expressions to describe it quantitatively are available in the literature [15]. Computations using DFT, mentioned above, are now routinely used to determine energy versus specific volume (or volume per atom) tables, and thereafter the zero-temperature pressure–volume relation. Results of such analyses are then used in semi-empirical expressions. We propose to use a four-parameter model, developed by Li et al. [16], which is expressed as:

$$E_{Li}(V) = -E_{coh}(1 + a + \delta a^3)e^{-a}, \quad \eta = \left(\frac{9B_0 V_{c0}}{E_{coh}}\right)^{1/2}, \quad x = \left(\frac{V}{V_{c0}}\right)^{1/3}, \tag{3}$$

$$P_{Li}(V) = 3B_0 \frac{(1-x)}{x^2}(1 - 3\delta a + \delta a^2)e^{-a}, \quad a = \eta(x-1), \quad \delta = \frac{B_0' - 1}{2\eta} - \frac{1}{3}. \tag{4}$$

The four parameters in the model are the specific volume V_{c0}, the bulk modulus B_0, its pressure derivative B_0', and the cohesive energy E_{coh} at zero temperature. These parameters occur in terms of dimensionless quantities η and δ, and a is related to the dimensionless length variable x. Furthermore, if energy E_{Li} is scaled with E_{coh} and pressure P_{Li} with B_0, then, these expressions are totally dimensionless—however, defined in terms of two parameters η and δ. The specific volume V_{c0} is slightly lower than the volume at ambient conditions ($T_0 = 300$ K and $P_0 = 1$ bar). In our approach, we adjust the value of V_{c0} such that the zero-temperature pressure together with thermal pressure of ions and electrons is just one bar at 300 K. The four-parameter model is a refinement over Rose equation [17] and Vinet equation [18], and is found to provide quite accurate descriptions of the zero-temperature energy and pressure over compressed volume up to $\sim V_{c0}/2$, which corresponds to about 100–150 GPa pressure, for about forty metals [16]. It also provides accurate representation for energy and pressure in the expanded volume up to about $\sim 2V_{c0}$. However, the formulation is inadequate in the region of extreme compression, as is evident from Equation (3), which shows that E_{Li} saturates as $V \to 0$. Theoretically, the zero-temperature energy and pressure should approach those of electron gas.

To rectify this problem, we use a procedure [12] to smoothly go over from the four-parameter model to the the quantum statistical model (QSM) [19], which is known to provide accurate descriptions of pressure and energy of electrons above few hundred GPa pressure. The QSM accounts for exchange and correlation effects in addition to corrections for electron density gradients [20]. Electron pressure in a compressed atom within the QSM model is expressed as:

$$P_{QSM}(V) = \frac{e^2}{5}\frac{a_0}{V^{5/3}}(3\pi^2)^{2/3}Z_n^{5/3}\exp[-\alpha - \beta], \tag{5}$$

$$\alpha = 0.3225 \, R_w \, Z_n^{[\,0.495 - 0.039\log_{10} Z_n\,]},$$

$$\beta = \frac{5}{3}R_w^2[\,0.068 + 0.078\log_{10} Z_n - 0.086(\log_{10} Z_n)^2\,].$$

Here, e is electron charge, Z_n is atomic number, a_0 is the Bohr radius, and R_w is the Wigner-Seitz cell radius in units of a_0. Specific internal energy E_{QSM} is obtained from pressure by integrating the thermodynamic relation $P = -dE/dV$ from a suitable initial volume, say V_{c0}.

Now, choose a value of V, say V_m, such that the four-parameter model $E_{Li}(V)$ is accurate for $V \geq V_m$. That is, we assume that the zero-temperature isotherm $E_{cold}(V) = E_{Li}(V)$ and $P_{cold}(V) = P_{Li}(V)$ for $V \geq V_m$. Then, for lower values of V, these are defined as

$$E_{cold}(V) = (E_{QSM}(V) - E_{QSM}(V_m))B_{int}(V) + E_{Li}(V_m), \quad V \leq V_m,$$
$$P_{cold}(V) = P_{QSM}(V)B_{int}(V) + (E_{QSM}(V) - E_{QSM}(V_m))B_{int}'(V), \quad V \leq V_m, \tag{6}$$
$$B_{int}(V) = (1 + b_1 V + b_2 V^{4/3} + b_3 V^{5/3}).$$

Here, $B_{int}(V)$ is a suitable interpolating function. Note that, by definition, $E_{cold}(V)$ is continuous at V_m. Now, the parameters b_k ($k = 1, 3$) in $B_{int}(V)$ are chosen such that $P_{cold}(V)$ and its first two derivatives are also continuous at V_m [12]. This procedure gives a smooth transition from the four-parameter model to the QSM. Plots of energy versus V for Cu using the two models are shown in Figure 1A with the choice $V_m = V_{c0}/1.4 = 0.07998$ cm^3/g.

As an application of the zero-temperature energy $E_{cold}(x)$, we use the lattice inversion method [21] and obtain an effective inter-particle potential between Cu atoms in the solid. We may imagine that the lattice is formed by assembling shells successively around a central atom. Then, the zero-temperature energy $E_{cold}(x)$ per atom, where $x = (V/V_{c0})^{1/3}$ is the scaled nearest-neighbor distance, can be readily written as a lattice sum involving the inter-particle potential $U(x)$. The inversion method provides a similar formula [22] for $U(x)$ in terms of $E_{cold}(x)$. For an FCC lattice, the direct formula for $E_{cold}(x)$ and the inverse formula for potential $U(x)$ are given by

$$E_{cold}(x) = \frac{1}{2}\sum_{n=1}^{N_s} r_n\, U(b_n x),$$ (7)

$$U(x) = 2\sum_{n=1}^{N_s} w_n E_{cold}(b_n x).$$

Here, n is the shell index, r_n is the number of atoms in the shell, b_n is the normalized radius of the shell, and w_n is the weight factor for the shell. This is a truncated formula, and the total number (N_s) of shells considered should be sufficiently large for convergence. The total potential (curve-1) and its repulsive (curve-2) and attractive (curve-3) components, as per the Weeks–Chandler–Andersen prescription [23], are shown in Figure 1B. For the sake of completeness, we have listed the constants r_n, b_n and w_n ($1 \le n \le N_s$) for FCC lattice in Table 1. The factors $1/2$ and 2 in Equation (7) arise because $U(x)$ is the energy for two atoms; while $E_{cold}(x)$ is the energy per atom.

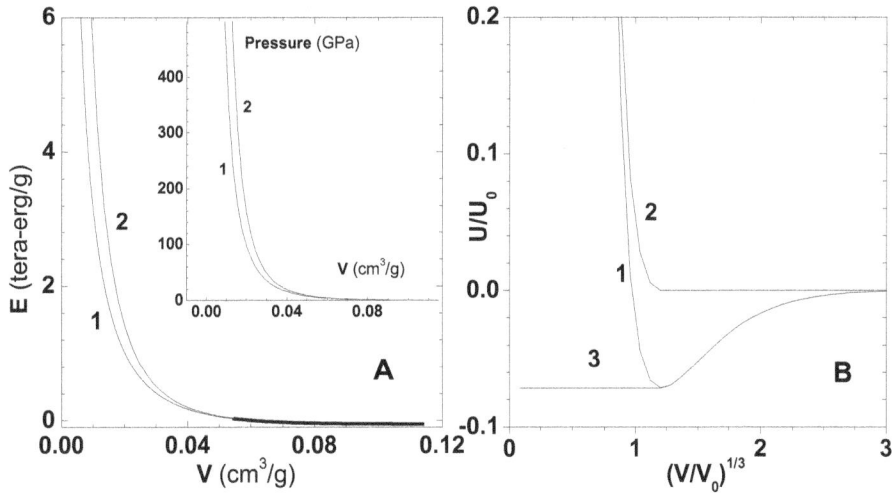

Figure 1. (A) Uncorrected energy $E_{Li}(V)$ for Cu (curve-1) according to the four-parameter model [16] and the zero-temperature energy $E_{cold}(x)$ corrected with QSM model (curve-2) as given in Equation (6). The insert figure shows pressure. (B) Effective inter-particle potential $U(x)$ for Cu (curve-1), obtained by inverting $E_{cold}(x)$, where $x = (V/V_{c0})^{1/3}$, using lattice inversion method. Curves 2 and 3 show, respectively, the repulsive and attractive components as per the WCA separation.

Table 1. Constants of inversion formula for FCC lattice.

n	r_n	b_n [1]	w_n	n	r_n	b_n [1]	w_n	n	r_n	b_n [1]	w_n
1	12	$\sqrt{1}$	$1/12$	6	8	$\sqrt{6}$	$1/9$	11	24	$\sqrt{11}$	$-1/6$
2	6	$\sqrt{2}$	$-1/24$	7	48	$\sqrt{7}$	$-1/3$	12	24	$\sqrt{12}$	$7/12$
3	24	$\sqrt{3}$	$-1/6$	8	6	$\sqrt{8}$	$1/32$	13	72	$\sqrt{13}$	$-1/2$
4	12	$\sqrt{4}$	$-1/16$	9	36	$\sqrt{9}$	$1/12$	14	0	$\sqrt{14}$	$-1/3$
5	24	$\sqrt{5}$	$-1/6$	10	24	$\sqrt{10}$	0	15	48	$\sqrt{15}$	$1/3$

[1] For FCC lattice $b_n = \sqrt{n}$. These are different for other lattices [21].

4. Ionic Thermal Component

The treatment of the ionic contribution to the EOS is best described by considering the compressed region and the expanded volume regions separately. In the former, we start with the low-temperature solid phase and go over to the melted fluid phase on increasing temperature. However, in the latter region, the material is always in the fluid phase, which also encompasses what is called the warm dense matter region. We consider different models for describing the EOS in these regions.

4.1. Compressed Region—Johnson's Model

The model for ionic thermal energy should describe the low-temperature properties of solids, fluid phase for temperatures above melting, and ideal gas behavior at higher temperatures. Thus, for a specified volume, the constant-volume molar specific heat of ions—denoted by C_{Vi}—must vary from the low temperature T^3 law to $3R$ above Debye's temperature and finally to $3R/2$. This feature is essential since shock compression of materials produce high temperatures and the solid melts and becomes a fluid after shock traversal. The mean field model used by Wang does not possess this crucial feature [24]. The parameters in the model used by Kormer et al. [25] need adjustments for every material.

So, we propose to use Johnson's ionic model [26] in the *compressed region*. In addition to the general constraints on C_{Vi} mentioned above, the model adds an extra contribution (3RT/Tm), linearly varying with T in the interval T_m to $1.2T_m$ to account for the heat of fusion. This corresponds to an increment of $0.6R$ in entropy, which has been determined from studies of several materials. Furthermore, the typical decrease of C_{Vi} from its value $3R$ at T_m to $9R/4$ at $5T_m$—and thereafter a linear variation in $\ln(T)$ to the ideal gas value $3R/2$—are built in to the model. Specific internal energy and pressure within the model (version-I) are thus given by [26]:

$$E_{ti}(V,T) = E_D + Nk_BT(E_0 + \epsilon_\psi). \tag{8}$$

$$P_{ti}(V,T) = \frac{\Gamma_i}{V}E_D + \frac{1}{V}(2\Gamma_i - 2/3)Nk_BT(E_0 + \epsilon_\psi), \tag{9}$$

$$E_D(V,T) = Nk_BT\left[\frac{9}{8}\frac{\theta_D}{T} + 9\left(\frac{T}{\theta_D}\right)^3 \int\limits_0^{\theta_D/T} \frac{z^3}{e^z - 1}\,dz\right].$$

Here, E_D is Debye's specific internal energy, $T_M(V)$ is melting temperature, $\theta_D(V)$ is Debye's temperature, k_B is Boltzmann's constant, N is number of atoms per gram, and $\psi = T/T_M$ is scaled temperature. Further, $\Gamma_i(V)$ is Grüneisen parameter for ions, to be defined below. The energy parameters E_0 and ϵ_ψ are fitted functions of ψ in order to account for the constraints on C_{Vi} mentioned above, and are given by

$$E_0 = -\frac{3}{2} + \frac{3}{2}a_4(\psi^{-3/2} - \psi^{-2}/2) + a_2\frac{(a_3y + \psi^{1-y})}{(\psi^y(a_3 + \psi^{1-y})^2)}, \quad 1 \le \psi < \infty,$$

$$\epsilon_\psi = \frac{3}{2}(\psi - 1/\psi), \quad 1 \le \psi \le 1.2, \tag{10}$$

$$\epsilon_\psi = 0.66/\psi, \quad 1.2 \le \psi < \infty.$$

Thus, Johnson starts with Debye's model in the region $T \le T_m$. The contribution to specific internal energy due to heat of fusion is given by ϵ_ψ; and E_0 describes the variation of C_{Vi} after melting. Note that E_0 varies from zero at T_m to $-3/2$ in the high temperature limit. An equally important feature is that the factor $(2\Gamma_i - 2/3)$ facilitates correct approach of the effective Γ_i to its ideal gas limit $(2/3)$. The constants a_2, a_3, a_4, which take care of the constraints, are given by

$$a_1 = -5.7 - \sum_i n_i \ln[(A_i/\bar{A})^{3/2}/n_i], \quad a_3 = 200,$$

$$y = \left(201[1600a_1^2 + 2398(4a_1 + 5)]^{1/2} - 40(5 - 197a_1)\right)/[3980(4a_1 + 5)], \quad (11)$$

$$a_2 = \frac{3}{2}(1 + a_3)^3/[a_3(1 - y)(a_3 y + 2 - y)],$$

$$a_4 = -\frac{8}{5}\left(a_1 + a_2/(1 + a_3)\right),$$

where, n_i, A_i, and $\bar{A}$ denote the number fraction, mass number, and its average, respectively. The summation in a_1, which contributes only for mixtures, accounts for ideal entropy mixing. Thus, the model is applicable to the case of compounds as well.

In a more elaborate method (version-II), region $1 \leq \psi < \infty$ is divided into three segments, viz., $1 \leq \psi \leq 5$, $5 \leq \psi \leq \zeta$, and $\zeta \leq \psi < \infty$. In the first region, C_{Vi} varies linearly in T; while in the the second, the variation is linear in $\ln(T)$. The value of ζ is determined so that entropy approaches ideal gas limit. Heat of fusion is added, as in the first version. Finally, specific energy and pressure are expressed as

$$E_{ti}(V, T) = E_D + Nk_B T(E_\psi + \epsilon_\psi), \quad (12)$$

$$P_{ti}(V, T) = \frac{\Gamma_i}{V}E_D + \frac{1}{V}(2\Gamma_i - 2/3)Nk_B T(E_\psi + \epsilon_\psi), \quad (13)$$

where ϵ_ψ is the same as that given in Equation (11), while E_ψ is given by

$$E_\psi = 3/16 - 3\psi/32 - 3/(32\psi), \quad 1 \leq \psi \leq 5,$$

$$E_\psi = -(3/4) + b\ln(\psi/5) - b + 5(b + 9/20)/\psi, \quad 5 \leq \psi \leq \zeta, \quad (14)$$

$$E_\psi = -(3/2) + 5(b + 9/20)/\psi - \zeta b/\psi, \quad \zeta \leq \psi < \infty.$$

The new constants b and ζ are given by

$$b = 9/\left(32(a_1 + 3/4 + 27\ln(5)/16)\right),$$

$$\zeta = 5\exp[-3/(4b)].$$

For illustration, we show in Figure 2 the variation of specific heat of Cu with temperature at normal volume V_0, using first version in graph-A and second version in graph-B. Both versions, thus, produce almost identical results. From the discussion summarized above, it is clear that Johnson's model is to be used in the compressed volume region. So, we next discuss a suitable model in the expanded volume region.

4.2. Expanded Region—Modified Soft-Sphere

The expanded volume region of the material, which covers the warm dense region and liquid–vapor transition, is important in several hydrodynamic applications. For instance, the material undergoing expansion induced via rarefaction wave is in this region. Similarly, highly porous materials (e.g., copper with porosity more than fifty percent) reach this state after shock-compression. It is necessary to have a separate model for this region as the physics here is mainly determined with excluded volume effect and a weak van der Waals type attractive interaction [27]. Young developed a soft-sphere model for liquid metals [28] using Monte Carlo simulations data for the thermal properties of particles interacting via soft-sphere potential, $\epsilon(\sigma/r)^n$. Here, ϵ, σ, and n define the parameters of the inverse power law potential. Together with the van der Waals attractive interaction, five parameters in the model were fitted to liquid–vapor co-existence data. This procedure is an extension of van der Waals theory of fluids, which treats attractive part as a zero temperature component of the EOS. We developed a modified soft-sphere model [29], wherein the use of simulation data was retained, however a generalized Lennard–Jones model was used as the attractive component. Instead of numerical fitting, all the parameters of the modified model were determined in terms of V_{c0}, E_{coh}, B_0, and B_0' at normal conditions.

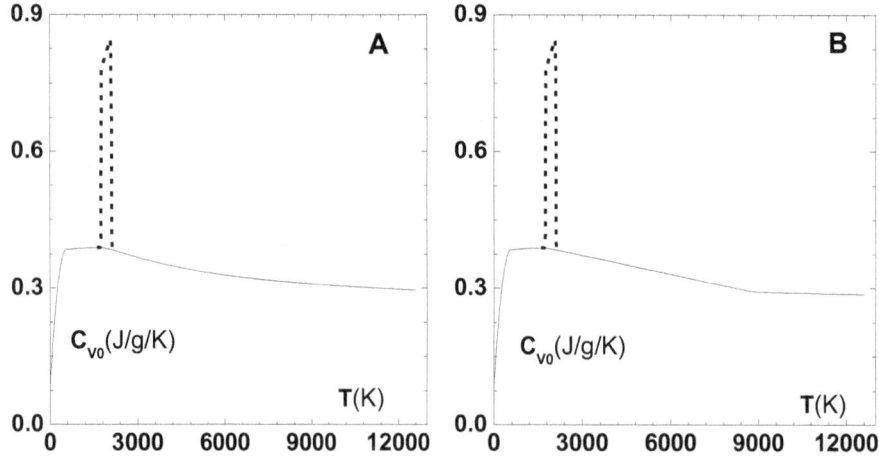

Figure 2. (**A**) Variation of ionic specific heat of Cu versus temperature at normal volume V_0 using Johnson's model, version-I. (**B**) Similar results using Johnson's model, version-II. Melting contribution is shown with dashed lines in both panels.

To discuss the formulation briefly, we start with a generalized Lennard–Jones pair potential between the atoms:

$$U_{LJ}(R) = \epsilon_{rep}\left(\frac{\sigma}{R}\right)^n - \epsilon_{att}\left(\frac{\sigma}{R}\right)^m. \tag{15}$$

Here, respectively, ϵ_{rep} and ϵ_{att} denote repulsive and attractive energy parameters, and σ is a spatial scale factor. This particular form is chosen because, as already mentioned above, extensive simulation data is available [27] for the repulsive component in Equation (15). Further, a proper choice of the exponent m will lead to an attractive part similar to that in van der Waals theory. The zero-temperature ionic energy per gram, $E_{LJ} = N\mathcal{E}_{LJ}$, where $\mathcal{E}_{LJ}$ is the energy per atom, can be expressed as

$$E_{LJ}(V) = N\frac{\epsilon_{rep}}{2} C_n\left(\frac{V_{c0}}{V}\right)^{n/3} - N\frac{\epsilon_{att}}{2} C_m\left(\frac{V_{c0}}{V}\right)^{m/3}, \tag{16}$$

where $V_{c0} = N\sigma^3/\sqrt{2}$ is the specific volume at zero-temperature, and C_n and C_m are lattice sums [30]. Now, on imposing the conditions that the zero-temperature pressure $P_{LJ} = -dE_{LJ}/dV$ vanishes at V_{c0} and the corresponding energy $E_{LJ}(V_{c0}) = E_{coh}$, we readily find that

$$\epsilon_{rep} = \frac{E_{coh}}{N} \frac{2}{C_n} \frac{m}{n-m}, \quad \epsilon_{att} = \frac{E_{coh}}{N} \frac{2}{C_m} \frac{n}{n-m}. \tag{17}$$

The expressions for E_{LJ} and P_{LJ} can be re-written as

$$E_{LJ}(V) = \frac{E_{coh}}{n-m}\left[m\left(\frac{V_{c0}}{V}\right)^{n/3} - n\left(\frac{V_{c0}}{V}\right)^{m/3}\right], \tag{18}$$

$$P_{LJ}(V) = \frac{E_{coh}}{n-m}\frac{n\,m}{3\,V_{c0}}\left[\left(\frac{V_{c0}}{V}\right)^{1+n/3} - \left(\frac{V_{c0}}{V}\right)^{1+m/3}\right]. \tag{19}$$

The exponents n and m are yet to be determined, however, bulk modulus B_0 and its pressure derivative B_0' can be computed from Equation (19) as

$$B_0 = E_{coh}\frac{n\,m}{9\,V_{c0}}, \quad B_0' = 2 + \frac{n}{3} + \frac{m}{3}. \tag{20}$$

Even though these relations imply that n and m are the roots of the quadratic equation $x^2 - 3(B_0' - 2)\,x + 9B_0V_{c0}/E_{coh} = 0$, experimental parameters generally lead to complex roots [31]. Therefore, in such situations, we can use only three parameters. Jiuxun has shown [32] that the spinodal condition $|B| \sim \sqrt{(P - P^*)}$, where P^* is the pressure corresponding to $B(P^*) = 0$, follows if we use the relation $m = (n-3)/2$. Then, the first relation in Equation (20) shows that n is determined from the quadratic equation $n^2 - 3n - 18B_0V_{c0}/E_{coh} = 0$. The positive root gives $n = 8.803$ and $m = 2.901$ for Cu. The accuracy of the generalized Lennard–Jones model can be assessed by comparing it with the four-parameter model [16] discussed earlier. In Figure 3A, we compare the repulsive component of the interparticle potential for Cu. The energy parameter ϵ_{rep} and the exponent n in the repulsive component are to be used in the model for ionic thermal energy. Figure 3B shows the generalized Lennard–Jones energy and pressure, which compare quite well.

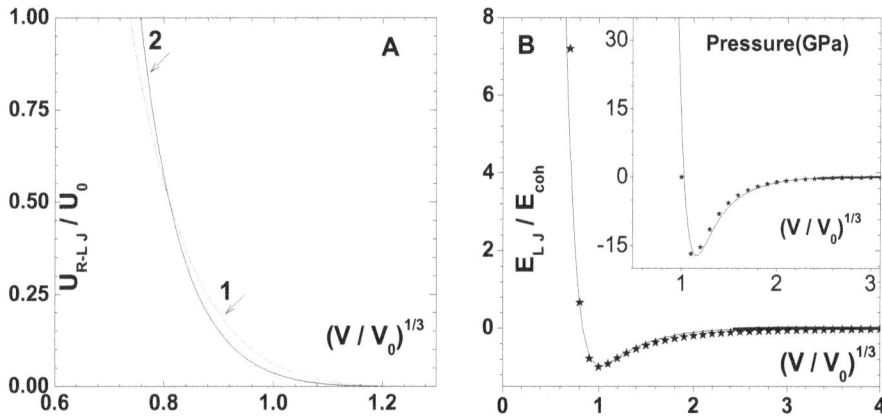

Figure 3. (**A**) Comparison of repulsive part of inter-particle potential for Cu versus scaled inter-particle distance $(V/V_0)^{1/3}$, using four-parameter model (curve-1) [16] and generalized Lennard–Jones model (curve-2). (**B**) Scaled cohesive energy and pressure using four-parameter model (solid lines) [16] and generalized Lennard–Jones model (symbols).

As explained in the beginning of this section, the ionic thermal energy of the modified soft-sphere model is precisely contributed by the repulsive part of the inter-particle potential. Monte Carlo simulation data of thermal pressure of soft particles, interacting via the potential $\epsilon_{rep}(\sigma/r)^n$, are accurately expressed in the parametric form [27]:

$$P = \frac{Nk_BT}{V}\left[1 + \frac{n}{3}\frac{C_n}{2}\left(\frac{\rho_{sc}}{\sqrt{2}}\right)^{n/3} + \frac{1}{18}n(n+4)\left(\frac{\rho_{sc}}{\sqrt{2}}\right)^{n/9}\right], \quad \rho_{sc} = \frac{N}{V}\sigma^3\left[\frac{\epsilon_{rep}}{k_BT}\right]^{3/n}. \tag{21}$$

We show the accuracy of this fit in Figure 4, where the scaled pressure $Z = PV/(Nk_BT)$ is plotted versus the scaled density, denoted as ρ_{sc}, for values of exponent $n = 4, 5, 6$. Similar results for the exponent $n = 7, 8, 9, 12$ are shown in Figure 5. Impressive agreement between the data and the parametric fit is evident in these figures. On adding this contribution to E_{LJ} and P_{LJ}, the ionic energy and pressure within the modified soft-sphere model are given by

$$E_{soft}(V,T) = Nk_BT\left[\frac{3}{2} + \frac{1}{6}(n+4)\left(\frac{V_{c0}}{V}\right)^{n/9}\left(\frac{\epsilon_{rep}}{Nk_BT}\right)^{1/3}\right] + E_{LJ}(V), \qquad (22)$$

$$P_{soft}(V,T) = \frac{Nk_BT}{V}\left[1 + \frac{1}{18}n(n+4)\left(\frac{V_{c0}}{V}\right)^{n/9}\left(\frac{\epsilon_{rep}}{Nk_BT}\right)^{1/3}\right] + P_{LJ}(V), \qquad (23)$$

where we have used the expression for zero-temperature reference volume, $V_{c0} = N\sigma^3/\sqrt{2}$, and the subscript $soft$ to denote the term 'soft-sphere'.

As mentioned earlier, the zero-temperature isotherms of the solid and the correction to ideal gas thermal energy (second term in brackets), determined from Monte Carlo simulation data, define the modified soft-sphere model. All the parameters in this model are determineda priori, so there is no need to fit its parameters.

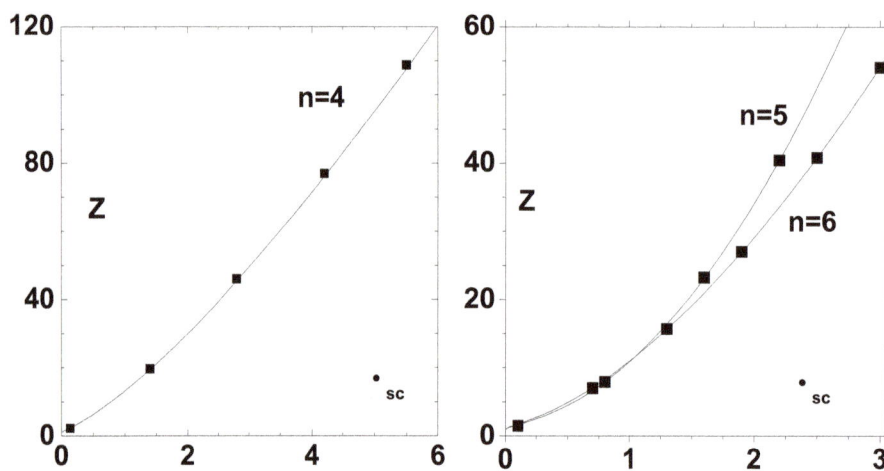

Figure 4. Scaled pressure (Z) versus scaled density (ρ_{sc}) for power law potentials with exponent $n = 4, 5, 6$. Lines correspond to Equation (21), while symbols denote Monte Carlo simulation data [27].

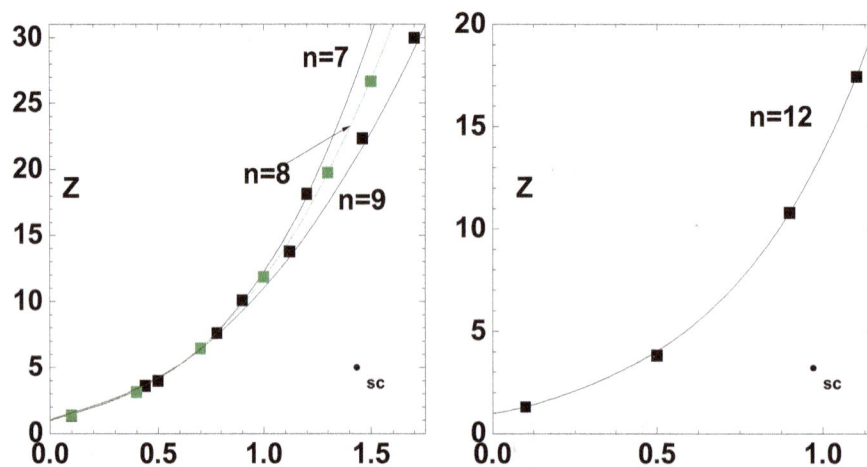

Figure 5. Scaled pressure (Z) versus scaled density (ρ_{sc}) for power law potentials with exponent $n = 7, 8, 9, 12$. Lines correspond to Equation (21) while symbols denote Monte Carlo simulation data [27].

4.3. Grüneisen Parameter

Ionic Grüneisen parameter $\Gamma_i(V)$, Debye temperature $\theta_D(V)$, and melting temperature $T_M(V)$ are needed in modeling ionic energy and pressure in the compressed volume region. While there are several empirical fits [13] for $\Gamma_i(V)$, the one due to Preston et al. [33], which has the correct asymptotic behavior for strong compression, is expressed as

$$\Gamma_i(V) = \frac{1}{2} + c_1 V^{1/3} + c_2 V^q. \tag{24}$$

The parameters c_1, c_2, and q are determined using experimental data on Γ_i at $T = 300$ K, at zero-pressure melting point and asymptotic ($V \to 0$) approximation for free electron states. These constraints define three nonlinear equations to determine the parameters. Expressions for $\theta_D(V)$ and $T_M(V)$, which follow from Debye–Grüneisen law and Lindemann's law, respectively, are given by

$$\theta_D(V) = \theta_0 \left(\frac{V_0}{V}\right)^{1/2} \exp[3c_1(V_0^{1/3} - V^{1/3}) + \frac{c_2}{q}(V_0^q - V^q)]. \tag{25}$$

$$T_M(V) = T_{M0} \left(\frac{V_r}{V}\right)^{1/3} \exp[6c_1(V_r^{1/3} - V^{1/3}) + \frac{2c_2}{q}(V_r^q - V^q)]. \tag{26}$$

Values of θ_0, T_{M0}, and V_r, which reference Debye temperature, melting temperature, and melting volume at $P = 0$, respectively, are tabulated for a variety of materials [33]. Further, the expression for $T_M(V)$ is found to compare well with experimental data, as shown below.

As an application of the models discussed in this section, we show the Grüneisen parameter for Cu in the compressed ($V \le V_0$) as well as the expanded ($V \ge V_0$) volume regions for 300 K (curve-1) and 5000 K (curve-2) in Figure 6A. The region to the left of the (vertical) dashed line in the figure is the compressed phase while the expanded region is on its right side. The thermodynamic definition for ionic Grüneisen parameter, $\Gamma_i = (1/V)(\partial P/\partial E)_V = (1/V)(\partial P/\partial T)_V (1/C_{Vi})$, can be readily used with analytical expressions for P and E, which are described in both volume regions (Johnson's model for $V \le V_0$; and the modified soft-sphere model for $V \ge V_0$). Experimental value 2.19 at 300 K is also shown (filled circle) in the figure. For a specified temperature, say 5000 K, the material goes over to the melted fluid region as volume is increased, and hence $\Gamma_i(V_0)$ is reduced to 1.96 from its value at 300 K. For the expanded volume region, the modified soft-sphere model shows explicit temperature dependent of $\Gamma_i(V)$.

In Figure 6B, we have also sketched $\theta_D(V)$ and $T_M(V)$, given in Equations (25) and (26). Experimental data on T_M taken from Preston et al. [33], and the value of $\theta_D(V_0)$ are also shown. Good comparison of T_M data, which are not used in fixing the parameters in Γ_i (c_1, c_2, and q in Equation (24)), shows the internal consistency of the procedure. In addition, we have shown θ_D (dashed line) derived from Slater's formula, $\Gamma_i = (1/6) - (1/2)(dB_S/dP)$ [11], where B_S is the zero-temperature bulk modulus obtained from Equation (4). Again, the agreement between θ_D obtained from two totally independent sources (Preston's and Slater's Γ_i) demonstrates its accuracy.

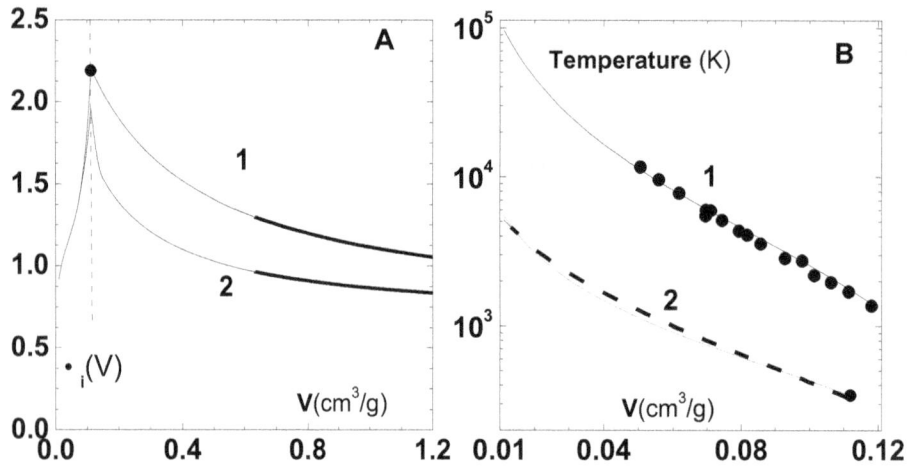

Figure 6. (**A**) Ionic Grüneisen parameter Γ_i versus volume V for Cu in the compressed (left side of dashed line) and expanded (right side of dashed line) regions for 300 K (curve-1) and 5000 K (curve-2). Experimental value 2.19 at 300 K is also shown. (**B**) Melting temperature T_M (curve-1) and Debye's temperature θ_D (curve-2) versus V for Cu in the compressed volume region, as given in Equations (26) and (25). Filled circles are experimental data on T_M [33]. Room temperature value 343 K of θ_D is also shown. The dashed line is Debye's temperature based on Slater model [11] for Γ_i.

5. Electronic Thermal Component

Electronic thermal component of energy and pressure is significant at temperatures reached in shock compression of porous materials. The EOS codes mentioned earlier uses the well-known Thomas–Fermi (TF) model [14] to describe electronic properties. However, the pressure and energy resulting from the approximations involved in this model need to be corrected in the low-temperature range. For example, the low-temperature specific heat constant predicted by the model differs from experimental values. Further, it is necessary to employ pressure and energy tables since in-line solutions of the TF equation are time consuming. Therefore, analytical fits [34] to results of Latter's calculations are sometimes employed [25] in high-pressure studies. However, it is important to note that this fit is valid in the compressed volume region, even though TF model as such may be applied even to an isolated atom.

So, we propose a somewhat different approach. First of all, following Atzeni et al [35], we assume that the Fermi gas model can be used to compute the thermal energy and pressure of electrons with a suitable average ionization degree Z^* of atoms, which depends on density and temperature. We use an excellent analytical fit for Z^* given by More [36] using results of Thomas–Fermi model:

$$
\begin{aligned}
R_5 &= \rho(g/cm^3)/(Z_n A), \quad tc = T(ev)/A^{4/3}, \quad t = tc/(1+tc), \\
a_5 &= 3.323 \times 10^{-3} \times tc^{0.971832} + 9.26148 \times 10^{-5} \times tc^{3.10165}, \\
b_5 &= -\exp[-1.763 + 1.43175 \times t + 0.315463 \times t^2], \\
c_5 &= -0.36667 \times t + 0.98333, \quad q_6 = a_5 \times R_5^{b_5}, \\
q_5 &= (\mathbf{factor} \times R_5^{c_5} + q_6^{c_5})^{1/c_5}, \quad x_5 = 14.3139 \times q_5^{0.6624}, \\
Z^* &= Z_n \times x_5/(1 + x_5 + [1 + 2x_5]^{1/2}),
\end{aligned}
\tag{27}
$$

where Z_n and A are atomic number and mass number, respectively. We have introduced a multiplicative correction term, termed '**factor**' in the equation above, which needs to be adjusted so that Z^* agrees with the experimental value of average ionization degree at ρ_0 and T_0. It takes value 0.079 for Cu, and the corrected and uncorrected variation of Z^* with temperature is shown in

Figure 7A. Note that Z^* is corrected only in the lower ranges of temperature, and it correctly saturates to Z_n at high temperature. The insert in this figure shows specific heat variation for three densities: (1) $\rho_0/10$, (2) ρ_0, and (3) $10\rho_0$. Similar correction factors for some other metals are given in Table 2.

Table 2. Z_0^* using More's formula [36].

Element	$\rho_0 (g/cc)$	$Z_{0UC}^*{}^1$	$Z_{0E}^*{}^1$	factor	$Z_{0C}^*{}^1$
Cu	8.93	4.38	1.01	0.079	1.02
Al	2.74	2.46	1.0	0.210	1.09
Fe	7.89	5.85	2.01	0.270	2.06
W	19.41	4.15	1.34	0.095	1.36

[1] Z_{0UC}^* (uncorrected), Z_{0E}^* (desired), Z_{0C}^* (corrected).

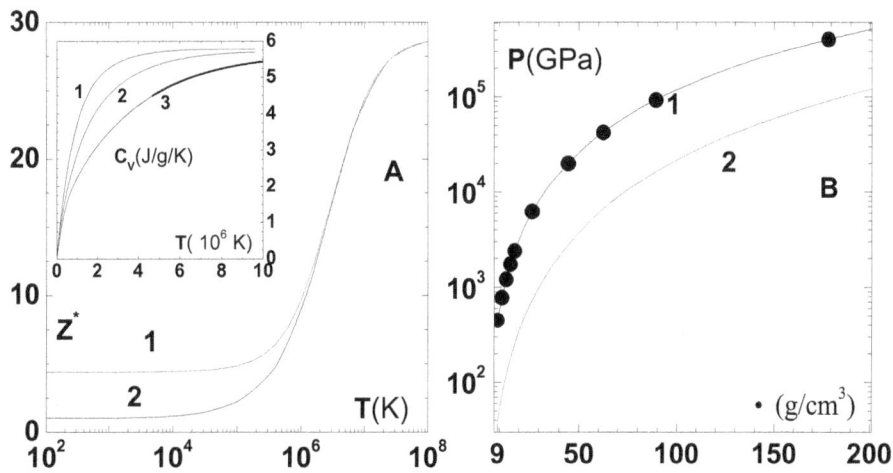

Figure 7. (**A**) Variation of effective charge Z^* versus temperature at density ρ_0—without correction (curve-1) and with correction (curve-2). Insert shows electron-specific heat at three densities: (1) $\rho_0/10$, (2) ρ_0, (3) $10\rho_0$. (**B**) Zero-temperature pressure versus density. Numerical results [37] employing TF theory (filled circles), results of free electron gas model without correction (curve-1) and with correction (curve-2) are shown.

To test this approach, we have plotted in Figure 7B the zero-temperature electron pressure $P = C_p(NZ^*/V)^{5/3}$, where $C_p = (3/\pi)^{2/3}(h^2/20m_e)$ versus density (ρ) using the uncorrected (curve-1) and corrected (curve-2) formula for Z^*. Numerical results [37] employing TF theory (filled circles) are also shown. It is interesting to note the TF results are accurately reproduced with the uncorrected Z^*, while the correction gives lower pressures. Finally, the thermal component of electron energy and pressure, within the free electron gas model, are given by

$$\frac{N}{V}Z^* = C_0(k_B T)^{3/2} I_{1/2}[E_F/k_B T]. \tag{28}$$

$$E_e[V,T] = C_0(k_B T)^{5/2} V\, I_{3/2}[E_F/k_B T] - \frac{3}{5}(NZ^*)E_F(V,0). \tag{29}$$

$$P_e[V,T] = \frac{2}{3}\frac{1}{V}E_e[V,T], \quad C_0 = \frac{4\pi}{h^3}(2m_e)^{3/2}. \tag{30}$$

Here, $E_F(V,T)$ is the Fermi energy, which is implicitly defined via Equation (28); and I_n are Fermi–Dirac integrals. Very accurate rational approximations for these integrals are now available [38]. The next level of improvement to the electron EOS is to add Coulomb interaction, exchange, and correlation energies, thereby accounting for all the terms in the uniform electron gas model [39].

6. Applications

In this section, we discuss three more applications of the EOS model. The first is its application to the calculation of the shock Hugoniot. Extensive data available in the shock wave database [40] is compared in Figure 8A with predictions of the model (curve-1). We find excellent agreement throughout the range of pressures obtained in the experiments. In the insert figure, we have shown the results without correcting the zero-temperature isotherm (curve-2), and without adding electronic terms (curve-3). Correction to the zero-temperature isotherm is found to be quite important as the four-parameter isotherm is valid up to 100–150 GPa. Addition of the electron component is found to improve the prediction even for pressures in the range of 200 GPa. Contrary to the common feeling, pressure on the Hugoniot is decreased when electron contribution is added. This is because the electron degrees of freedom reduces temperature, and consequently the pressure, for a specified volume on the Hugoniot. Temperature along the Hugoniot (curve-1) and the melting temperature (curve-2) displayed in Figure 8B show that melting occurs around 300 GPa pressure. Therefore, proper accounting of the melting transition is important even though the transition is not evident in the pressure–volume Hugoniot.

Next, we compare the liquid–vapor phase diagram of Cu employing the modified soft-sphere model, which was briefly considered in our earlier work [29]. It is well known that the attractive and thermal components of energy finely balance to produce the van der Waals loops in the isotherms in the vapor–liquid co-existence region. The critical point parameters we have obtained, via Maxwell's construction, ($\rho_c = 2.246$ g/cm^3, $T_c = 8345$ K, and $P_c = 0.8935$ GPa) are very well within the range quoted in the literature [41]. The phase diagram (curve-1) is shown in Figure 9A, and compared with simulation data (filled circles) [42]. These data were obtained via molecular dynamics simulations using an effective pair potential deduced from DFT calculations of energy–volume curve in the compressed and expanded volume regions. We find good comparison except in the liquid region of the phase diagram. The spinodal curve (curve-2), which is the locus of points where isothermal compressibility diverges; and the diameter (curve-3), which is the average of liquid and vapor phase densities on the phase diagram, are also shown in the figure.

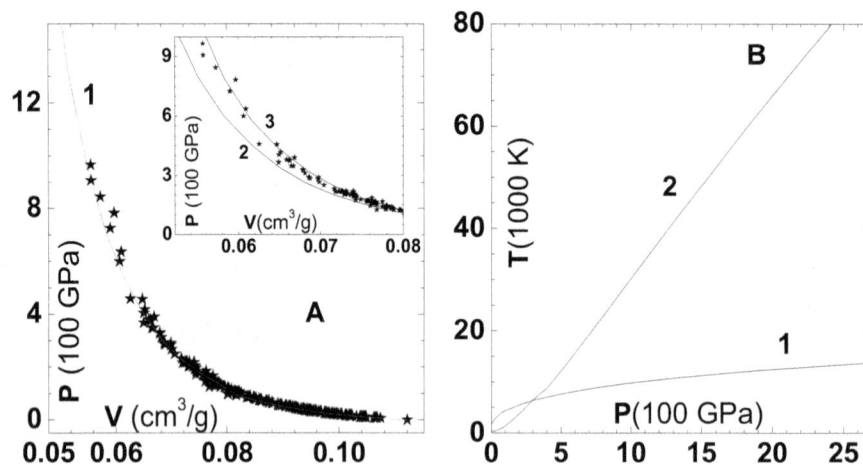

Figure 8. (**A**) Pressure–volume Hugoniot of Cu. Symbols are experimental data taken from the shock wave database [40], while curve-1 is based on the EOS model. The insert shows the effect of using uncorrected zero-temperature isotherm (curve-2) and without electronic contribution (curve-3). (**B**) Temperature along the Hugoniot (curve-1) and melting temperature (curve-2) versus pressure.

In Figure 9B, we have compared results of the EOS model for the enthalpy (H) of Cu versus temperature, along the 1-bar isobar, with experimental data taken from Trainor et al. [43]. Starting from 50 K, the material expands from 0.1105 to 0.1268 (cm^3/g), where it crosses the melting line around

1310 K. This results in a jump in enthalpy, and thereafter a smooth increase, as seen from the figure. The model for Γ_i employed in Johnson's model is inappropriate in this volume region. Therefore, we have taken a constant value $\Gamma_i = 2.19$, which is the experimental value at normal conditions, and corresponding expressions for $\theta_D(V)$ and $T_M(V)$ in lieu of Equations (25) and (26). Furthermore, we find that the heat of fusion to be added at the melting point is about 225 kJ/kg, in good agreement with the experimental value of 205 kJ/kg.

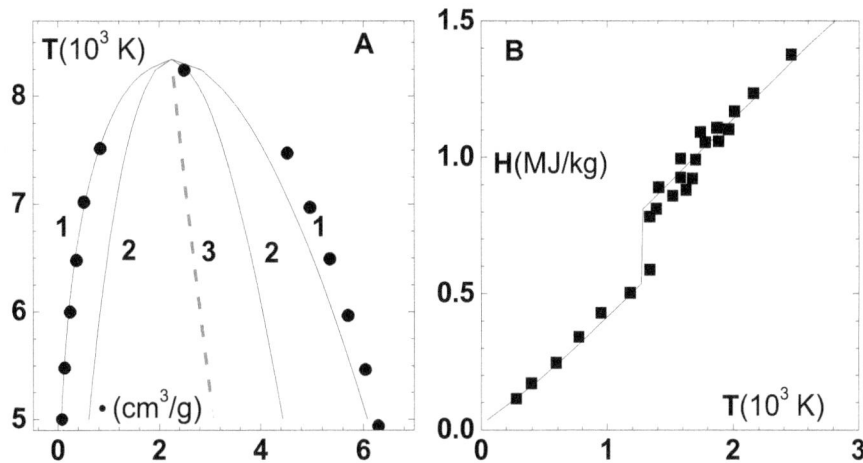

Figure 9. (**A**) Liquid–vapor phase diagram (curve-1) of Cu obtained using the modified soft-sphere model. Symbols are simulation results [42]. The spinodal line (curve-2) and the diameter (curve-3) are also shown. (**B**) Enthalpy of Cu versus temperature along the isobar at 1 bar. Symbols denote data taken from Trainor [43] while the line is based on the EOS model. The jump in enthalpy is due to the melting transition.

Finally, we mention that the EOS model described above (except the correction term for the zero-temperature isotherm and the free electron gas model) has been applied very successfully for the development and application of enthalpy-based approach to describe shock wave propagation in porous materials. There, we started with the computation of enthalpy-parameter, defined as $\chi = P(\partial V/\partial H)_P$, and developed the method treating P as independent variable. First, we showed [5] (using analytical fits [34] to results of TF theory) the correct approach to incorporate electronic effects explicitly. Then, we developed the modified soft-sphere model to properly evaluate the Hugoniot of highly porous materials, as their final shocked states are in the fluid region [29]. In another development, we used the EOS to show that the enthalpy-based approach can be implemented [44] in hydrodynamic simulations to describe shock wave propagation in solid as well as porous materials.

Applications of the EOS model to other materials can be easily carried out using the experimental/theoretical values of the parameters [16,33] employed in the different components of the model. A list of these parameters for Cu, Al, Fe, and W is given in our earlier publication [44].

7. Summary

The main aim of this paper is to discuss the basic components of an EOS model for metals for high-pressure physics applications. Thus, we started with the division of the EOS in to three parts: Zero-temperature isotherm, thermal ionic component, and thermal electronic component. This division is convenient as there are theories of different levels of sophistication dealing with them. We mentioned that results of DFT-based electronic structure calculations can be suitably fitted into a functional form, as is done in the case of the four-parameter model. A method to correct this model—which is only valid in the lower pressure range, so as to approach the results of the quantum statistical model—was discussed. Similarly, in lieu of detailed lattice dynamic calculations for density of states of lattice vibration modes, we have used the simple Debye model with a suitable form for volume dependence of Debye temperature. However, we have stressed the need for incorporating the energy of fusion due to melting in the thermal ionic component. A model for this component, which incorporates continuous temperature dependence of constant volume-specific heat, was discussed. It is the valence electrons in the metal which contribute to the thermal electronic component, which we modeled using the results of the Fermi gas model. However, we have shown that the degree of ionization, and hence the free electron density, can be determined from the Thomas–Fermi model. A simple correction to get the experimental electron density at normal conditions was also discussed. Finally, for the purpose of demonstration, the model was applied to calculate the shock-Hugoniot, liquid-vapor phase diagram, and isobaric expansion of Cu. We hope that the model described here can be used to generate tables for hydrodynamic applications of impact experiments, shock wave studies, and above all, design and analysis of high-energy-density systems.

Author Contributions: S.V.G.M.—conceptualization, methodology, formal analysis, investigation, software, validation, writing-review and editing; B.N.—investigation, validation and editing.

Funding: This research received no external funding.

Acknowledgments: The authors would like to thank the reviewers of *Condensed Matter Journal* for critical reviews and suggestions to improve the presentation of the paper.

Conflicts of Interest: The authors declare no conflict of interest.

References

1.　Zeldovich, Y.B.; Raizer, Y.P. *Physics of Shock Waves and High-Temperature Hydrodynamic Phenomena*; Academic: New York, NY, USA, 1967.

2.　Tillotson, J.H. *Metallic Equations of State for Hypervelocity Impact*; Report GA-3216; General Atomic Division of General Dynamics: San Diego, CA, USA, 1962; Unpublished.

3.　Rice, M.H.; Walsh, J.M. Equation of State of Water to 250 Kilobars. *J. Chem. Phys.* **1957**, *26*, 824. [CrossRef]

4.　Wu, Q.; Jing, F. Thermodynamic equation of state and application to Hugoniot predictions for porous materials. *J. Appl. Phys.* **1996**, *80*, 4343. [CrossRef]

5.　Nayak, B.; Menon, S.V.G. Explicit accounting of electronic effects on the Hugoniot of porous materials. *J. Appl. Phys.* **2016**, *119*, 125901. [CrossRef]

6.　More, R.M.; Warren, K.H.; Young, D.A.; Zimmerman, G.B. A new quotidian equation of state (QEOS) for hot dense matter. *Phys. Fluid* **1988**, *31*, 3059. [CrossRef]

7.　Young, D.A.; Corey, E.M. A new global equation of state model for hot, dense matter. *J. Appl. Phys.* **1995**, *78*, 3748. [CrossRef]

8.　Baroni, S.; Giannozzi, P.; Isaev, E. Thermal properties of materials from ab-initio quasi-harmonic phonons. *arXiv* **2011**, arXiv:1112.4977.

9.　Belmonte, D. First Principles Thermodynamics of Minerals at HP–HT Conditions: MgO as a Prototypical Material. *Minerals* **2017**, *7*, 183. [CrossRef]

10.　Chisolm, E.D.; Crockett, S.D.; Wallace, D.C. Test of a theoretical equation of state for elemental solids and liquids. *Phys. Rev. B* **2003**, *68*, 104103. [CrossRef]

11.　Otero-de-la-Roza, A.; Abbasi-Pérez, D.; Luaña, V. Gibbs2: A new version of the quasiharmonic model code. II. Models for solid-state thermodynamics, features and implementation. *Comp. Phys. Commun.* **2011**, *182*, 2232. [CrossRef]

12.　Kerley, G.I. *User's Manual for PANDA II- A Computer Code for Calculating Equation of State*; Sandia Report, SAND88-229.UC-405; Sandia National Labs.: Albuquerque, NM, USA, 1991.

13.　Godwal, B.K.; Sikka, S.K.; Chidambaram, R. Equation of state theories of condensed matter up to about 10 TPa. *Phys. Rep.* **1983**, *102*, 121. [CrossRef]

14. Latter, R. Thermal behavior of Thomas-Fermi statistical model of atoms. *Phys. Rev.* **1955**, *99*, 1854. [CrossRef]

15. Holzapfel, W.B. Physics of solids under strong compression. *Rep. Prog. Phys.* **1996**, *59*, 29. [CrossRef]

16. Li, J.H.; Liang, S.H.; Guo, H.B.; Liu, B.X. Four-parameter equation of state and determination of the thermal and mechanical properties of metals. *J. Alloys Comp.* **2007**, *431*, 23. [CrossRef]

17. Rose, J.H.; Smith, J.R.; Guinea, F.; Ferrante, F. Universal features of the equation of state of metals. *Phys. Rev. B* **1984**, *29*, 2963. [CrossRef]

18. Vinet, P.; Ferrante, F.; Smith, J.R.; Rose, J.H. Universal equation of state for solids. *J. Phys. Condens. Matter* **1989**, *1*, 1941. [CrossRef]

19. Kalitkin, N.N.; Kuz'mina, L.V. Curves of cold compression at high pressures. *Sov. Phys. Solid State* **1972**, *13*, 1938.

20. More, R.M. Quantum-statistical model for high-density matter. *Phys. Rev. A* **1979**, *19*, 1234. [CrossRef]

21. Chen, N.X.; Chen, Z.D.; Shen, Y.N.; Liu, S.J.; Li, M. 3D inverse lattice problems and Mobius inverssion. *Phys. Lett. A* **1994**, *184*, 347. [CrossRef]

22. Mookerjee, A.; Chen, N.X.; Kumar, V.; Satter, M.A. *Ab initio* pair potentials for FCC metals: an application of the method of Mobius transformation. *J. Phys. Condens. Matter* **1992**, *4*, 2439. [CrossRef]

23. Weeks, J.D.; Chandler, D.; Andersen, H.C. Role of Repulsive Forces in Determining the Equilibrium Structure of Simple Liquids. *J. Chem. Phys.* **1971**, *54*, 5237. [CrossRef]

24. Wang, Y.; Ahuja, A.; Johansson, B. Calculated Hugoniot curves of porous metal: Copper, nickel, and Molybdenum. *AIP Conf. Proc.* **2002**, *620*, 67.

25. Kormer, S.B.; Funtikov, A.I.; Urlin, V.D.; Kolesnikova, A.N. Dynamic compression of porous metals and equation of state with variable specific heat at high temperatures. *Sov. Phys. JETP* **1962**, *15*, 477.

26. Johnson, J.D. A generic model for the ionic contribution to the equation of state. *High Press. Res.* **1991**, *6*, 277. [CrossRef]

27. Hoover, W.H.; Stell, G.; Goldmark, E.; Degani, G.D. Generalized van der Waals equation of state. *J. Chem. Phys.* **1975**, *63*, 5434. [CrossRef]

28. Young, D. *A Soft-Sphere Model for Liquid Metals*; UCRL-Report UCRL-52353; University of California: Oakland, CA, USA, 1977.

29. Nayak, B.; Menon, S.V.G. Enthalpy-based equation of state for highly porous materials employing modified soft sphere fluid model. *Phys. B Phys. Condens. Matter* **2017**, *529*, 66. [CrossRef]

30. Mohazzabi, P.; Behroozi, F. A re-examination of the continuum approach to the calculation of lattice sums. *Phys. Stat. Sol. B* **1987**, *144*, 459. [CrossRef]

31. Stacey, F.D. High pressure equations of state and planetary interiors. *Rep. Prog. Phys.* **2005**, *68*, 341. [CrossRef]

32. Jiuxun, S. A modified Lennard-Jones-type equation of state for solids strictly satisfying the spinodal condition. *J. Phys. Condens. Matter* **2005**, *17*, L-103. [CrossRef]

33. Burakovsky, L.; Preston, D.L. Analytic model of the Grüneisen parameter for all densities. *J. Phys. Chem. Solids* **2004**, *65*, 1581. [CrossRef]

34. McCloskey, D.J. *An Analytic Formulation of Equations of State*; Memorandum RM-3905-PR; RAND Corporation: Santa Monica, CA, USA, 1964.

35. Atzeni, S.; Caruso, A.; Pais, V.A. Model equation-of-state for any material in conditions relevant to ICF and to stellar interiors. *Laser Part. Beams* **1986**, *4*, 393. [CrossRef]

36. More, R.M. Pressure Ionization, Resonances, and the Continuity of Bound and Free States. In *Advances in Atomic and Molecular Physics*; Academic Press, Inc.: San Diego, CA, USA, 1985; Volume 21, p. 305.

37. Perrot, F. Zero-temperature equation of state of metals in the statistical model with density gradient correction. *Phys. A* **1979**, *98*, 555. [CrossRef]

38. Antia, H.M. Rational function approximations for fermi-Dirac integrals. *Astrophys. J. Suppl. Ser.* **1993**, *84*, 101. [CrossRef]

39. Ichimaru, S.; Iyetomi, H.; Tanaka, S. Statiatical physics of dense plasmas: Thermodynamics, transport coefficients and dynamic correlations. *Phys. Rep* **1987**, *149*, 91. [CrossRef]

40. Shock Hugoniot Database. Available online: http://www.ihed.ras.ru/rusbank/ (accessed on 1 March 2019).

41. Levashov, P.R.; Fortov, V.E.; Khishchenko, K.V.; Lomonosov, I.V. Equation of state for Liquid metals. *AIP Conf. Proc.* **2000**, *505*, 89–92.

42. Sai Venkata Ramana, A. Molecular Dynamics Simulation of liquid–vapor Phase Diagrams of Metals Modeled Using Modified Empirical Pair Potentials. *Fluid Phase Equilib.* **2014**, *361*, 181. [CrossRef]
43. Trainor, K.S. Construction of a wide-range tabular equation of state for copper. *J. Appl. Phys.* **1983**, *54*, 2372. [CrossRef]
44. Nayak, B.; Menon, S.V.G. Numerical solution of Euler equations employing enthalpy-based equation of state for simulating shock wave propagation in porous materials. *Mater. Res. Express* **2019**, *6*, 055514. [CrossRef]

3

Effect of Phase Errors on a Quantum Control Protocol Using Fast Oscillations

Francesco Petiziol [1,2] ⓘ and **Sandro Wimberger** [1,2,*] ⓘ

1   Dipartimento di Scienze Matematiche, Fisiche e Informatiche, Università di Parma,
     Parco Area delle Scienze 7/A, 43124 Parma, Italy; francesco.petiziol@studenti.unipr.it
2   Italian Institute for Nuclear Physics (INFN), Sezione di Milano Bicocca, Gruppo Collegato di Parma,
     Parco Area delle Scienze 7/A, 43124 Parma, Italy
*   Correspondence: sandromarcel.wimberger@unipr.it

Abstract: It has been recently shown that fast oscillating control fields can be used to speed up an otherwise slow adiabatic process, making the system always follow an instantaneous eigenvector closely. In applying this method though, one typically assumes perfect phase relations among the control fields. In this work, we discuss the effect of potential static phase errors. We show that the latter can in some cases produce higher fidelities, leading to an unexpected improvement of the method. This is shown numerically and explained via a perturbative expansion of the error produced by the control strategy. When high-precision phase control is accessible, the results suggest that the phases of the control field can be used as free parameters whose optimization can be beneficial for the control protocol.

Keywords: quantum control theory; shortcuts to adiabaticity; fast oscillations; phase offsets; nonlinear phenomena; Landau–Zener; Jaynes–Cummings model

---

1. Introduction

The capability of preparing and manipulating with high precision the state of a quantum system is essential for the study of quantum phenomena, both for the investigation of fundamental aspects and for the development of technological applications. One of the most interesting control methodologies relies on a fundamental result of Quantum Mechanics, namely the adiabatic theorem [1]. This states that, given a time-dependent Hamiltonian, if the rate of the time evolution is sufficiently slow and there are no level crossings, then a system initially in an instantaneous eigenstate always remains in the corresponding instantaneous eigenstate at future times.

Although the adiabatic method is simple and robust by construction, for practical high-precision applications it often requires too long total evolution timescales, which are ultimately limited by dissipation and decoherence effects. For this reason, many strategies have been recently proposed for speeding up the quantum adiabatic dynamics, which are referred to as "shortcuts to adiabaticity" [2]. These methods can produce the desired evolution exactly in arbitrarily short time, provided one is able to implement sufficiently strong control fields. However, the implementation of the accelerated protocol is often experimentally challenging. This is so since it typically requires time-dependent tuning of matrix elements which are not controllable in the original Hamiltonian [2,3], therefore demanding new resources and modifications of the experimental setup. Indeed, experimental proofs have been limited to few level systems and typically needed adaptations [4–7]. The problem can be circumvented by working in carefully selected interaction pictures and relaxing the constraint of following the adiabatic path at all times, i.e., by asking the evolution to match the adiabatic eigenstates only at the beginning and at the end of the protocol, see e.g., [8–10]. This, however, results in a completely nonadiabatic method, thus arguably dropping the advantage of using adiabatic strategies.

Recently, it was shown that it is possible to achieve both effective instantaneous following of the adiabatic states and no necessity to control new matrix elements at the same time [3]. This can be done by introducing fast oscillations in the parameters of the initial Hamiltonian which mimic the action of a full shortcut-to-adiabaticity Hamiltonian. Therefore, the necessity of new control knobs is traded for a high degree of controllability of those initially available and for an approximate rather than exact driving. This methodology has been theoretically exemplified in a precise circuit quantum electrodynamics (QED) experimental context in Ref. [11], where the experimental realizability is discussed in detail.

An interesting feature recognized in previous work [3,11] is that this scheme requires precise phase relations between oscillations in different Hamiltonian matrix elements, which may not be easily guaranteed experimentally. Nonetheless, it was also noticed that small phase offsets could interestingly lead to an unpredicted improvement of the method. For this reason, in this work we elaborate in more detail on this phenomenon, studying how small biases in the relative phases of the control fields affect the performance of the protocol. We confirm that improvements can be obtained and we explain their origin in relation to the procedure for constructing the accelerating control fields. This new understanding allows us to speculate on the potential use of the phases of the control elements as additional parameters for optimizing the control method, also opening a connection with optimal control theory.

After reviewing the general method in Section 2, we derive general system-independent expressions for the corrections to the final fidelity in Section 3, which are then applied to the case study of a two-level system in Section 4, connecting our results to previous work. In particular, it is explained that, in some cases, small phase shifts can be exploited to improve the control method without adding higher harmonics to the oscillating control functions. Our central result is Equation (26) highlighting the role of the phase biases on the efficiency of the protocol. In Section 5, it is then shown numerically that these mechanisms are present also in multilevel systems, by considering the case of a sped-up adiabatic protocol which produces entanglement between two qubits dispersively coupled to a resonator [11].

2. Effective Counterdiabatic Driving

In this section, the method introduced in Ref. [3] is reviewed. Let us begin by summarizing the main features. The fundamental idea is that of introducing fast oscillations and use them to perturbatively counteract nonadiabatic transitions. Technically, this is done by decomposing the evolution into time steps determined by the small fundamental period of oscillation T of the control functions. Then, the propagator is written down at the end of each time step as a Magnus (exponential) [12] perturbation expansion in small T. The time integration over a full period of the oscillating control functions averages out the oscillations, leaving an effective expression for the propagator which depends only on the oscillation amplitudes and on the phases. These parameters are then used as free parameters to enforce equality of the actual dynamics, to first orders in T, with the desired adiabatic one. In the following, the details are provided by formulating the problem in the instantaneous adiabatic frame and including potential phase shifts among the control functions.

General Setup

Let the Hamiltonian of the finite-dimensional system be $H(t) = \sum_{n=0}^{N} E_n(t) |n_t\rangle \langle n_t|$, where $E_n(t)$ are the instantaneous eigenvalues, assumed to be never crossing, while $|n_t\rangle$ are the instantaneous eigenvectors. The Hamiltonian $H(t)$ is assisted by a correcting control Hamiltonian $H_c(t)$. The time dependent Schrödinger equation governing the evolution of the system is then ($\hbar = 1$)

$$i\partial_t U(t) = [H(t) + H_c(t)]U(t). \tag{1}$$

Let us assume that both $H(t)$ and $H_c(t)$ can be written as a linear combination of the same M time-independent Hamiltonians $\{H_k\}_{k=1,\ldots,M}$: this assumption guarantees that H_c can be formally absorbed into H, i.e., that H_c does not include initially unavailable matrix elements. The matrices $\{H_k\}$ are assumed to be linearly independent and they determine the matrix structure of H:

$$H(t) = \sum_{n=1}^{M} h_n(t) H_n, \qquad H_c(t) = \sum_{n=1}^{M} F_n(t) H_n. \tag{2}$$

The correcting Hamiltonian H_c is also assumed to include only fast oscillating terms with fundamental frequency Ω and a limited—small, in principle—number of harmonics $\pm \ell \Omega$ with $1 \leq \ell \leq L$ and L a natural number. Let us indicate with $\mathbb{Z}_L = \{-L, \ldots, -1, 1, \ldots, L\}$ the set of integer numbers from $-L$ to L with zero excluded. One can thus write the control functions $F_n(t)$ in terms of a truncated Fourier series:

$$F_n(t) = \sqrt{\Omega} \sum_{l \in \mathbb{Z}_L} f_{n,l} e^{il\Omega t + i\phi_{n,l}}. \tag{3}$$

Most relevant for the present work are the potentially different phases $\phi_{n,l}$ occurring in different control functions. The proportionality with Ω is chosen since we have in mind applications to avoided crossing scenarios; as will become clear in Section 4, for this kind of problems this choice allows to counteract nonadiabatic effects to order T. In fact, it permits one to enhance second order (in T) terms in the expansion of the propagator which can then be used to counteract first order undesired transitions.

For studying nonadiabatic effects, it is now convenient to move to a time-dependent frame defined by the instantaneous adiabatic basis, that is the basis of instantaneous eigenvectors multiplied by the dynamical and geometric phase factors [13],

$$A(t) = \sum_{n=0}^{N} e^{-i \int_0^t E_n(t') dt' - \int_0^t \Theta_n(t') dt'} |n_t\rangle \langle n_0|, \tag{4}$$

where $\Theta_n(t) = \langle n_t | \partial_t n_t \rangle$ is the effective vector potential which produces the geometric phase [14]. In this frame, the Hamiltonian can be written like $H^{(A)}(t) = H_c^{(A)}(t) + H_{\text{na}}^{(A)}(t)$ and the Schrödinger Equation (1) is thus

$$i \partial_t U_A = H^{(A)} U_A, \tag{5}$$

where $U_A = A^\dagger U$ and, introducing $\omega_{nm} \equiv E_n(t) - E_m(t) - i[\Theta_n(t) - \Theta_m(t)]$ and using Equations (2) and (3),

$$H_c^{(A)} = A^\dagger H_c A = \sum_{n,m}^{1,N} \sum_{k=1}^{M} \sum_{l \in \mathbb{Z}_N} f_{k,l} e^{il\Omega t + i\phi_{k,l}} e^{i \int_0^t \omega_{nm}(t') dt'} \langle n_t | H_k | m_t \rangle |n_0\rangle \langle m_0|, \tag{6}$$

$$H_{\text{na}}^{(A)} = i \partial_t A^\dagger A = i \sum_{\substack{n,m \\ n \neq m}}^{1,N} e^{i \int_0^t \omega_{nm}(t') dt'} \langle \partial_t n_t | m_t \rangle |n_0\rangle \langle m_0|. \tag{7}$$

Recall that the controllable matrices $\{H_k\}$ were introduced in Equation (2). Using Equation (2) and the relation $\langle \partial_t n_t | m_t \rangle = \frac{\langle n_t | \partial_t H(t) | m_t \rangle}{E_{nm}(t)}$ with $E_{nm}(t) = E_n(t) - E_m(t)$, which can be obtained by taking the time derivative of the eigenvalue equation $H(t) |n_t\rangle = E_n(t) |n_t\rangle$, Equation (7) can be further rewritten like

$$H_{\text{na}}^{(A)} = \sum_{\substack{n,m \\ n \neq m}}^{1,N} \sum_{k=1}^{M} e^{i \int_0^t \omega_{nm}(t') dt'} \frac{\partial_t h_k(t)}{E_{nm}(t)} \langle n_t | H_k | m_t \rangle |n_0\rangle \langle m_0|. \tag{8}$$

Let us now consider an evolution time equal to one period $T = 2\pi/\Omega$. A formal solution of Equation (5) can be written in exponential form as a Magnus expansion [12] (assuming that T is sufficiently small for the expansion to converge): $U_A(T) = \exp[\sum_n \mathcal{M}^{(n)}(t)]$. The first terms are

$$\mathcal{M}^{(1)}(T) = -i \int_0^T H^{(A)}(t_1) dt_1 \tag{9a}$$

$$\mathcal{M}^{(2)}(T) = \frac{(-i)^2}{2} \int_0^T dt_1 \int_0^{t_1} dt_2 \left[H^{(A)}(t_1), H^{(A)}(t_2) \right] \tag{9b}$$

$$\mathcal{M}^{(3)}(T) = \frac{i}{3!} \int_0^T dt_1 \int_0^{t_1} dt_2 \int_0^{t_2} dt_3 \left(\left[H^{(A)}(t_1), \left[H^{(A)}(t_2), H^{(A)}(t_3) \right] \right] \right. \tag{9c}$$

$$\left. + \left[H^{(A)}(t_3), \left[H^{(A)}(t_2), H^{(A)}(t_1) \right] \right] \right). \tag{9d}$$

The terms of the Magnus expansion generated by $H_c^{(A)}$ alone will be denoted as $\mathcal{M}_c^{(k)}$ with k relating to the number of nested commutators involved. Similarly, those generated by $H_{na}^{(A)}$ will be denoted as $\mathcal{M}_{na}^{(k)}$, while terms whose integrands are mixed commutators of $H_c^{(A)}$ and $H_{na}^{(A)}$ will be denoted by $\mathcal{M}_{mix}^{(k)}$. As an example, $\mathcal{M}_{mix}^{(0)} = 0$ and

$$\mathcal{M}_{mix}^{(1)} = -\frac{1}{2} \int_0^T dt_1 \int_0^{t_1} dt_2 \left\{ \left[H_c^{(A)}(t_1), H_{na}^{(A)}(t_2) \right] + \left[H_{na}^{(A)}(t_1), H_c^{(A)}(t_2) \right] \right\}. \tag{10}$$

Let us remark that, since our objective is that of achieving an expansion in T for the propagator, if the solution of the equation $i\partial_t U_c^{(A)} = H_c^{(A)} U_c^{(A)}$ is known in closed form then it would be convenient to further move to the frame defined by $U_c^{(A)}$; this would remove from Equation (9) the $\mathcal{M}_c^{(k)}$ terms, being re-summed, leaving an expansion in T as desired. However, this is typically not the case in general. The recipe of the control method here discussed is to compute the expressions (9) while keeping terms up to a certain desired order in T. Then, the amplitudes $\{f_{n,k}\}$ of the oscillations in Equation (3) are to be used to make vanish as many terms as possible for increasing powers of T.

3. System-Independent Expressions

3.1. Magnus Terms

Using the expressions of Equations (6) and (7) into Equation (9), one can compute the general expressions for the first order terms of the expansion in T of the propagator at the end of a time step. Since it is assumed that the oscillations in H_c are much faster than the timescales of the system, we will assume that all the quantities except the exponentials in Equation (6) are constant within the time interval T. This formally amounts to consider an expansion at $T/2$ for all of them, where only the value evaluated at $T/2$ is kept. This approximation allows us to compute the integrals. Recalling that $H_c \propto \sqrt{\Omega}$ (see Equation (3)) the results up to order $\Omega^{-3/2}$ are of the form

$$\mathcal{M}_c^{(0)}(T) = m_c^{(0)} \Omega^{-3/2} + o(\Omega^{-5/2}); \qquad \mathcal{M}_c^{(1)} = m_c^{(1)} \Omega^{-1} + o(\Omega^{-2});$$

$$\mathcal{M}_c^{(2)}(T) = m_c^{(2)} \Omega^{-3/2} + o(\Omega^{(-5/2)}); \qquad \mathcal{M}_{na}^{(0)} = m_{na}^{(0)} \Omega^{-1} + o(\Omega^{-2}),$$

where

$$m_c^{(0)} = -2\pi i \sum_{n,m}^{1,N} \sum_{k=1}^M \sum_{l \in \mathbb{Z}_L} \frac{f_{k,l}}{l} e^{i\phi_{k,l}} \omega_{nm} \langle n_t | H_k | m_t \rangle |n_0\rangle \langle m_0| \tag{11a}$$

$$m_{\mathrm{c}}^{(1)} = -2\pi i \sum_{n,m}^{1,N} \sum_{k,k'}^{1,M} \sum_{l=1}^{L} \frac{f_{k,l} f_{k',-l}}{l} e^{i(\phi_{k,l} + \phi_{k',-l})} \langle n_t | [H_k, H_{k'}] | m_t \rangle | n_0 \rangle \langle m_0 | \tag{11b}$$

$$m_{\mathrm{c}}^{(2)} = -\frac{2\pi i}{3} \sum_{n,m}^{1,N} \sum_{k,k',k''}^{1,M} \left\{ \sum_{l \in \mathbb{Z}_L} \sum_{l'=1}^{L} \frac{3}{ll'} f_{k,l} f_{k',l'} f_{k'',-l'} e^{i(\phi_{k'',-l'} + \phi_{k',l'} + \phi_{k,l})} \right.$$

$$\left. + \sum_{l,l'} f_{k,l} f_{k',l'} f_{k'',-l-l'} e^{i(\phi_{k,l} + \phi_{k',l'} + \phi_{k'',-l-l'})} \frac{1}{l(l+l')} \right\} \langle n_t | [H_k, [H_{k'}, H_{k''}]] | m_t \rangle | n_0 \rangle \langle m_0 | \tag{11c}$$

$$m_{\mathrm{na}}^{(0)} = 2\pi \sum_{\substack{n,m \\ n \neq m}}^{1,N} \sum_{k=1}^{M} \frac{h_k'(T/2)}{E_{nm}} \langle n_t | H_k | m_t \rangle | n_0 \rangle \langle m_0 | . \tag{11d}$$

Notice that U_{A} of Equation (5) can thus be written as

$$U_{\mathrm{A}} = \exp \left[\frac{m_{\mathrm{c}}^{(1)} + m_{\mathrm{na}}^{(0)}}{\Omega} + \frac{m_{\mathrm{c}}^{(0)} + m_{\mathrm{c}}^{(2)}}{\Omega^{3/2}} + o(\Omega^{-2}) \right] . \tag{12}$$

Let us remark that the first order in $\mathcal{M}_{\mathrm{mix}}^{(1)}$ is $o(\Omega^{-5/2})$.

3.2. Infidelity

Since our objective is that of studying the effect of the phases $\{\phi_{k,j}\}$ on the efficiency of the method, we want to derive an expression which quantifies the deviations from adiabaticity as a function of such phases. In order to do so, it is convenient to study the probability that the system has deviated from the instantaneous eigenstate at the end of the time step T. This is defined by

$$\mathbb{I}_{\mathrm{T}} = 1 - |\langle \psi_{\mathrm{tg}} | \psi(T) \rangle|^2,$$

and we call it the infidelity between the system state $|\psi(t)\rangle$ and target state $|\psi_{\mathrm{tg}}(t)\rangle$ at the end of one time step T. In order to express $\mathbb{I}_{\mathrm{T}}$ in our adiabatic frame, let us note that, given the initial system state $|\psi_0\rangle$, it holds that $|\psi_{\mathrm{tg}}(t)\rangle = A(t) |\psi_0\rangle$, with $A(t)$ defined in Equation (4), while for the evolving state $|\psi(t)\rangle$ it holds $|\psi(t)\rangle = U(t) |\psi_0\rangle$, with $U(t)$ defined by Equation (1). Therefore $\mathbb{I}_{\mathrm{T}}$ can be written like

$$\mathbb{I}_{\mathrm{T}} = 1 - |\langle \psi_0 | U_{\mathrm{A}}(T) | \psi_0 \rangle|^2, \tag{13}$$

which is consistent with the fact that the adiabatic evolution in the adiabatic frame is by definition time independent.

In order to compute an expansion of Equation (13) for small T, let us consider the propagator U_{A} written in exponential form, and let us write the exponent as a power series in T with half-integer powers, $k = 1, 3/2, 2, \ldots$. The half-integer powers are necessary due to the proportionality of H_{c} with $\sqrt{\Omega}$, see Equations (2) and (3). Thus,

$$U_{\mathrm{A}}(T) = \exp \left(\sum_k u_k T^k \right) .$$

From Equation (12), one can immediately identify

$$u_1 = m_{\mathrm{c}}^{(1)} + m_{\mathrm{na}}^{(0)}; \quad u_{\frac{3}{2}} = m_{\mathrm{c}}^{(0)} + m_{\mathrm{c}}^{(2)}. \tag{14}$$

For easing up the notation, let $\langle \cdot \rangle$ denote the average value $\langle \psi_0 | \cdot | \psi_0 \rangle$. An expansion of $|\langle U_A(T) \rangle|^2$ can be obtained by expanding the matrix exponential for small T, taking the average value termwise and then compute the modulus squared. Let us remark that, since all the Magnus terms are skew-hermitian, the expectation value $\langle \mathcal{M}^k \rangle$ is imaginary and thus $\Re \langle \mathcal{M}^{(k)} \rangle = 0$ for all k. This implies in turn that $\Re \langle u_k \rangle = 0$ for all k. Up to order T^3, the infidelity can then be written like

$$
\begin{aligned}
\mathbb{I}_T = &- \left[\Re \left\langle u_1^2 \right\rangle + |\langle u_1 \rangle|^2 \right] T^2 \\
&- \left[2\Re \left(\langle u_1 \rangle^* \left\langle u_{\frac{3}{2}} \right\rangle \right) + \Re \left\langle \{u_1, u_{\frac{3}{2}}\} \right\rangle \right] T^{5/2} \\
&- \left[\Re \left\langle \{u_1, u_2\} \right\rangle + \Re \left\langle u_{\frac{3}{2}}^2 \right\rangle + \left| u_{\frac{3}{2}} \right|^2 + 2\Re \langle u_1 \rangle^* \langle u_2 \rangle \right] T^3.
\end{aligned}
\tag{15}
$$

Let us now assume that the system is initially in an instantaneous eigenstate, say $|\psi_0\rangle = |j_0\rangle$. Then, using Equations (11), one can note that $\left\langle m_c^{(1)} \right\rangle = 0$ and $\left\langle m_{na}^{(1)} \right\rangle = 0$, so from Equation (14) it holds that $\langle u_1 \rangle = \left\langle m_c^{(1)} \right\rangle$, $\left\langle u_{\frac{3}{2}} \right\rangle = \left\langle m_c^{(2)} \right\rangle$.

3.3. Real Hamiltonians

For real Hamiltonians, the expression (15) for the infidelity can be further simplified. Let us indicate with $\mathrm{ad}^{(k)}(X)$ the nested commutator $[X(t_{i_1}), [\ldots, [X(t_{i_k}), X(t_{i_{k+1}})]] \ldots]$ involving k commutators, which appear in the Magnus expansion. Denoting with $\mathcal{M}_{mix}^{(n)}(\text{even})$ the commutators in $\mathcal{M}_{mix}^{(n)}$ which contain an even number of $H_c^{(A)}$, we can then state

Observation 1. *For real $H(t)$ and for an initial state $|\psi_0\rangle = |j_0\rangle$, it holds that*

(i) $\left\langle \mathcal{M}_c^{(2n-1)} \right\rangle = 0$ *for $n \in \mathbb{N}$;*
(ii) $\left\langle \mathcal{M}_{mix}^{(n)}(\text{even}) \right\rangle = 0.$

Proof. (i) Due to the properties of the commutator under unitary transformations

$$
\langle j_0 | \mathrm{ad}^{(2n-1)} \left[H_c^{(A)} \right] |j_0\rangle = \langle j_0 | A^\dagger \mathrm{ad}^{(2n-1)} [H_c] A |j_0\rangle = \langle j_t | \mathrm{ad}^{(2n-1)} [H_c] |j_t\rangle.
\tag{16}
$$

Since $H(t)$ is real, the matrices $\{H_k\}$ are real symmetric and so is H_c. Then, the commutators $\mathrm{ad}^{(2k-1)}[H_c]$ are real skew-symmetric and are thus orthogonal, with respect to the Hilbert–Schmidt inner product, to the whole subspace spanned by the $\{H_k\}$. From this it follows, since $|j_t\rangle$ is an eigenstate of a combination of $\{H_k\}$, that $\langle j_t | \mathrm{ad}^{(2n-1)} |j_t\rangle = 0$ and thus that $\left\langle \mathcal{M}_c^{(2n-1)} \right\rangle = 0$.

(ii) For (real) symmetric $H(t)$, $A(t)$ of Equation (4) is a real orthogonal matrix. Therefore $-i\partial_t A A^\dagger = A(i\partial_t A^\dagger A)A^\dagger$ is imaginary skew-symmetric and is thus orthogonal to the subspace spanned by the real symmetric matrices $\{H_k\}$. Then, it also holds that $H_{na}^{(A)} = i\partial_t A^\dagger A$ is orthogonal to the subspace spanned by the matrices $\{A^\dagger H_k A\}$. Since the commutator of two matrices is symmetric only if one of the two matrices is symmetric and the other is skew-symmetric, the commutators in $\mathcal{M}_{mix}^{(n)}(\text{even})$ are of the form $A^\dagger X A$ with X skew-symmetric, and they are thus orthogonal to the subspace spanned by the matrices $\{A^\dagger H_k A\}$. Their average value is then zero. $\square$

Furthermore, we also have $\Re \left\langle \{u_1, u_{\frac{3}{2}}\} \right\rangle = 0$ since u_1 is purely real skew-symmetric while $u_{3/2}$ is purely imaginary skew-symmetric. In conclusion, the infidelity of Equation (15) reduces to

$$\mathbb{I}_T = -\left[\Re\left\langle u_1^2\right\rangle\right] T^2 - \left[\Re\left\langle u_{\frac{3}{2}}^2\right\rangle + \Im^2\left\langle u_{\frac{3}{2}}\right\rangle\right] T^3 \tag{17}$$

$$= -\left[\Re\left\langle \left(m_c^{(1)} + m_{na}^{(0)}\right)^2\right\rangle\right] T^2 - \left[\Re\left\langle \left(m_c^{(0)} + m_c^{(2)}\right)^2\right\rangle + \Im^2\left\langle m_c^{(2)}\right\rangle\right] T^3. \tag{18}$$

4. Effect of Phase Shifts: Two-Level System

Using the general expressions reported in Section 3, we study in this section the effects of phase shifts $\phi_{k,l}$ in the control functions in a simple yet important scenario, namely that of a two-level system. In particular, the case of a two-level avoided crossing problem is considered in detail, being this a fundamental phenomenon where non-adiabaticity mechanisms become of paramount importance [15]. An avoided crossing occurs when two levels whose energy difference is varied would cross if uncoupled, but, if they are coupled instead, they repel each other after reaching a minimal gap. This phenomenon was first described theoretically by Landau [16], Zener [17] and Majorana [18], and has been studied thoroughly in a variety of experiments, see e.g., [19–23]. Many complex nonadiabatic phenomena can also be reformulated by means of multiple single-avoided-crossing scenarios [15,24–27]. Our aim is to show that slight phase offsets $\{\phi_{k,l}\}$ can lead to an improvement of the fast adiabatic method based on fast oscillations discussed in the previous sections, and to explain this effect.

Let $\sigma = \{\hat{\sigma}_1, \hat{\sigma}_2, \hat{\sigma}_3\}$ be the Pauli matrices. Let us consider a two-state problem ($N = 2$) with initial Hamiltonian $H(t) = \boldsymbol{H}(t) \cdot \sigma$ with $\boldsymbol{H}(t) = \{h_1(t), 0, h_3(t)\}$ ($M = 2$). That is, we assume that H is controllable only along the $\hat{\sigma}_1$ and $\hat{\sigma}_3$ directions, while the $\hat{\sigma}_2$ direction is not directly available due to experimental constraints. For instance, this can happen in planar superconducting circuits such as the transmon [28] where it is not easy to control both real and imaginary parts of the couplings between bare levels [7]. Note that the Hamiltonian $H(t)$ is real-valued, and we are thus in the case of Section 3.3. The correcting Hamiltonian H_c, being a linear combination of the same control matrices $\hat{\sigma}_1$ and $\hat{\sigma}_3$, can be written in the form $H_c(t) = \boldsymbol{F}(t) \cdot \sigma$ with $\boldsymbol{F}(t) = \{F_1(t), 0, F_3(t)\}$. As it has been suggested in Ref. [3] a first-order cancellation method can be achieved by choosing control functions of the form $F_1(t) = A\sqrt{\Omega}\cos(\Omega t)$ and $F_3(t) = B\sqrt{\Omega}\sin(\Omega t)$ and this will be explained in more detail in the following. Here, since we want to study "dephasing" effects between the control fields, we allow for the presence of phase offsets ϕ_1 and ϕ_3 by taking

$$F_1(t) = A\sqrt{\Omega}\cos(\Omega t + \phi_1); \qquad F_3(t) = B\sqrt{\Omega}\sin(\Omega t + \phi_3). \tag{19}$$

This choice, relative to Equation (3), implies that only the fundamental frequency is present, $L = 1$, and it moreover holds that $\phi_{k,l} = -\phi_{k,-l} \equiv \phi_k$, $f_{1,1} = f_{1,-1} = A/2$ while $f_{3,1} = -f_{3,-1} = B/(2i)$.

In order to compute an expression for the infidelity of Equation (18) to the first two orders (T^2 and T^3), one can use Equation (11). Moreover, one can use the exact expressions for the eigenvectors of $H(t)$, which, since $H(t)$ is real-valued, can be written in the parametrized form

$$|1_t\rangle = \{-\sin\chi, \cos\chi\}, \quad |2_t\rangle = \{\cos\chi, \sin\chi\}, \tag{20}$$

with $2\chi = \arctan(h_1(t)/h_3(t))$ the azimuthal angle between $\boldsymbol{H}(t)$ and the z axis. In this way, one determines, using Equation (11),

$$m_c^{(0)} = \frac{4\pi i}{\Omega^{3/2}}\left[Ah_3\sin\phi_1 + Bh_1\cos\phi_3\right]\hat{\Sigma}_2, \tag{21}$$

$$m_c^{(1)} = \frac{2\pi i}{\Omega}AB\cos\Phi\hat{\Sigma}_2,$$

$$m_c^{(2)} = \frac{4\pi i}{\Omega^{3/2}}\frac{AB}{\sqrt{h_1^2 + h_3^2}}\left[(A\sin\phi_1 h_3 + B\cos\phi_3 h_1)\hat{\Sigma}_3 + (A\sin\phi_1 h_1 - B\cos\phi_3 h_3)\hat{\Sigma}_1)\right]\cos\Phi, \tag{22}$$

$$m_{na}^{(0)}(T) = \frac{2\pi i}{\Omega}\eta\hat{\Sigma}_2, \tag{23}$$

where we have introduced the relative phase $\Phi = \phi_1 - \phi_3$, the matrices

$$\hat{\Sigma}_1 = |1_0\rangle\langle 2_0| + |2_0\rangle\langle 1_0|, \quad \hat{\Sigma}_2 = -i(|1_0\rangle\langle 2_0| - |2_0\rangle\langle 1_0|), \quad \hat{\Sigma}_3 = |1_0\rangle\langle 1_0| - |2_0\rangle\langle 2_0|$$

and the nonadiabatic coupling $\eta = \langle 1_t|\partial_t 2_t\rangle = \frac{\langle 1_t|\partial_t H(T/2)|2_t\rangle}{E_1 - E_2}$. The latter can be computed from the expressions in Equation (20) to be $\eta = \frac{h_1\partial_t h_3 - h_3\partial_t h_1}{2(h_3^2 + h_1^2)}$. Thus, the first constraint equation, $u_1(T) = m_c^{(2)} + m_{na}^{(1)} = 0$, is

$$(iAB\cos\Phi + i\eta)\sigma_2 = 0 \implies AB\cos\Phi = -\eta. \tag{24}$$

If one was not aware of the presence of the phase offset Φ, a solution for $\Phi = 0$ could be taken to be

$$A = \sqrt{|\eta|}; \quad B = -\text{sign}(\eta)\sqrt{|\eta|}, \tag{25}$$

as was done in Ref. [3]. This would be a good choice unless $\sqrt{|\eta(t)|}$ was not a smooth function, in which case different solutions may be more convenient. Insertion of Equation (22) into Equation (18) gives the infidelity at the end of one time step T. Assuming that the system is initially in the ground state, $|\psi_0\rangle = |1_0\rangle$, the result is

$$\mathbb{I}_T = \eta^2(1 - \cos\Phi)^2 T^2 + \frac{2}{\pi}\left\{(Ah_3\sin\phi_1 + Bh_1\cos\phi_3)^2 + \frac{\eta^2\cos^2\Phi}{h_1^2 + h_3^2}(Ah_1\sin\phi_1 - Bh_3\cos\phi_3)^2\right\}T^3. \tag{26}$$

With this expression at hand, the effect of phase shifts in the initial control functions can now be discussed. For very small T, one can focus on the first order, T^2, in Equation (26). This determines the general behaviour; the infidelity oscillates for varying Φ with a period of 2π, reaching the maximum value for $\Phi = \pi + 2k\pi, k \in \mathbb{Z}$—that is, when the oscillating part of the control functions of Equation (19) are in perfect phase match. On the other hand, the first order term is minimum for $\Phi = 0$, i.e., for control function with totally off-phase oscillating part as chosen in Equation (19). Interestingly, the first order term turns out to be perfectly symmetric with respect to sign of Φ, so, if higher-order terms are neglected, no benefit can come for the presence of phase offsets, since in such a case it always holds that $\mathbb{I}_T(\Phi) \geq \mathbb{I}_T(0)$.

Although the first term of the expansion of Equation (26) determines the global behaviour of $\mathbb{I}_T$, more details can be obtained by inspecting the second term, of order T^3. First of all, one can see that this is no more symmetric with respect to the sign of Φ, in general. In fact, a term $\propto \sin\Phi$ appears, breaking the symmetry unless $\phi_1 = 0$. This asymmetry has important consequences on the efficiency of the method; indeed, for certain values of $\Phi \neq 0$ the infidelity can be become smaller than $\mathbb{I}_T(0)$ due to a compensation of the terms in Equation (26), leading eventually to an improvement of the fidelity.

Secondly, the $o(T^3)$ term depends explicitly on ϕ_1 and ϕ_3, so when this term becomes relevant the infidelity does not depend any more on the relative phase Φ only. This means that, even once the relative phase is fixed, one may be able to tune ϕ_1 and ϕ_3 so to further minimize $\mathbb{I}_T$.

Thirdly, the $o(T^3)$ term depends also on the choice of A and B as solutions of the constraint Equation (24). This induces a dependence on the choice of the adiabatic sweep function, via the nonadiabatic coupling η.

As a final remark, it must be noted that, for very small Φ, no information can in general be obtained from Equation (26) if Φ enters a regime in which its magnitude is comparable with that of T. In this case, in fact, Equation (26) would no longer be a valid expansion, and more terms would need to be considered.

In order to corroborate the previous observations, we treat a specific example in the following section.

4.1. Application: Avoided Level Crossing

The standard scenario where nonadiabatic effects become of great importance is when an avoided crossing of energy levels occurs. A simple model to describe this phenomenon is to consider a two-level Hamiltonian of the form

$$H(t) = h_3(t)\hat{\sigma}_3 + h_1\hat{\sigma}_1 \tag{27}$$

with h_1 constant in time while $h_3(t)$ is a sweep function such that $h_3(t_c) = 0$—that is, such that in absence of the coupling h_1 the levels would cross at $t = t_c$. Depending on the rate of the sweep, which determines the speed of the evolution, adiabaticity may be satisfied or violated. The simplest choice for $h_3(t)$ is a linear sweep $h_3(t) = \alpha t$, as proposed by Landau, Zener and Majorana [15–18]. For this case, an exact solution of the Schrödinger equation for t going from $-\infty$ to $+\infty$ can be computed, for the system starting in the ground state, giving an asymptotic probability at time $+\infty$ that the system has jumped to the excited state. This is exponentially small in a certain adiabatic parameter which quantifies the speed of the evolution—namely $\exp(-\tau)$ with $\tau = \pi h_1^2/\alpha$. In practical finite-time problems though, more refined choices for $h_3(t)$ are much preferable if one wants to achieve the best fidelity at the end of the sweep in a certain total time (while always remaining as close as possible to true adiabaticity, i.e., to being in an instantaneous eigenstate). For this reason, here we will work with the specific choice

$$h_3(t) = h_1 \tan\left[c_1\left(1 - 2\frac{t}{t_f}\right)\right], \tag{28}$$

with $c_1 = \arctan(\Delta/h_1)$ where 2Δ is the initial energy gap, t_f is the total duration time of the sweep, and $t \in [0, t_f]$. This kind of sweep function was introduced in Ref. [11], following general recipes from Refs. [29,30], and guarantees that the instantaneous rate of the evolution satisfies an instantaneous adiabatic condition for all times.

Using the expressions (19) and (25), a correcting Hamiltonian for this problem is

$$H_c(t) = \sqrt{\Omega|\eta_t|}[-\text{sign}(\eta_t)\cos(\Omega t)\hat{\sigma}_1 + \sin(\Omega t)\hat{\sigma}_3], \tag{29}$$

with $\eta_t = -h_1\partial_t h_3/2(h_1^2 + h_3^2)$.

Introducing the dimensionless quantities $\tau = t_f h_1, \omega = \Omega/h_1, \xi = t_f\eta$ and rescaling the physical time like $s = t/t_f, s \in [0,1]$, the Schrödinger equation can be written in the dimensionless form:

$$i\partial_s U(s) = \tau[H(s) + H_c(s)]U(s), \tag{30a}$$

$$H(s) = \tan[c_1(1 - 2s)]\hat{\sigma}_3 + \hat{\sigma}_1, \tag{30b}$$

$$H_c(s) = \sqrt{\frac{\omega|\xi|}{\tau}}[-\text{sign}(\xi)\cos(\omega s\tau)\hat{\sigma}_1 + \sin(\omega s\tau)\hat{\sigma}_3]. \tag{30c}$$

The numerical results discussed in detail in the following are obtained with the values of the dimensionless parameters $\Delta/h_1 = 10, \tau = 0.1$. The instantaneous energy levels and the shape of the sweep function of Equation (28) for these parameters are plotted in Figure 1a, while the evolution given by the control method, for $\omega/2\pi = 15$ is compared with the adiabatic dynamics in Figure 1b.

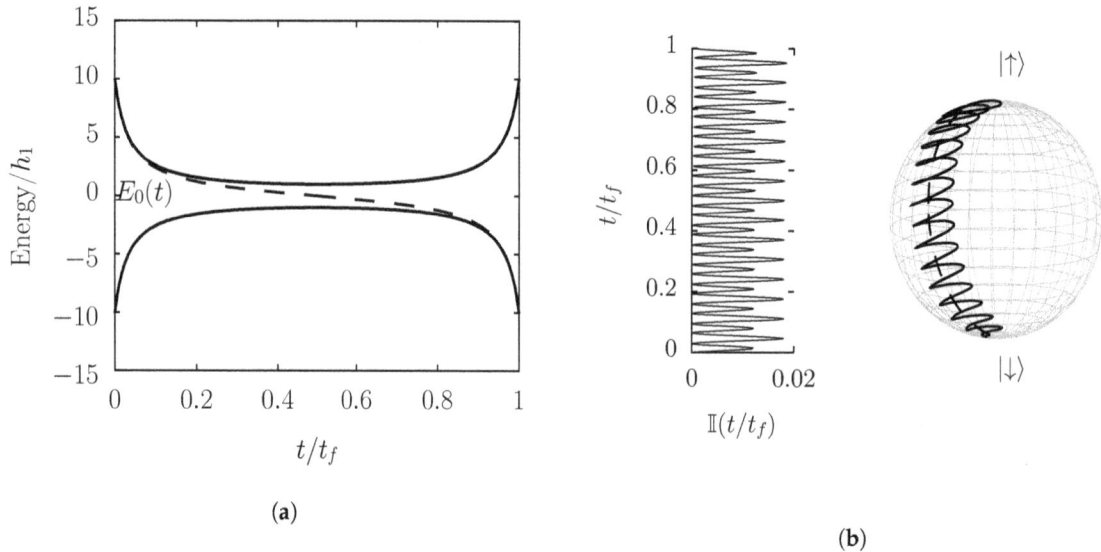

(a)

(b)

Figure 1. (a) The blue solid lines represent the time evolution of the two energy levels, measured in units of the coupling h_1, of the two-level Hamiltonian of Equation (27) with a sweep function of the form in Equation (28). The latter is represented, again in units of the coupling h_1, by the black dashed line. (b) Infidelity as a function of time and representation on the Bloch sphere of the evolution given by the effective counterdiabatic driving for a two-level avoided crossing problem. The dashed black line shows the instantaneous ground state, which is initially close to the $\hat{\sigma}_3$ eigenstate $|\downarrow\rangle$ while at the end is close to $|\uparrow\rangle$. The solid blue line shows the evolution given by the control method with Hamiltonian of Equation (30) and parameters $\Delta/h_1 = 10, \tau = 0.1, \omega = 2\pi \times 15$. The controlled dynamics oscillates around the target evolution.

Figures 2 and 3 show the dependence of the infidelity on the phase offset Φ for a sweep function of the form of Equation (28) and for fixed $\phi_3 = 0$ (so $\Phi = \phi_1$). Numerical data are indicated with symbols. In Figure 2, the general behaviour for an ample range of phase errors $-\pi \leq \Phi \leq \pi$ is shown. It is appreciable how this is very well described by the oscillating term $\propto (1 - \cos \Phi)^2$ of Equation (26). Nonetheless, such a description becomes ineffective for values of Φ closer to zero. This can be seen from Figure 3a, where the same general behavior for $-\pi \leq \Phi \leq \pi$ is shown for different values of T, in semilogarithmic scale; the blue dashed line represents the $o(T^2)$ term in Equation (26), which is a good description only when Φ is not small. The behavior near zero is captured by taking into account also the $o(T^3)$ term (red solid line). The latter introduces an asymmetry around zero with respect to the sign of Φ which becomes evident for small Φ: this is shown in Figure 3c for different values of T, with the blue line produced by considering both terms in Equation (26). As T increases, so does the asymmetry, due to the increasing importance of higher-order terms. From Figure 3c, it is also interesting to note that the asymmetry can produce a minimum which can be smaller than the value for $\Phi = 0$, witnessing again the fact that small phase offsets can be beneficial for the control method. Once the relative phase Φ is fixed, one has still freedom to choose the phase $\phi_1 = \phi_3$; although the first term is invariant with respect to this choice, the second one is not. This leaves room for optimization of ϕ_1 for further minimizing the infidelity in each time step. As an example, in Figure 3b, the infidelity is shown for $\Phi = 0$ and varying ϕ_1. It is evident that some choices of ϕ_1 can strongly reduce the infidelity with respect to a null ϕ_1. Numerical simulations show that performing such an optimization in each time step can decrease the infidelity at the end of the whole protocol almost by one order of magnitude.

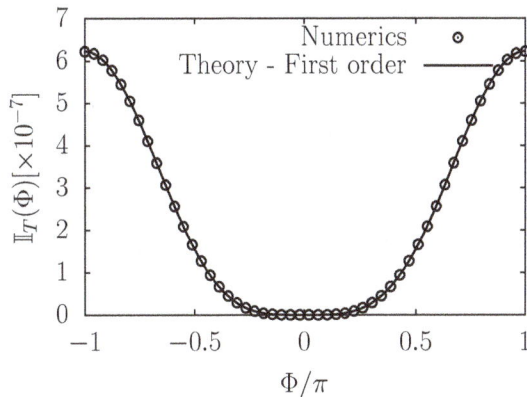

Figure 2. General behaviour of the infidelity $\mathbb{I}_T(\Phi)$ at the end of one time step $T \sim 7 \times 10^{-4}$ as a function of the phase error Φ, in units of π, for the accelerated adiabatic protocol described by Equation (30). The blue circles indicate the result of numerical simulations, while the solid black line indicates the prediction given by the first term of Equation (26).

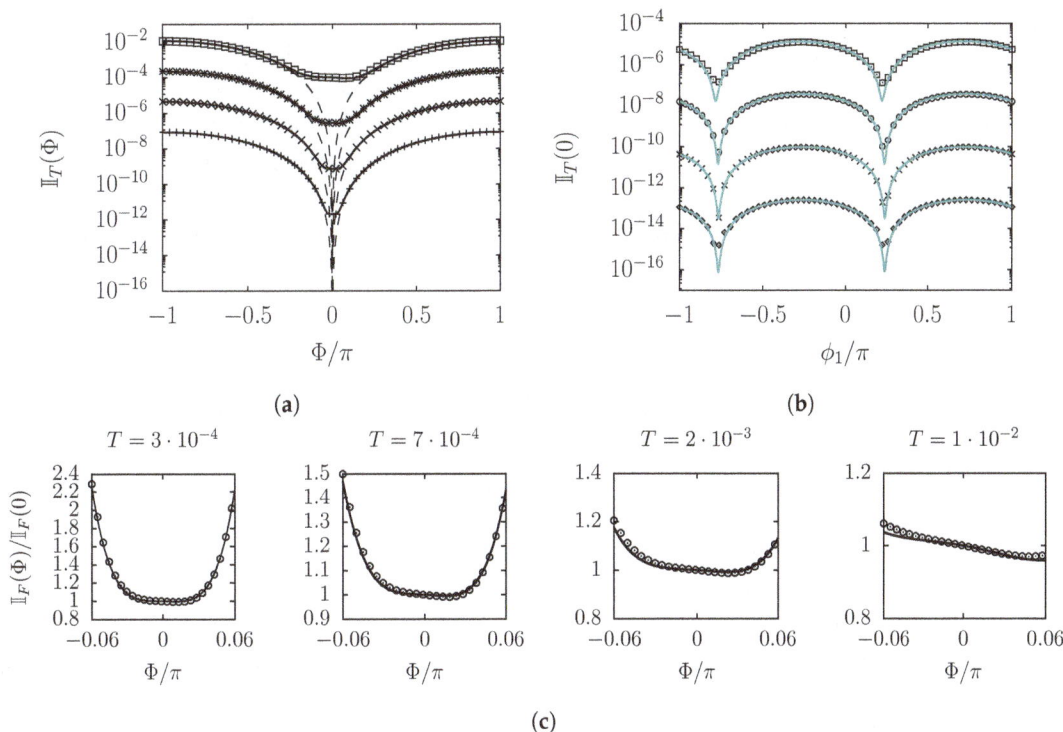

Figure 3. Behavior of the infidelity at the end of a time step T, for an avoided crossing problem with sweep function of the form in Equation (28), as a function of phase shifts. (**a**) Dependence on the relative phase Φ between the control functions of Equation (19) for different values of T (and so of Ω): $T = 10^{-4}, 7 \times 10^{-4}, 5 \times 10^{-3}, 4 \times 10^{-2}$. Symbols represent the result of numerical simulations, while the dashed blue curve represents the description given by the first term of Equation (26) and the solid red line is obtained by including both terms of Equation (26). Far from zero the first term is a sufficiently good description, and this is also true in the vicinity of zero when T decreases. The behavior near zero requires to take into account also the second term. The latter introduces asymmetries with respect to the sign of Φ, which are more pronounced for smaller T, as can be seen from panel (**c**). The second term also introduces an explicit dependence on the specific phases ϕ_1, ϕ_3: this is evident from panel (**b**), where the infidelity for null relative phase $\Phi = 0$ is shown as a function of $\phi_1 = \phi_3$. This interestingly shows that, once the first term is made to vanish by choosing $\Phi = 0$, the second term can be minimized by a suitable choice of phases $\phi_1 = \phi_3$ in each time step. Again, the symbols represent numerical data while the solid green-blue line is the prediction of Equation (26).

5. Beyond Two Levels: Two Atoms in a Cavity

In Ref. [11] an effective counterdiabatic scheme was proposed for speeding up an adiabatic entangling protocol between two superconducting qubits coupled to a transmission line resonator. Due to the complexity of the full multi-level problem, exact analytical expressions in the general case are hardly accessible. Therefore, we discuss the data produced by numerical simulation in the light of the results of the previous section, showing that an insight into the dependence on the phase shift can be obtained also in this scenario. The Hamiltonian of the problem in the rotating wave approximation is of Jaynes–Cummings form:

$$H = \omega_r \hat{a}^\dagger \hat{a} + \sum_{k=1}^{2} \frac{\omega_k}{2} \hat{\sigma}_3^{(k)} + \sum_{k=1}^{2} g_k [\hat{\sigma}_+^{(k)} \hat{a} + \hat{\sigma}_-^{(k)} \hat{a}^\dagger], \qquad (31)$$

where ω_r and ω_k—$k = 1, 2$—are the transition frequencies of the resonator and of the two qubits, respectively, while g_k denotes the coupling between qubit k and the resonator. The idea of the entangling protocol is that of exploiting an avoided crossing which emerges between states $|n \uparrow\downarrow\rangle$ and $|n \downarrow\uparrow\rangle$, where n is the number of photons while the $|\uparrow\rangle$ denote the states of the qubits, when the two qubits get in resonance with each other far in frequency from ω_r. This avoided crossing results in the dispersive regime $\omega_k - \omega_r \gg g_k$ from the emergence of an effective exchange interaction mediated by the exchange of virtual photons through the resonator. If the system starts in the state $|n \uparrow\downarrow\rangle$, for instance, driving the system adiabatically into qubit resonance produces the entangled singlet state $(|n \uparrow\downarrow\rangle - |n \downarrow\uparrow\rangle) / \sqrt{2}$. In order to bring the two qubits into resonance, one drives their transition frequencies, so that the sweep function is $f(t) = (\omega_1(t) - \omega_2(t))/2$. For the system starting in $|1 \uparrow\downarrow\rangle$ and for $g_1 = g_2 \equiv g$, the correcting Hamiltonian H_c, derived in Ref. [11], requires a time-dependent driving of the couplings g. It reads

$$H_c = \sqrt{2\Omega\eta_{2,3}} \left\{ \sin(\Omega t) [\hat{\sigma}_+^{(1)} \hat{a} + \hat{\sigma}_-^{(1)} \hat{a}^\dagger] + \cos(\Omega t) [\hat{\sigma}_+^{(2)} \hat{a} + \hat{\sigma}_-^{(2)} \hat{a}^\dagger] \right\}, \qquad (32)$$

where $\eta_{2,3} = |\langle 2_t | \partial_t 3_t \rangle|$ and $|2_t\rangle$ and $|3_t\rangle$ are the second and third excited instantaneous eigenvectors of the whole system. As opposed to the two-level case, the nonadiabatic coupling $\eta_{2,3}(t)$ requires numerical evaluation. This correcting Hamiltonian counteracts unwanted transitions between the two levels involved in the crossing, neglecting effects in the rest of the spectrum, since these are the major obstacle to the desired state transfer.

Let us now suppose that a phase shift is introduced such that, in the above Equation (32), $\cos(\Omega t) \longrightarrow \cos(\Omega t + \Phi)$. Figure 4 shows the dependence of the infidelity at the end of one time step T, in units of the unperturbed infidelity $\mathbb{I}_T(0)$, as a function of the phase error Φ. The blue symbols are the results of numerical simulations with parameters $\omega_1(0)/2\pi = 6.01$ GHz, $f(0)/2\pi = 10$ MHz, $\Omega/2\pi = 7$ GHz, $g_1/2\pi = g_2/2\pi \equiv g/2\pi = 10$ MHz and a sweep function of the form

$$f(t) = g_0 \tan[\alpha(1 - t/t_f)],$$

with $\alpha = \arctan[f(0)/g_0]$ and $2g_0$ the minimal width of the anticrossing extrapolated from numerical diagonalization. This is similar to the one in Equation (28), but it stops ($f(t_f) = 0$) at the crossing. One can recognize from Figure 4 the same general oscillatory behaviour found in Section 4, which is well captured, far from its nodes, by the black solid line: this line represents a fit with functional dependence as in the first term of Equation (26), namely $k[1 - \cos(\Phi)]^2$ with k free parameter. In the vicinity of the values $\Phi = \pm 2n\pi$ though, higher order effects again come into play. Focusing in the neighborhood of zero, an asymmetry with respect to the sign of Φ is present, so that for small positive Φ the infidelity becomes lower than the unperturbed value. We thus see once again that small phase offsets can lead to an accidental improvement of the control protocol. The systematics of this phenomenon further suggest that it can be exploited for optimizing the method.

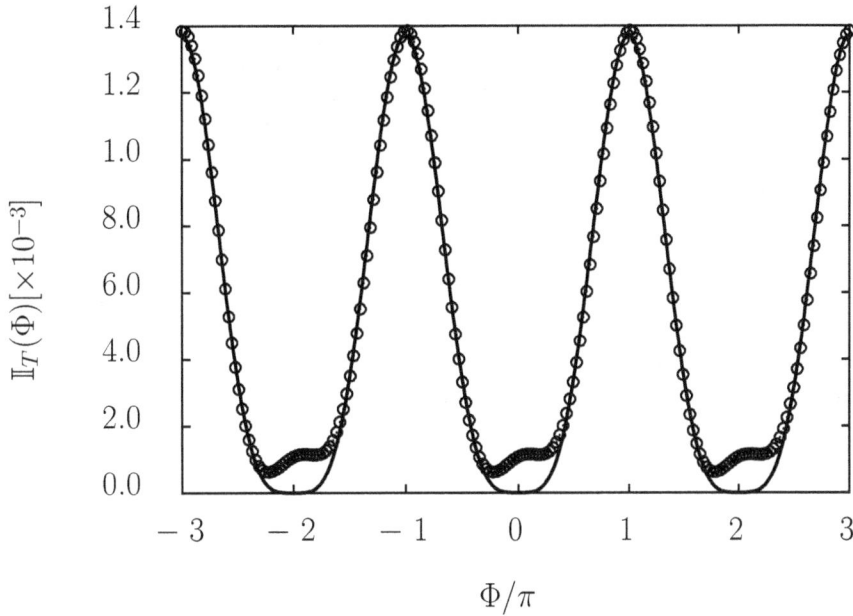

Figure 4. Infidelity at the end of one time step T as a function of the phase error Φ. The blue symbols indicate the result of a numerical simulation, while the solid black line represents a fit according to the dependence $k[1 - \cos(\Phi)]^2$, suggested by the first term in Equation (26), with k a free parameter. The fit well describes the data far from the nodes of the fitting function, well reproducing their oscillatory pattern. Closer to $\pm 2n\pi$, higher order effects in T show up, inducing an asymmetry which can lead to smaller infidelities.

In conclusion of this analysis, the parameter ranges should be discussed in which these effects are appreciably detectable. This is strictly related to the presence of other secondary phenomena which compete in slightly affecting the value of the infidelity on the same order of magnitude. For the specific framework treated in this section, the essential phenomenon which must be particularly taken into account is that of transitions towards levels other than the two involved in the crossing. We refer to Ref. [11] for a detailed discussion. The strength of this transition channel depends on the total duration of the protocol and on how good the dispersive-regime approximation is. The latter in turn depends on how large is the ratio $(\omega_k - \omega_r)/g$. The effects due to the phase shift are related instead to the strength of second-highest-order term in the expansion of the infidelity in the time step T, see Equation (26), which in turn depends on the system parameters and especially on how small T is. Assuming that T is sufficiently small for the method to converge at the end of each time step, these effects become visible for larger T, while they are reduced as T diminishes as discussed in Section 4.1. In Figure 5 we show how the dependence of the infidelity on the phase offset Φ behaves as the parameter g is varied, thus affecting the dispersive-regime approximation. For facilitating the comparison, all the curves are renormalized in such a way that the highest value reached in the interval considered lies at ordinate 1. It is immediately clear the general $\sim (1 - \cos\Phi)^2$ oscillating behaviour is robust even if g becomes large (larger values would challenge the validity of the Hamiltonian (31) in the experimental platform considered here, due to effects beyond the strong-coupling regime between qubits and resonator [31]). For small $g \sim 3$ MHz, transitions to other levels can be neglected and the time step T is sufficiently small for the $\sim (1 - \cos\Phi)^2$ behaviour to be completely dominant. For larger values of g, asymmetries in the vicinity of zero become more visible. For larger values of g the whole pattern start to shift with respect to $\Phi = 0$, indicating the emergence of further effects.

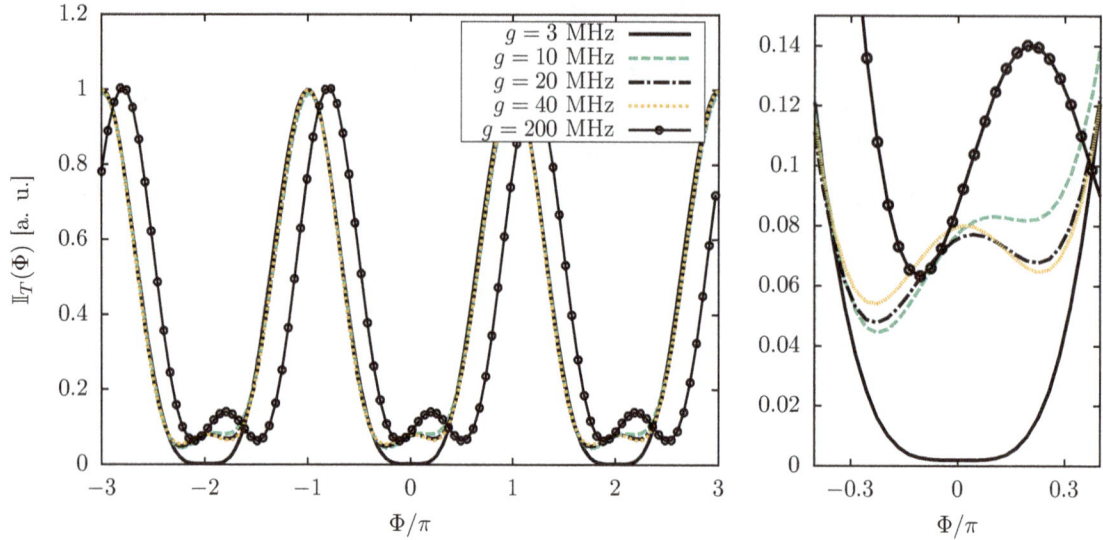

Figure 5. The left panel shows the behaviour of the infidelity at the end of one time step as a function of the phase offset Φ, for different values of the qubits-resonator couplings g. The values are rescaled in such a way that the highest value in the interval lies at ordinate 1 for all curves (all nonscaled maximal values lie below 2×10^{-2}). All curves present a general $\sim (1 - \cos \Phi)^2$ pattern, as predicted using Equation (26). For small $g = 3$ MHz, this pattern is dominant (black solid curve), while for larger values the contributions in the vicinity of $\Phi = \pm 2k\pi$ become relevant, showing substantial asymmetries with respect to Φ. The right panel shows an inset for Φ close to zero where asymmetries with respect to the sign of Φ are evident.

6. Discussion

We have discussed the effect of phase errors in the control functions for a control method which performs effective counterdiabatic driving by using fast oscillations. We have derived general expressions for the first order terms of the error produced by the protocol. These have been applied to the case of an avoided crossing problem, well explaining the behaviour found in numerical simulations. The results show that, on the scale of large values of the phase error Φ (as compared to π), the fidelity oscillates as a function of Φ and the protocol typically becomes sensibly less efficient. Nonetheless, it turns out to be rather stable against small values of Φ. Remarkably, small values of Φ can also produce higher fidelities. This is due to the small phase error leading to an accidental compensation of the first terms in an expansion of the infidelity. As a result, nonadiabatic transitions are further suppressed with respect to the basic protocol and the method turns out to be more efficient. This mechanism, well understood in the case of a two-level system, is confirmed numerically also for a more complex multilevel system. Also in this scenario, in fact, the infidelity oscillates for large Φ, while it can diminish for small Φ.

These results are a good evidence that a fine tuning of the phases of the control field can be used for an optimization of the control protocol, without needing to include more expensive resources, such as new control fields or higher harmonics.

Author Contributions: Project design: F.P. and S.W.; theory and data production: F.P.; both authors contributed to the writing of the manuscript.

Funding: This research received no external funding.

Acknowledgments: The authors are very grateful to the organizers of the conference SuperFluctuations 2018 for their invitation. F.P. thanks Florian Mintert for hospitality at Imperial College London during the early development of this work.

Conflicts of Interest: The authors declare no conflict of interest.

References

1. Messiah, A. *Quantum Mechanics*; Dover Publications: Mineola, NY, USA, 1961.

2. Torrontegui, E.; Ibàñez, S.; Martínez-Garaot, S.; Modugno, M.; del Campo, A.; Guéry-Odelin, D.; Ruschhaupt, A.; Chen, X.; Muga, J.G. Shortcuts to Adiabaticity. *Adv. At. Mol. Opt. Phys.* **2013**, *62*, 117–169. [CrossRef]

3. Petiziol, F.; Dive, B.; Mintert, F.; Wimberger, S. Fast adiabatic evolution by oscillating initial Hamiltonians. *Phys. Rev. A* **2018**, *98*, 043436. [CrossRef]

4. Bason, M.G.; Viteau, M.; Malossi, N.; Huillery, P.; Arimondo, E.; Ciampini, D.; Fazio, R.; Giovannetti, V.; Mannella, R.; Morsch, O. High-fidelity quantum driving. *Nat. Phys.* **2012**, *8*, 147–152. [CrossRef]

5. Du, Y.X.; Liang, Z.T.; Li, Y.C.; Yue, X.X.; Lv, Q.X.; Huang, W.; Chen, X.; Yan, H.; Zhu, S.L. Experimental realization of stimulated Raman shortcut-to-adiabatic passage with cold atoms. *Nat. Commun.* **2016**, *7*, 12479. [CrossRef]

6. Vepsäläinen, A.; Danilin, S.; Paraoanu, G.S. Optimal superadiabatic population transfer and gates by dynamical phase corrections. *Quantum Sci. Technol.* **2018**, *3*, 024006. [CrossRef]

7. Vepsäläinen, A.; Danilin, S.; Paraoanu, G.S. Superadiabatic population transfer in a three-level superconducting circuit. *Sci. Adv.* **2019**, *5*, eaau5999. [CrossRef]

8. Ibáñez, S.; Chen, X.; Torrontegui, E.; Muga, J.G.; Ruschhaupt, A. Multiple Schrödinger Pictures and Dynamics in Shortcuts to Adiabaticity. *Phys. Rev. Lett.* **2012**, *109*, 100403. [CrossRef]

9. Martínez-Garaot, S.; Torrontegui, E.; Chen, X.; Muga, J.G. Shortcuts to adiabaticity in three-level systems using Lie transforms. *Phys. Rev. A* **2014**, *89*, 053408. [CrossRef]

10. Baksic, A.; Ribeiro, H.; Clerk, A.A. Speeding up Adiabatic Quantum State Transfer by Using Dressed States. *Phys. Rev. Lett.* **2016**, *116*, 230503. [CrossRef]

11. Petiziol, F.; Dive, B.; Carretta, S.; Mannella, R.; Mintert, F.; Wimberger, S. Accelerating adiabatic protocols for entangling two qubits in circuit QED. *arXiv* **2019**, arXiv:1901.07344.

12. Blanes, S.; Casas, F.; Oteo, J.A.; Ros, J. The Magnus expansion and some of its applications. *Phys. Rep.* **2009**, *470*, 151–238. [CrossRef]

13. Sakurai, J.J.; Napolitano, J. *Modern Quantum Mechanics*, 2nd ed.; Cambridge University Press: Cambridge, UK, 2017. [CrossRef]

14. Berry, M.V. Quantal phase factors accompanying adiabatic changes. *Proc. R. Soc. A* **1984**, *392*, 45–57. [CrossRef]

15. Shevchenko, S.; Ashhab, S.; Nori, F. Landau-Zener-Stückelberg interferometry. *Phys. Rep.* **2010**, *492*, 1–30. [CrossRef]

16. Landau, L.D. A theory of energy transfer II. In *Collected Papers of L. D. Landau*, 1st ed.; Ter Haar, D., Ed.; Pergamon: Oxford, UK; 1965; pp. 63–66

17. Zener, C. Non-Adiabatic Crossing of Energy Levels. *Proc. R. Soc. A* **1932**, *137*, 696–702. [CrossRef]

18. Majorana, E. Atomi orientati in campo magnetico variabile. *Nuovo Cimento* **1932**, *9*, 43–50. [CrossRef]

19. Sillanpää, M.; Lehtinen, T.; Paila, A.; Makhlin, Y.; Hakonen, P. Continuous-Time Monitoring of Landau-Zener Interference in a Cooper-Pair Box. *Phys. Rev. Lett.* **2006**, *96*, 187002. [CrossRef]

20. Cao, G.; Li, H.-O.; Tu, T.; Wang, L.; Zhou, C.; Xiao, M.; Guo, G.-C.; Jiang, H.-W.; Guo, G.-P. Ultrafast universal quantum control of a quantum-dot charge qubit using Landau-Zener-Stückelberg interference. *Nat. Commun.* **2013**, *4*, 1401. [CrossRef]

21. Quintana, C.M.; Petersson, K.D.; McFaul, L.W.; Srinivasan, S.J.; Houck, A.A.; Petta, J.R. Cavity-Mediated Entanglement Generation Via Landau-Zener Interferometry. *Phys. Rev. Lett.* **2013**, *110*, 173603. [CrossRef]

22. Zhou, J.; Huang, P.; Zhang, Q.; Wang, Z.; Tan, T.; Xu, X.; Shi, F.; Rong, X.; Ashhab, S.; Du, J. Observation of Time-Domain Rabi Oscillations in the Landau-Zener Regime with a Single Electronic Spin. *Phys. Rev. Lett.* **2014**, *112*, 010503. [CrossRef]

23. Silveri, M.P.; Kumar, K.S.; Tuorila, J.; Li, J.; Vepsäläinen, A.; Thuneberg, E.V.; Paraoanu, G.S. Stückelberg interference in a superconducting qubit under periodic latching modulation. *New J. Phys.* **2015**, *17*, 043058. [CrossRef]

24. Brundobler, S.; Elser, V. S-matrix for generalized Landau-Zener problem. *J. Phys. A Math. Gen.* **1993**, *26*, 1211–1227. [CrossRef]

25. Zhu, L.; Widom, A.; Champion, P.M. A multidimensional Landau-Zener description of chemical reaction dynamics and vibrational coherence. *J. Chem. Phys.* **1997**, *107*, 2859–2871. [CrossRef]

26. Dodin, A.; Garmon, S.; Simine, L.; Segal, D. Landau-Zener transitions mediated by an environment: Population transfer and energy dissipation. *J. Chem. Phys.* **2014**, *140*, 124709. [CrossRef]

27. Theisen, M.; Petiziol, F.; Carretta, S.; Santini, P.; Wimberger, S. Superadiabatic driving of a three-level quantum system. *Phys. Rev. A* **2017**, *96*, 013431. [CrossRef]

28. Koch, J.; Yu, T.M.; Gambetta, J.; Houck, A.A.; Schuster, D.I.; Majer, J.; Blais, A.; Devoret, M.H.; Girvin, S.M.; Schoelkopf, R.J. Charge-insensitive qubit design derived from the Cooper pair box. *Phys. Rev. A* **2007**, *76*, 042319. [CrossRef]

29. Roland, J.; Cerf, N.J. Quantum search by local adiabatic evolution. *Phys. Rev. A* **2002**, *65*, 042308. [CrossRef]

30. Rezakhani, A.T.; Kuo, W.J.; Hamma, A.; Lidar, D.A.; Zanardi, P. Quantum Adiabatic Brachistochrone. *Phys. Rev. Lett.* **2009**, *103*, 080502. [CrossRef]

31. Gu, X.; Kockum, A.F.; Miranowicz, A.; Liu, Y.; Nori, F. Microwave photonics with superconducting quantum circuits. *Phys. Rep.* **2017**, *718–719*, 1–102. [CrossRef]

4

Evaluating Superconductors through Current Induced Depairing

Milind N. Kunchur

Department of Physics and Astronomy, University of South Carolina, Columbia, SC 29208, USA;
kunchur@sc.edu

Abstract: The phenomenon of superconductivity occurs in the phase space of three principal parameters: temperature T, magnetic field B, and current density j. The critical temperature T_c is one of the first parameters that is measured and in a certain way defines the superconductor. From the practical applications point of view, of equal importance is the upper critical magnetic field B_{c2} and conventional critical current density j_c (above which the system begins to show resistance without entering the normal state). However, a seldom-measured parameter, the depairing current density j_d, holds the same fundamental importance as T_c and B_{c2}, in that it defines a boundary between the superconducting and normal states. A study of j_d sheds unique light on other important characteristics of the superconducting state such as the superfluid density and the nature of the normal state below T_c, information that can play a key role in better understanding newly-discovered superconducting materials. From a measurement perspective, the extremely high values of j_d make it difficult to measure, which is the reason why it is seldom measured. Here, we will review the fundamentals of current-induced depairing and the fast-pulsed current technique that facilitates its measurement and discuss the results of its application to the topological-insulator/chalcogenide interfacial superconducting system.

Keywords: vortex; vortices; theory

The phenomenon of superconductivity has a long and rich history: from the initial discovery in 1911 of superconductivity in mercury at liquid-helium temperature [1] to the recent discovery of room-temperature superconductivity in lanthanum superhydride [2,3]. Numerous parameters, probed by a variety of techniques, are used to characterize the superconducting state. However, the mixed-state upper critical field B_{c2}, reflective of the coherence length ξ, and the penetration depth λ_L, reflective of the superfluid density $\rho_s = 1/\lambda^2$, are two crucial measurements that are amongst the first to be performed. There are multiple techniques for determining such parameters, each of which has its own advantages and limitations. Our group has developed some uncommon, and in some cases unique, experimental techniques that investigate superconductors at ultra-short time scales, and under unprecedented and extreme conditions of current density j, electric fields E, and power density $p = \rho j^2$ (where ρ is the resistivity). These techniques have led to the discovery or confirmation of several novel phenomena and regimes in superconductors and in addition provide an alternative method to glean information on fundamental superconducting parameters, which in some cases may be hard to obtain by other methods. These methods and approaches are highly relevant in the search for new superconducting materials and in developing an understanding of their fundamental properties. This article discusses the physical meaning of j_d and its interrelationships with other basic parameters of the superconducting state, as well as the technical challenges in measuring this important critical parameter. We discuss our results from this approach in the study of the topological-insulator/chalcogenide interfacial superconducting system.

1. Introduction

Attractive interactions between charge carriers cause them to condense by pairs into a coherent macroscopic quantum state below some transition temperature T_c. The formation of this state is governed principally by a competition between four energies: condensation, magnetic-field expulsion, thermal, and kinetic. The order parameter Δ, which describes the extent of condensation and the strength of the superconducting state, is reduced as the temperature T, magnetic field B, and electric current density j are increased. In type-II superconductors, there is partial flux entry at B values above the lower critical magnetic field B_{c1} and complete destruction of superconductivity above the upper critical field B_{c2} (type-I superconductors can be viewed as a special case where the thermodynamic critical field $B_c = B_{c1} = B_{c2}$). The boundary in the T-B-j phase space that separates the superconducting and normal states is where Δ vanishes, and the three parameters attain their critical values $T_{c2}(B, j)$, $B_{c2}(T, j)$, and $j_d(T, B)$. j_d sets the intrinsic upper limiting scale for supercurrent transport in any superconductor, and for $j > j_d$, the system attains its normal-state resistivity ρ_n. j_d should be distinguished from the conventional critical value j_c (related to extrinsic characteristics such as the depinning of vortices) above which there is partial resistivity $\rho < \rho_n$.

The resistivity ρ in the superconducting state is usually less than its normal-state value ρ_n. The reason for the presence of resistance at all in the superconducting state is because of fluctuations, percolation through junctions (in the case of granular superconductors), and the motion of magnetic flux vortices. For singly-connected superconductors not very close to T_c, only the last mechanism dominates as the cause of resistance. In the magnetic field region between B_{c1} and B_{c2}, a type II superconductor enters a "mixed state" with quantized magnetic flux vortices, each containing an elementary quantum of flux $\Phi_0 = h/2e$. Under the Lorentz driving force of an applied current, $j \times \Phi_0$, vortices move transverse to j, leading to a flux-flow resistivity:

$$\rho_f \sim \rho_n B / B_{c2} \tag{1}$$

in the free-flux-flow (large driving force) limit.

Two length scales characterize the superconducting state [4]. One is the coherence length:

$$\xi = v_F \tau_\Delta \simeq \hbar v_F / \pi \Delta \tag{2}$$

which is the characteristic length scale for spatial modulations in Δ (here, v_F is the Fermi velocity and τ_Δ is the order-parameter relaxation time). The normal core of a flux vortex has an approximate effective radius of ξ. The destruction of the superconducting state occurs when these normal cores overlap, corresponding to the condition:

$$B_{c2} = \frac{\Phi_0}{2\pi \xi^2} = \frac{\Phi_0}{2\pi \xi_1 \xi_2} \tag{3}$$

where ξ is the coherence length perpendicular to B; the single ξ is replaced by the geometric mean $\sqrt{\xi_1 \xi_2}$ in cases where the plane perpendicular to B is characterized by two anisotropic values.

The other characteristic length scale in a superconductor is the magnetic-field penetration depth λ, whose London value is given by:

$$\lambda_L = \sqrt{\frac{m^*}{\mu_0 n_s e^2}} \tag{4}$$

where m^* is the effective electronic mass and n_s is the density of superconducting electrons. The theory behind this important quantity and its relationship to j_d is described below. Figure 1 shows the profile of the magnetic field as it gets screened from the interior of a superconductor.

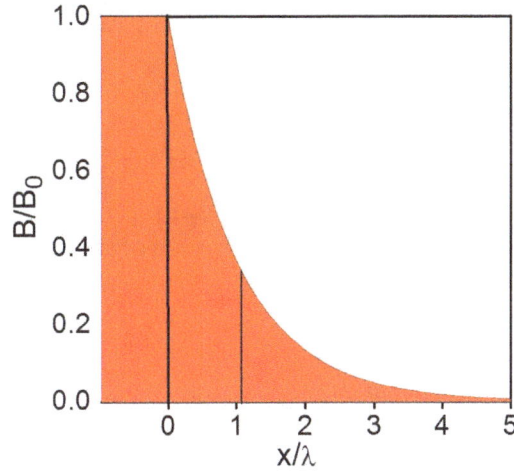

Figure 1. An externally-applied magnetic field B_0 is screened from the interior of a superconductor in the Meissner state over a characteristic length scale, which is the magnetic-field penetration depth λ: $B \sim B_0 e^{-x/\lambda}$. The circulating screening current is of roughly the depairing magnitude j_d, so that $\lambda \propto 1/j_d$.

1.1. Superfluid Density

For clean metallic superconductors, $n_s \to n$ as $T \to 0$, where n is the concentration of carriers in the normal state. Non-local effects and other corrections lead to deviations in λ from its London value. Hence, the superfluid density ρ_s can be conveniently and more completely defined as:

$$\rho_s \equiv 1/\lambda^2 \tag{5}$$

which includes the effective mass and other corrections to the effective n_s, rather than the simpler definitions $\rho_s = n_s$ or $\rho_s = n_s/m^*$ that are sometimes used in the literature. ρ_s, a quantity of central importance in superconductivity, characterizes the phase stiffness of the condensate [5] and its effectiveness at screening out magnetic fields and feeds into expressions for the transition temperature (such as the Uemura relation $T_c \propto \rho_s(0)$ that applies to the underdoped cuprate superconductors [6–8]).

Traditionally, many common methods for obtaining ρ_s do so by directly or indirectly measuring λ through its effect on a superconducting sample's magnetic-field profile and consequent magnetic susceptibility. This category includes methods such as reflection of spin-polarized slow neutrons [9], mutual inductance altered by an intervening superconducting film [10,11], changes in the self-inductance of a coil that is part of an LC resonating circuit [12,13], muon-spin rotation [14], magnetic force microscopy [15], microwave cavity resonance [16], and measurements of the lower critical field. These measurements are understandably affected if the material's internal magnetic field is altered, for example by a large paramagnetic background as in the case of the $Nd_{2-x}Ce_xCuO_4$ superconductor because of its Nd^{3+} magnetic moments.

Another approach to obtaining ρ_s is by measuring the inertia of the superfluid (kinetic inductance) during its ballistic acceleration phase [17–20]. This method requires the sample to be patterned into very high aspect ratio meanders for the highest accuracy.

A measurement of j_d provides an alternative to the above approaches for obtaining ρ_s. It requires a minimal amount of material (typically just a microbridge or nanobridge), does not require the complicated meander patterning needed for a kinetic-inductance measurement, and is unaffected by a material's normal-state magnetism that affects inductive measurements of ρ_s as discussed above. This immunity to material magnetism was used to good advantage for directly obtaining ρ_s in the $Nd_{2-x}Ce_xCuO_4$ superconductor for the first time [21]. Furthermore, unlike some of the methods for measuring λ that do not provide an accurate absolute value but only provide the temperature variation

$\lambda(T)/\lambda(0)$, j_d does provide the absolute value of λ and ρ_s and can hence provide information on the total carrier concentration n.

1.2. Normal-State Resistivity

Another valuable byproduct of measuring j_d is that it provides a direct measurement of the normal state resistivity ρ_n for temperatures below T_c. One of the starting points in developing an understanding of any newly-discovered superconductor is to understand the underlying normal state: the type of carriers and their concentration, their band-related properties, and the relevant scattering mechanisms and rates. At low applied B and j, $\rho(T)$ drops precipitously below T_c, thereby obscuring how $\rho_n(T \ll T_c)$ would have behaved if the superconductivity had not set in. The most common method for measuring $\rho_n(T < T_c)$ utilizes high magnetic fields $B > B_{c2}$ to drive the system normal below T_c; however, this measurement is subject to magnetoresistance (typically $R(B) \neq$ const) and may require prohibitively high magnetic fields ($B_{c2} > 100$ T for some superconductors).

One alternative is to use the core of a magnetic flux vortex as a window to the normal state. From Equation (1), a measurement of ρ_f elucidates ρ_n [22]. However, this extraction of ρ_n requires interpretation and modeling, since Equation (1) usually holds only approximately except for very high driving forces, and the exact prefactor depends on the detailed regime of flux flow [23–25].

Current-induced depairing provides an especially clean method for destroying superconductivity and accessing $\rho_n(T \ll T_0)$. Like the method of applying $B > B_{c2}$ to drive the system normal, applying $j > j_d$ is also free of interpretation and modeling, unlike flux-flow dissipation measurements. On the other hand, unlike the potential errors in the B_{c2}-based measurement due to normal-state magnetoresistance, the j_d method is immune to this issue because the normal-state electroresistance is quite negligible (i.e., $R(E) \simeq$ const) under the electric fields that arise at depairing conditions.

Thus, besides the investigation of interesting phenomena and regimes in superconductivity related directly to current-induced depairing itself, the study of j_d provides information on the important parameters of the superconducting state such as $\rho_s(T)$ and $\rho_n(T)$.

2. Relationship between the Depairing Current and Other Parameters

In a microscopic theory such as the Bardeen–Cooper–Schrieffer (BCS) theory, experimental quantities are calculated from microscopic parameters such as the strength of the effective attractive interaction that leads to Cooper pair formation and the density of states at the Fermi level. Often, these microscopic parameters are not sufficiently well known. In the London and Ginzburg–Landau (GL) phenomenological theories, connections are made between the different observables from constraints based on thermodynamic principles and electrodynamical properties of the superconducting state, leading to an adequate estimation of the depairing current. These phenomenological formulations are described next [4,26].

2.1. London Formulation

The London theory [4,27] of superconductivity provides a description of the observed electrodynamical properties by supplementing the basic Maxwell equations by additional equations that constrain the possible behavior to reflect the two hallmarks of the superconducting state: perfect conductivity and the Meissner effect. Note that these properties hold only partially when vortices are present.

An ordinary metal (normal conductor) requires a driving electric field E to maintain a constant current against resistive losses. In the simple Drude picture, this produces Ohm's law behavior, $j = \sigma E$, with a conductivity given by $\sigma = ne^2\tau/m^*$. A superconductor can carry a resistanceless current, and so, an electric field is not required for maintaining a persistent current. Instead, E in a perfectly-conducting state causes a ballistic acceleration of charge so that:

$$E = \left(\frac{m^*}{n_s e^2}\right)\frac{\partial j}{\partial t} \tag{6}$$

This is the first London equation, which reflects the dissipationless acceleration of the superfluid.

The second property that needs to be accounted for is the expulsion of magnetic flux by a superconductor. The magnetic field is exponentially screened from the interior following a spatial dependence:

$$\nabla^2 B = B/\lambda_L^2 \tag{7}$$

Together with the Maxwell equation $\nabla \times B = \mu_0 j$, this implies the following condition between B and j:

$$B = -\mu_0\lambda_L^2(\nabla \times j) \tag{8}$$

This is the second London equation, which describes the property of a superconductor to exclude magnetic flux from its interior. Taken together with the Maxwell equation $\nabla \times E = -\partial B/\partial t$, Equations (6) and (8) yield the expression for λ_L of Equation (4).

Besides the London equations themselves, a third ingredient needed for the estimation of j_d in this framework is the thermodynamic critical field B_c and its relationship to the Helmholtz free energy density f. When flux is expelled, the free energy density is raised by the amount $B^2/2\mu_0$. The critical flux expulsion energy (for the ideal case of a type-I superconductor with a non-demagnetizing geometry and dimensions large compared to the penetration depth) corresponds to the condition:

$$f_c = f_n - f_s = \frac{B_c^2}{2\mu_0} \tag{9}$$

where the L.H.S. of the equation represents the condensation energy density, which is the difference in free energy densities $f_n - f_s$ between the normal and superconducting states. j_d represents the condition when the kinetic energy density equals the condensation energy density: $\frac{1}{2}n_s m^* v_s^2 = \frac{m^* j_d^2}{2n_s e^2} = \frac{B_c^2}{2\mu_0}$, where v_s is the superfluid speed. Substituting for λ_L (Equation (4)) gives the London estimate for the depairing current density:

$$j_d \leq \frac{B_c}{\mu_0\lambda_L} \tag{10}$$

The inequality reflects the fact that n_s does not remain constant, but diminishes as j approaches j_d.

2.2. Ginzburg–Landau Formulation

There are situations where a system's quantum wavefunction cannot be solved for by usual means because the Hamiltonian is unknown or not easily approximated. The GL formulation [28] is a clever construction that allows useful information and conclusions to be extracted in such a situation where one cannot solve the problem quantum mechanically. For describing macroscopic properties, such as j_d that we are about to calculate, the GL theory is in fact more amenable than the microscopic theory [4,29].

The idea is to introduce a complex phenomenological order parameter (pseudo wavefunction) $\psi = |\psi|e^{i\varphi}$ to represent the superconducting state. $|\psi(r)|^2$ is assumed to represent the order parameter Δ introduced earlier, and to approximate the local density of paired superconducting charge carriers (Cooper pairs), which in turn is half the density of superconducting electrons n_s.

The free energy density f_s of the superconducting state is then expressed as a reasonable function of $\psi(r)$ plus other energy terms. A "solution" to $\psi(r)$ is now obtained by the minimization of free energy rather than through quantum mechanics. The unknown parameters of the theory are then solved in terms of measurable physical quantities, thereby providing constraints between the different quantities of the superconducting state.

Close to the phase boundary, $|\psi|^2$ is small, and so, f_s can be expanded keeping the lowest two orders of $|\psi|^2$. First, let us consider the simplest situation where there are no currents, gradients in $|\psi|$, or magnetic fields present. Then, we have:

$$f_s = f_n + \alpha|\psi|^2 + \frac{\beta}{2}|\psi|^4, \tag{11}$$

where α and β are temperature-dependent coefficients whose values are to be determined in terms of measurable parameters. The coefficients can be determined as follows. First of all, for the solution of $|\psi|^2$ to be finite at the minimum free energy, β must be positive. Second, for the solution of $|\psi|^2$ to be non-zero, α must be negative. Since $|\psi|^2$ vanishes above T_c, α must change its sign upon crossing T_c. The minimum in f_s occurs at:

$$|\psi|^2 = -\alpha/\beta. \tag{12}$$

Substituting this back in Equation (11) and using the definition of B_c (Equation (9)), Equation (12) can be written as:

$$f_c = \frac{B_c^2}{2\mu_0} = \frac{\alpha^2}{2\beta} \tag{13}$$

giving one of the connections between α and β and a measurable quantity (B_c). A second connection can be obtained by noting that n_s in Equation (4) can be replaced by $2|\psi|^2$, taking its equilibrium value from Equation (12):

$$\lambda^2 = \frac{m^*}{2\mu_0|\psi|^2 e^2} = \frac{-\beta}{\alpha}\left(\frac{m^*}{2\mu_0 e^2}\right) \tag{14}$$

Solving Equations (13) and (14) simultaneously gives the GL coefficients:

$$\alpha = -\frac{2e^2 B_c^2 \lambda^2}{m^*} \quad\quad \text{and} \quad\quad \beta = \frac{4\mu_0 e^4 B_c^2 \lambda^4}{m^{*2}} \tag{15}$$

Note that e and m^* refer to single-carrier values and not pair values.

To calculate j_d, we include the effect of a current in Equation (11) by adding a kinetic energy term $\frac{1}{2}n_s m^* v_s^2 = |\psi|^2 m^* v_s^2$ to it:

$$f_s = f_n + \alpha|\psi|^2 + \frac{\beta}{2}|\psi|^4 + |\psi|^2 m^* v_s^2. \tag{16}$$

For zero j and v_s, we saw earlier (Equation (12)) that the equilibrium value of $|\psi|^2$ that minimizes the free energy is $|\psi_{j=0}|^2 = -\alpha/\beta$. For a finite j and v_s, minimization of Equation (16) gives the value of $|\psi|^2$ when it is suppressed by a current:

$$|\psi_{j\neq0}|^2 = \frac{-\alpha}{\beta}\left(1 - \frac{m^* v_s^2}{|\alpha|}\right) = |\psi_{j=0}|^2\left(1 - \frac{m^* v_s^2}{|\alpha|}\right) \tag{17}$$

The corresponding supercurrent density is then:

$$j = 2e|\psi_{j\neq0}|^2 v_s = \frac{-2e\alpha}{\beta}\left(1 - \frac{m^* v_s^2}{|\alpha|}\right)v_s \tag{18}$$

The maximum possible value of this expression can now be identified with j_d:

$$j_d(T) = \frac{-4e\alpha}{3\beta}\left(\frac{|\alpha|}{3m^*}\right)^{1/2} = \left(\frac{2}{3}\right)^{3/2}\frac{B_c(T)}{\mu_0\lambda(T)} \tag{19}$$

where the GL-theory parameters were replaced by their expressions in terms of the physical measurables B_c and λ through Equation (15). As anticipated at the end of Equation (10) for the

London derivation for j_d, that simpler estimate is indeed larger than this more rigorous GL derivation by the factor $(3/2)^{3/2} = 1.84$.

The approximate temperature dependence of j_d can be obtained by inserting the generic empirical temperature dependencies $B_c(T) \approx B_c(0)[1 - (T/T_c)^2]$ and $\lambda(T) \approx \lambda(0)/\sqrt{[1 - (T/T_c)^4]}$, giving:

$$j_d(T) \approx j_d(0)[1 - (T/T_c)^2]^{\frac{3}{2}}[1 + (T/T_c)^2]^{\frac{1}{2}} \tag{20}$$

which close to T_c reduces to:

$$j_d(T) \approx \sqrt{2}j_d(0)[1 - (T/T_c)^2]^{\frac{3}{2}}. \tag{21}$$

where $j_d(0)$ is given by Equation (19) by setting $T = 0$ (for high scattering "dirty" superconductors, the $\sqrt{2}$ prefactor can be smaller or absent [29,30]).

Since B_c is not an easy quantity to measure directly, the relation:

$$B_c = \sqrt{\frac{\Phi_0 B_{c2}}{4\pi\lambda^2}} \tag{22}$$

can be used along with Equation (19) to write the expression for $j_d(0)$:

$$j_d(0) = \sqrt{\frac{2\Phi_0 B_{c2}(0)}{27\pi\mu_0^2\lambda^4(0)}} \tag{23}$$

that has the more easily measurable B_{c2}. Since both B_{c2} and j_d can be obtained from transport measurements, this becomes a convenient way to obtain λ and, hence, ρ_s.

2.3. Microscopic Formulations and Generalizations

Various authors have calculated $j_d(T)$ from a microscopic basis [29,31,32]. For arbitrary temperatures and mean free paths, one must use the Gorkov equations as the starting point. Kupriyanov and Lukichev [33] have derived $j_d(T)$ from the Eilenberger equations, which are a simplified version of the Gorkov equations. This derivation is beyond the scope of the present review, but a nice shortened version can be found in [30]. The microscopic calculation confirms the overall temperature dependence predicted by GL, and the two normalized curves differ only slightly from each other (e.g., see Figure 4 of [30]). Thus, the GL theory can be applied over the entire temperature range down to $T \ll T_c$. The previous equations relating j_d to B_c and λ are expected to hold in the case of multiple bands and other gap symmetries, as long as one uses the actual empirical temperature dependencies of B_c and λ, which account for modifications in these unconventional cases. This was experimentally demonstrated in the case of MgB_2 [26], which was recognized as a multi-gap superconductor tuned by strain and doping in the early part of this century; in fact, MgB_2 showed superconductivity near a Lifshitz transition as in iron-based superconductors [34–36].

3. Pulsed Measurement Technique

Depairing current densities in superconductors is extremely high: on the order of $j_d(T = 0) = 10^{11}$–10^{13} A/m^2. If the cross-section of the sample is even as narrow as just 1 mm^2, the current required would reach a value of $I = jA \sim 10^6$ A. Such a magnitude of current would be exceedingly difficult to produce and control. There are three steps to overcoming this dilemma: (1) Fabricate samples with very narrow cross-sectional areas. This can be achieved by growing nanowires and nanorods or by depositing very thin films and using lithography to pattern narrow bridges (alternatively, the films can be deposited onto nanowires or carbon nanotubes). (2) The next step is by pulsing the current at very low duty cycles so that large values of I can be handled while reducing the time-averaged current and time-averaged power dissipation to manageable levels. (3) The last step is limiting the measurement of j_d to the regime close to T_c. From Equation (21), it would seem that j_d can be made arbitrarily small by making T very close to T_c;

however, the $T - T_c$ distance needs to be large compared to the transition width for the measurement to be meaningful. Even for this near-T_c measurement of j_d, the current usually will have to be pulsed to avoid significant sample heating. Furthermore, the near-T_c measurement will only measure ρ_s in that region, and its zero-T value will have to be extrapolated using theory. While this is better than nothing, it will not shed light on any abnormal temperature dependence of ρ_s over the entire range, which could be of special interest if the superconductor has some exotic behavior.

Thus, the experimental ingredients needed to conduct a j_d measurement are: a superconducting sample with a very narrow cross-section; a means to control the temperature, i.e., a cryostat; and a method for sourcing pulsed signals (current or voltage) and detecting the consequent complementary signal (voltage or current). There are numerous methods for sample fabrication, which vary widely with the different superconducting materials. Some deposition systems for preparing superconducting films can be bought off the shelf. Cryostats also represent standard equipment that can be bought off the shelf. The principal distinguishing the experimental capabilities of our work center on the pulsed electrical measurements. Therefore, the rest of the experimental section will be devoted to describing this unique measurement setup.

Figure 2 shows the overall configuration and functional schematic. The pulsed current/voltage source puts out a time-varying current and voltage. This signal flows through a standard impedance, usually a resistor R_{std} (although an inductor is preferable in some situations) and the superconducting sample of resistance R that are in series. The initial signal can be taken directly from the output of a standard pulse generator (one of the models used was a Wavetek Model 801). These signal generators will typically have an output impedance of $Z_{out} = 50\ \Omega$. If a lower Z_{out} is desirable (to allow for constant voltage control), the signal generator's output can be passed through any standard buffer amplifier (e.g., a transistor-emitter-follower-based circuit, a power-operational-amplifier-based circuit, or an off-the-shelf audio amplifier). If a higher voltage than the signal generator's output is desirable, its output can be passed through any standard voltage amplifier (fast high quality audio amplifiers can serve this purpose as well). Combining a higher voltage signal with a large series resistor (which can be the R_{std} itself or an additional series resistor) can provide a relatively constant current. In general, the measurement will be in current-controlled or voltage-controlled mode depending on whether the combination of the final Z_{out} (after the amplifier if any) plus R_{std} is greater than or less than R. If the current needs to be held constant to a high accuracy (for example, if a series of R vs. T resistive transition curves needs to be traced out at various constant currents, as will be seen later), then it is better to follow the pulse generator with a transconductance amplifier, which converts the generator's voltage pulse into a constant current pulse. The transconductance amplifier is able to hold the current constant by electronic circuitry instead of needing an enormous series resistance. While conducting a pulsed current-voltage (IV) curve, which is usually done manually, it is preferable to have the voltage-controlled mode. The reason for this is that as the current and voltage are pushed higher, the sample's resistance will increase, and at some point, the sample will be driven to normal as j_d is exceeded. In constant-current mode, the power dissipation $P = I^2 R$ rises as R rises, causing an increase in heating and a further rise in R. This can lead to a run-away condition, which can destroy the sample. On the other hand, the constant-voltage mode is self-stabilizing since in this case, $P = V^2/R$ decreases as R rises, thus reducing heating and controlling the situation.

Once the current pulse flows through the sample and R_{std}, the corresponding time-varying voltages, $V(t)$ and $V_{std}(t)$, will be developed across them respectively. These must be observed and quantified using an oscilloscope. A digital storage oscilloscope (DSO) allows multiple pulses to be averaged. Since the signal is exactly repetitive, because the DSO is triggered off of the pulse generator's sync signal, the averaging effectively suppresses random uncorrelated noise. As long as the sample condition (T, B, etc.) is stable, a very high number of averages can be taken to improve the signal-to-noise ratio (SNR) vastly. Coaxial cables with 50-Ohm characteristic impedance are used between all connection points, including the wiring within the cryostat. Where possible, the originating and/or terminating points at the ends of the cables need to have matching 50-Ohm values to avoid reflections. Multiple ground connections to the circuit must be avoided to prevent ground loops. This means the two channels of the DSO cannot be simultaneously connected to both the sample and R_{std}; either a differential instrumentation preamplifier (Princeton Applied Research and Stanford Research Systems are two brands that make instrumentation amplifiers) must be used between the DSO channels and the sample and R_{std}, or only one of the two must be measured at a time.

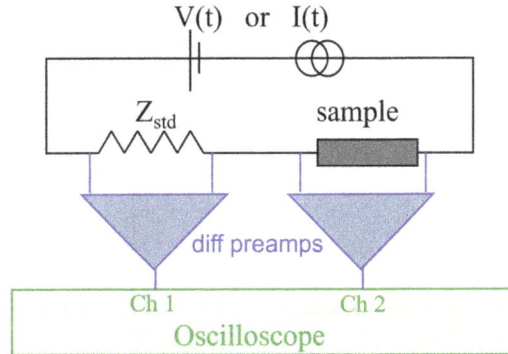

Figure 2. The overall configuration and functional schematic of the pulsed-signal measurement system. The differential preamplifiers (diff preamps) convert the time-varying potential differences across standard impedance (Z_{std}) and the sample, $V_{std}(t)$ and $V(t)$ respectively, into ground referenced single-ended signals that can be fed to the inputs of a digital storage oscilloscope.

Figure 3 shows the pair of time-varying current $I(t) = V_{std}(t)/R_{std}$ and sample-voltage $V(t)$ signals that results. The topmost trace is the scaled calculated resistance $50R(t) = 50V(t)/I(t)$. Note that the pulses reach constant plateaus after their initial transients. R, V, and I are defined by taking the plateau values of the individual quantities. The thermal rise in a sample because of Joule heating involves several processes: thermal diffusion occurs within the sample essentially instantaneously; on the time scale of nanoseconds, phonons transfer heat across the interface between the film and substrate; heat then diffuses within the substrate in a matter of microseconds and finally into the heat sink in milliseconds. For those processes that have time scales comparable to or longer than the pulse duration, there will be a visible rise in $V(t)$, causing the pulse to be distorted. Thus, as long as the $V(t)$ pulse is flat, slow causes of heating that influence the $V(t)$ shape can be assumed to be negligible. The work in [37] discusses a method to evaluate a sample's thermal resistance for pulsed signals quantitatively.

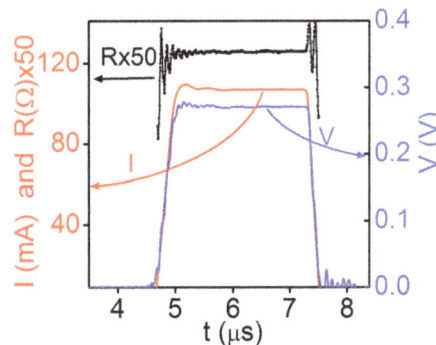

Figure 3. The measured oscilloscope traces of the sample voltage $V(t)$, current $I(t)=V_{std}(t)/R_{std}$, and scaled calculated resistance $50R(t) = 50V(t)/I(t)$ for an MgB$_2$ bridge at 42 K (normal state just above T_c). The resistance rises from 10–90% of its final value in about 50 ns (adapted from Reference [26]).

Figure 4 shows an example [26] of a set of IV curves at various fixed temperatures (in zero magnetic field), where each data point represents a pulsed measurement (plateau values) as described above. As the temperature is increased, j_d is reduced, and hence, the "jump" occurs at a lower value of I. Notice that the resistance (the V/I slope) jumps from zero (dissipationless superconducting state) to a constant finite value (normal-state) as the current crosses its depairing value. This is one direct way of obtaining ρ_n below T_c. In this particular material, high impurity scattering dominates over electron-phonon scattering at all temperatures, leading to a relatively flat $\rho(T)$. A more interesting application of this technique for elucidating a variable $\rho(T)$ is described in a later section.

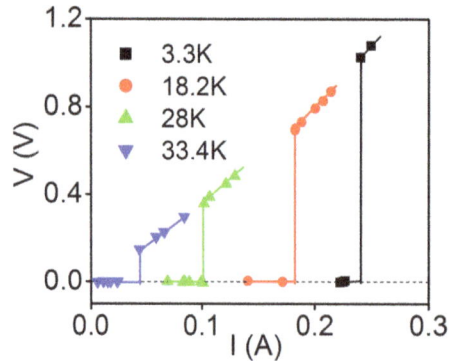

Figure 4. Current-voltage curves in a MgB_2 bridge at various fixed temperatures. As the temperature is increased, j_d is reduced, and hence, the "jump" occurs at a lower value of I. The sloped portions above each jump represent the normal-state resistance $R_n = \Delta V / \Delta I$ (adapted from from Reference [26]).

Figure 5a represents a set of pulsed constant-current $R(T)$ curves in zero magnetic field [38]. As the current is increased, the transition is progressively pushed down in temperature. Figure 5b plots these midpoint transition temperatures (T_{c2}) versus I, and they are seen to follow a $I^{2/3}$ law as per Equation (21). From this measured slope and Equation (21), one can estimate $j_d(0)$ without requiring the application of this enormous value of current. This is especially useful for systems (e.g., cuprate high-temperature superconductors) that have a very high $j_d(0)$ value.

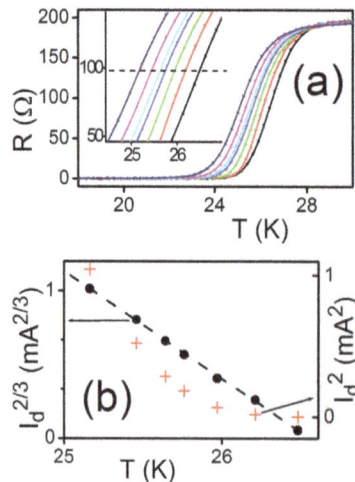

Figure 5. (a) Resistive transitions in a $Sr_{1-x}La_xCuO_2$ superconducting film bridge in zero magnetic field at various transport current values (right to left): 12.9, 132, 258, 426, 533, 721, and 1020 μA. The lowest current is continuous DC; the remaining currents are pulsed. The inset shows a magnified view of the midpoint region. (b) Left axis (circle symbols): two-thirds power of the depairing current versus the midpoint transition temperature corresponding to the depairing law. Right axis (plus symbols): square of the depairing current versus the midpoint transition temperature corresponding to Joule heating (adapted from Reference [38]).

Figure 6 shows the companion very low DC-current $R(T)$ curves at various constant B values. Here, the shift occurs because of the $B_{c2}(T)$ boundary; the current level is small enough for its depairing to be negligible. Unlike $j_d(T)$, $B_{c2}(T)$ has a linear dependence near T_c, and that slope can be related to $B_{c2}(0)$ through the WHH (Werthamer, Helfand, and Hohenberg) theory [39] and its variations [40] by relationships such as $B_{c2}(0) \simeq 0.7 dB_{c2}/dT$.

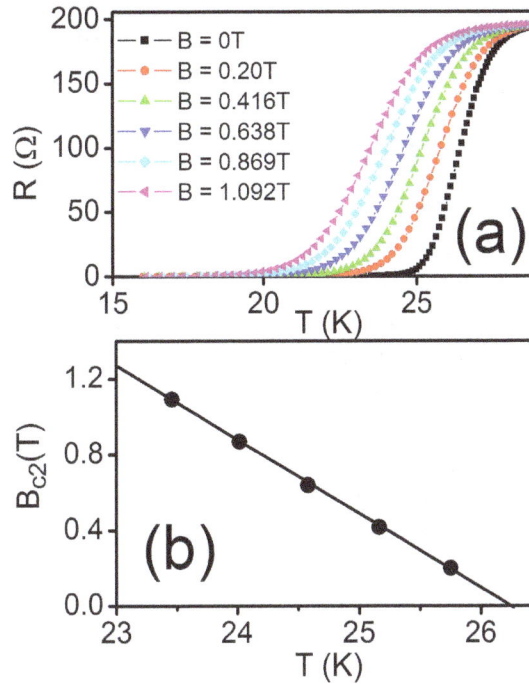

Figure 6. (**a**) Resistive transitions of a $Sr_{1-x}La_xCuO_2$ superconducting film bridge at a constant current of I = 13 μA in various perpendicular magnetic field values as indicated in the key. (**b**) Upper critical magnetic field versus the midpoint transition temperature, extracted from the curves in (a) (adapted from Reference [38]).

The measurements represented by Figures 5 and 6 together with Equation (23) are the key to obtaining ρ_s through relatively straightforward transport measurements. We now look at one recent example of a current-induced depairing study of an exotic superconducting system.

4. Investigations in a Topological Insulator/Chalcogenide Interfacial Superconductor

4.1. Background

The interface between the Bi_2Te_3 topological insulator and the FeTe chalcogenide provides a fascinating 2D superconducting system, in which neither Bi_2Te_3, nor FeTe are superconducting by themselves [41]. While the exact origin of the superconductivity is not known, it has been suggested that the robust topological surfaces states may be doping the FeTe and suppressing the antiferromagnetism in a thin region close to the interface, thus inducing the observed 2D superconductivity. These surface states represent a conducting system with very high normal conductivity because of protection against time-reversal-invariant scattering mechanisms. Therefore, it is of great interest to understand the nature and origin of the charge carriers that underlie this interfacial superconductivity, and in particular, to see if the topologically-protected surface states might be a source of the normal carriers. The relevance of this question is broader than the specific system studied here, since it has been recently proposed that interfacial superconductivity may even play a role in cuprates: for example, in the interface located between charge density wave nanoscale puddles [42] and between oxygen-rich grains where the interface is made of a filamentary network with hyperbolic geometry [43,44]. We describe below how the current-induced depairing approach was used to answer these questions to elucidate the nature of the normal state in the $Bi_2Te_3/FeTe$ system.

4.2. Samples and Experimental Information

The $Bi_2Te_3/FeTe$ samples consisted of a ZnSe buffer layer (50 nm) deposited on a GaAs (001) semi-insulating substrate, followed by a deposition of 220 nm thick FeTe, which was then capped with a 20 nm-thick Bi_2Te_3 layer. Upper-critical-field measurements [41] and vortex-explosion measurements [45] showed that the superconductivity occurred within an interfacial layer of thickness $d = 7$ nm, which was much thinner than both the FeTe and Bi_2Te_3 layers. Projection photolithography followed by argon-ion milling was used to pattern narrow microbridges optimized for the high current-density pulsed four-probe measurements. Two bridges were studied: Sample A with lateral dimensions of width $w = 11.5$ μm and length $l = 285$ μm and Sample B with $w = 12$ μm and $l = 285$ μm. The onset T_c (defined as the intersection of the extrapolation of the normal-state portion and the extrapolation of the steep transition portion of the $R(T)$ curve) for both bridges, was 11.7 K. Details about sample preparation are provided in [41]. All measurements were made in zero applied magnetic field. While the very low reference curves at $I \leq 60$ μA were measured using continuous DC signals, the main electrical transport measurements were made with pulsed signals. Contact resistances (<1 Ω) were much lower than the normal resistance R_n of the bridge, and heat generated at contacts did not reach the bridge within the time duration t of each pulse, since the thermal diffusion distance ($\sqrt{Dt} \sim 10$ μm) was much shorter than the contact-to-bridge distance (>1 mm); here, D is the diffusion constant.

4.3. Results and Discussion

The normal-state resistivity $\rho_n(T)$ and depairing current density $j_d(T)$ in the $Bi_2Te_3/FeTe$ samples were extracted over the entire temperature range [46], by driving the system normal with high pulsed currents using the methods described earlier and illustrated in Figures 4 and 5. Figure 7 shows the raw depairing current results. The dashed horizontal lines in Panels (a) and (b) provide the values $I_d(T \rightarrow 0) \geq 0.131$ A and 0.136 A for Samples A and B, respectively.

In order to obtain more accurate intrinsic j_d and ρ_n of the 7 nm-thick superconducting interfacial layer itself, we needed to subtract the small parallel current through the normally conductive underlying FeTe layer. For this purpose, a separate measurement of pure FeTe deposited on ZnSe/GaAs, without the Bi_2Te_3 top layer, was conducted [46]. With this subtraction, the previous raw $I_d(T \rightarrow 0)$ values gave a corrected $j_d(T = 0)$ of 1.5×10^8 A/cm^2 for both samples (which is a typical value: j_d ranges 10^7–10^9 A/cm^2 for most superconductors), and the correction gave the intrinsic $\rho_n(T)$ for the two samples, as shown in Figure 8. This absolute value of $\rho_n(T \rightarrow 0) \sim 200$ nΩ cm represents an extraordinarily conductive normal state for a superconducting system, as most superconductors are poor conductors in the normal state. This information will be analyzed below within the framework of an anisotropic Ginzburg–Landau (GL) approach [46], to obtain information on the superfluid density, carrier concentration, and scattering rate, as well as their implications for the nature of the normal-state.

From the previously-published measurements of He et al. [41], we have the following orientation-dependent values of B_{c2}: perpendicular-to-interface $B_{c2}^{\perp}(0) \approx 17$ T and parallel-to-interface $B_{c2}^{\parallel}(0) \approx 40$ T. The corresponding coherence lengths from Equation (3) are: in-plane $\xi_{\parallel}(0) \approx 4.4$ nm and perpendicular $\xi_{\perp}(0) \approx 1.9$ nm. Using Equation (23) together with this $B_{c2}^{\perp}(0)$ and our measured in-plane $j_d^{\parallel}(0)$ gave $\lambda_{\parallel}(0) = 124$ nm and a corresponding $\rho_s(0) = 1/\lambda_{\parallel}^2(0)$. From $\rho_s(0) = \mu_0 n_s(0)e^2/m^* \approx \mu_0 ne^2/m$ applicable in the clean limit at $T = 0$, we get $n \approx 1.8 \times 10^{21}$ per cm^3, approximating $m^* \approx m$. This effective single-band value of n evaluated above is similar to n characteristic of high temperature superconductors and about two orders of magnitude lower than n in highly-conductive metals such as copper.

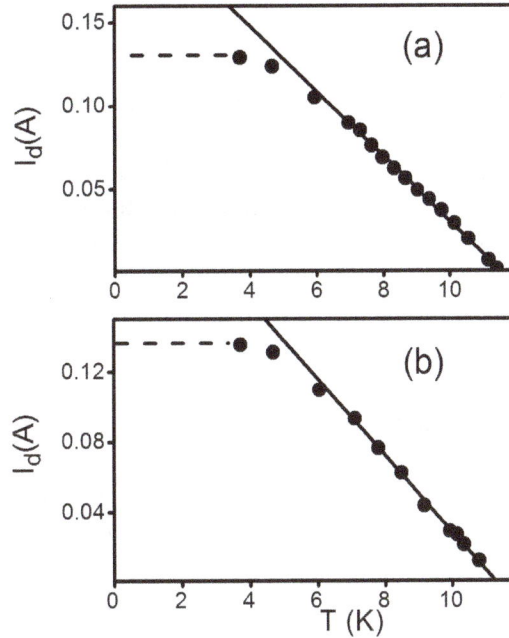

Figure 7. Raw depairing current versus temperature for $Bi_2Te_3/FeTe$ bridges: **(a)** Sample A and **(b)** Sample B. The dashed horizontal lines provide the values $I_d(T \to 0) \geq 0.131$ A and 0.136 A for Samples A and B, respectively (adapted from Reference [46]).

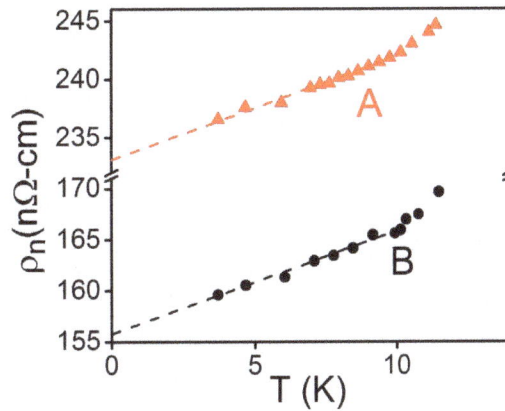

Figure 8. Intrinsic normal-state resistivity of $Bi_2Te_3/FeTe$ interfacial superconducting bridges, Samples A and B (adapted from Reference [46]).

The low value of n together with the very high normal conductivity implies a rather long scattering time τ and mean-free-path l. The Fermi wave number for this n computes to $k_F(3D) = m^*v_F/\hbar = (3\pi^2n)^{1/3} = 3.8 \times 10^9$ m^{-1} and $k_F(2D) = (2\pi nd)^{1/2} = 9.0 \times 10^9$ m^{-1} in three and two dimensions, respectively. In both cases, the Fermi wavelength $\lambda_F = 2\pi/k_F \ll d$, validating the continuum approximation for states along the perpendicular direction and justifying the anisotropic 3D treatment of the normal state. Then, from the Drude relationship $\rho \approx m/ne^2\tau$, we get $\tau \approx 10$ ps, which agrees well with the scattering rates ($\sim \hbar/0.05$ meV = 13 ps) measured by Pan et al. [47] using spin- and angle-resolved photoemission spectroscopy. Combining this value of τ with the Fermi velocity $v_F = \hbar k_F/m \approx 440$ km/s, we get $l = v_F\tau = 4.2$ μm. The very long l, which well exceeds the superconducting layer thickness d, indicates that scattering from the faces that bound the superconducting layer was of a specular nature. This surprising dramatically low scattering indeed supports the possible role of the topological surface states in the formation of the normal state that underlies this exotic interfacial superconducting system.

5. Concluding Remarks

Fast pulsed signals of short duration and low duty cycle make it possible to study transport behavior in superconductors at extreme current densities, power densities, and electric fields. In this article, we focused on the use of this technique in the measurement of one of the fundamental critical parameters of the superconducting state, the depairing current j_d. It was shown how through j_d, one can obtain information on various other key parameters of the superconducting state, in particular the penetration depth and consequent superfluid density, which cast light on the normal state. As an example and illustration of this procedure, we described a recent study of the superconducting system formed at the interface between a topological insulator and a chalcogenide. We hope that the information provided here will encourage other groups to utilize this approach.

Funding: This work was supported by the U.S. Department of Energy through Grant Number DE-FG02-99ER45763.

Acknowledgments: The following are acknowledged for useful discussions and other assistance: Charles L. Dean, Manlai Liang, Gabriel F. Saracila, James M. Knight, Luc Fruchter, Ziang Z. Li, Qing Lin He, Hongchao Liu, Jiannong Wang, Rolf Lortz, Iam Keong Sou, Alex Gurevich, Richard A. Webb, Ken Stephenson, and David K. Christen.

Conflicts of Interest: The author declares no conflict of interest.

References

1. Onnes, K.H. Further experiments with liquid helium. C. On the change of electric resistance of pure metals at very low temperatures, etc. IV. The resistance of pure mercury at helium temperatures. In *Through Measurement to Knowledge*; Springer: Dordrecht, The Netherlands, 1911; pp. 261–263. [CrossRef]
2. Somayazulu, M.; Ahart, M.; Mishra, A.K.; Geballe, Z.M.; Baldini, M.: Meng, Y.; Struzhkin, V.V.; Hemley, R.J. Evidence for Superconductivity above 260 K in Lanthanum Superhydride at Megabar Pressures. *Phys. Rev. Lett.* **2019**, *122*, 027001–027004. [CrossRef] [PubMed]
3. Drozdov, A.P.; Kong, P.P.; Minkov, V.S.; Besedin, S.P.; Kuzovnikov, M.A.; Mozaffari, S.; Balicas, L.; Balakirev, F.; Graf, D.; Prakapenka, V.B.; et al. Superconductivity at 250 K in lanthanum hydride under high pressures. *Nature* **2019**, *569*, 528–531. [CrossRef] [PubMed]
4. Tinkham, M. *Introduction to Superconductivity*, 2nd ed.; Dover Publications: Mineola, NY, USA, 2004; ISBN-10: 0486435032, ISBN-13: 978-0486435039.
5. Emery, V.J.; Kivelson, S.A. Importance of phase fluctuations in superconductors with small superfluid density. *Nature* **1995**, *374*, 434–437. [CrossRef]
6. Uemura, Y.J.; Luke, G.M.; Sternlieb, B.J.; Brewer, J.H.; Carolan, J.F.; Hardy, W.N.; Kadono, R.; Kempton, J.R.; Kiefl, R.F.; Kreitzman, S.R.; et al. Universal Correlations between T_c and n_s/m^* (Carrier Density over Effective Mass) in High-T_c Cuprate Superconductors. *Phys. Rev. Lett.* **1989**, *62*, 2317–2320. [CrossRef] [PubMed]
7. Uemura, Y.J.; Le, L.P.; Luke, G.M.; Sternlieb, B.J.; Wu, W.D.; Brewer, J.H.; Riseman, T.M.; Seaman, C.L.; Maple, M.B.; Ishikawa, M.; et al. Basic similarities among cuprate, bismuthate, organic, Chevrel-phase, and heavy-fermion superconductors shown by penetration-depth measurements. *Phys. Rev. Lett.* **1991**, *66*, 2665–2668. [CrossRef] [PubMed]
8. Hardy, W.N.; Bonn, D.A.; Morgan, D.C.; Liang, R.; Zhang, K. Precision measurements of the temperature dependence of λ in $Y_1Ba_2Cu_3O_7$: Strong evidence for nodes in the gap function. *Phys. Rev. Lett.* **1993**, *70*, 3999–4002. [CrossRef] [PubMed]
9. Felcher, G.P.; Kampwirth, R.T.; Gray, K.E.; Felici, R. Polarized-Neutron Reflections: A New Technique Used to Measure the Magnetic Field Penetration Depth in Superconducting Niobium. *Phys. Rev. Lett.* **1984**, *52*, 1539–1542. [CrossRef]
10. Claassen, J.H.; Evetts, J.E.; Somekh, R.E.; Barber, Z.H. Observation of the superconducting proximity effect from kinetic-inductance measurements. *Phys. Rev. B* **1991**, *44*, 9605–9608. [CrossRef]
11. Yong, J.; Lee, S.; Jiang, J.; Bark, C.W.; Weiss, J.D.; Hellstrom, E.E.; Larbalestier, D.C.; Eom, C.B.; Lemberger, T.R. Superfluid density measurements of $Ba(Co_xFe_{1-x})_2As_2$ films near optimal doping. *Phys. Rev. B* **2011**, *83*, 104510–104514. [CrossRef]
12. Boghosian, C.; Meyer, H.; Rives, J.E. Density, Coefficient of Thermal Expansion, and Entropy of Compression of Liquid Helium-3 under Pressure below 1.2 K. *Phys. Rev.* **1966**, *146*, 110–119. [CrossRef]

13. Van Degrift, C.T. Tunnel diode oscillator for 0.001 ppm measurements at low temperatures. *Rev. Sci. Instr.* **1975**, *46*, 599. [CrossRef]

14. Sonier, J.E. Muon spin rotation studies of electronic excitations and magnetism in the vortex cores of superconductors. *Rep. Prog. Phys.* **2007**, *70*, 1717–1756. [CrossRef]

15. Luan, L.; Auslaender, O.M.; Lippman, T.M.; Hicks, C.W.; Kalisky, B.; Chu, J.-H.; Analytis, J.G.; Fisher, I.R.; Kirtley, J.R.; Moler, K.A. Local measurement of the penetration depth in the pnictide superconductor $Ba(Fe_{0.95}Co_{0.05})_2As_2$. *Phys. Rev. B* **2010**, *81*, 100501–100504. [CrossRef]

16. Shibauchi, T.; Kitano, H.; Uchinokura, K.; Maeda, A.; Kimura, T.; Kishio, K. Anisotropic penetration depth in $La_{2-x}Sr_xCuO_4$. *Phys. Rev. Lett.* **1994** *72*, 2263–2266. [CrossRef]

17. Lee, J.Y.Y.; Lemberger, T.R. Penetration depth $\lambda(T)$ of $Y_1Ba_2Cu_3O_7$ films determined from the kinetic inductance. *Appl. Phys. Lett.* **1993**, *62*, 2419–2421. [CrossRef]

18. Saracila, G.F.; Kunchur, M.N. Ballistic acceleration of a supercurrent in a superconductor. *Phys. Rev. Lett.* **2009**, *102*, 077001–077004. [CrossRef] [PubMed]

19. Diener, P.; Leduc, H.G.; Yates, S.J. Design and Testing of Kinetic Inductance Detectors Made of Titanium Nitride. *J. Low Temp. Phys.* **2012**, *167*, 305–310. [CrossRef]

20. Jochem, B. Kinetic Inductance Detectors. *J. Low Temp. Phys.* **2012**, *167*, 292–304. [CrossRef]

21. Kunchur, M.N.; Dean, C.; Liang, M.; Moghaddam, N.S.; Guarino, A.; Nigro, A.; Grimaldi, G.; Leo, A. Depairing current density of $Nd_{2-x}Ce_xCuO_4$ superconducting films. *Physica C* **2013** *495*, 66–68. [CrossRef]

22. Kunchur, M.N.; Ivlev, B.I.; Christen, D.K.; Phillips, J.M. Metallic Normal State of $Y_1Ba_2Cu_3O_7$. *Phys. Rev. Lett.* **2000**, *84*, 5204–5207. [CrossRef] [PubMed]

23. Larkin, A.I.; Ovchinnikov, Y.U.N. Nonlinear conductivity of superconductors in the mixed state. *Sov. Phys.—JETP* **1976**, *41*, 960–965.

24. Blatter, G.; Feigel'man, M.V.; Mikhail, V.B.; Larkin, A.I.; Valerii, V.M. Vortices in high-temperature superconductors. *Rev. Mod. Phys.* **1994**, *66*, 1125–1388. [CrossRef]

25. Kunchur, M.N. Unstable flux flow due to heated electrons in superconducting films. *Phys. Rev. Lett.* **2002**, *89*, 137005–137008. [CrossRef] [PubMed]

26. Kunchur, M.N. Current-induced pair breaking in Magnesium Diboride. *J. Phys. Condens. Matter.* **2004**, *16*, R1183–R1204. [CrossRef]

27. London, F.; London, H. The electromagnetic equations of the supraconductor. *Proc. Roy. Soc.* **1935**, *A149*, 71–88. [CrossRef]

28. Ginzburg, V.L.; Landau, L.D. On the Theory of superconductivity. *Zh. Eksperim. Teor. Fiz.* **1950**, *20*, 1064–1082.

29. Bardeen, J. Critical Fields and Currents in Superconductors. *Rev. Mod. Phys.* **1962**, *34*, 667–681. [CrossRef]

30. Romijn, J.; Klapwijk, T.M.; Renne, M.J.; Mooij, J.E. Critical pair-breaking current in superconducting aluminum strips far below T_c. *Phys. Rev. B* **1982**, *26*, 3648–3655. [CrossRef]

31. Maki, K. On Persistent Currents in a Superconducting Alloy. II. *Progr. Theor. Phys.* **1963**, *29*, 333–340. [CrossRef]

32. Ovchinnikov, Y.U.N. Critical current of thin films for diffuse reflection from the walls. *Sov. Phys.—JETP* **1969**, *29*, 853–860.

33. Kupriyanov, M.Y.; Lukichev, V.F. Temperature dependence of pair-breaking current in superconductors. *Sov. J. Low Temp. Phys.* **1980**, *6*, 210.

34. Bauer, E.; Paul, C.; Berger, S.; Majumdar, S.; Michor, H.; Giovannini, M.; Saccone, A.; Bianconi, A. Thermal conductivity of superconducting MgB_2. *J. Phys. Condens. Matter* **2001**, *13*, L487–L493. [CrossRef]

35. Agrestini, S.; Metallo, C.; Filippi, M.; Simonelli, L.; Campi, G.; Sanipoli, C.; Liarokapis, E.; De Negri, S.; Giovannini, M.; Saccone, A.; et al. Substitution of Sc for Mg in MgB_2: Effects on transition temperature and Kohn anomaly. *Phys. Rev. B* **2004**, *70*, 134514–134518. [CrossRef]

36. Kagan, M.Y.; Bianconi, A. Fermi-Bose Mixtures and BCS-BEC Crossover in High-T_c Superconductors. *Condens. Matter* **2019**, *4*, 51. [CrossRef]

37. Kunchur, M.N.; Christen, D.K.; Klabunde, C.E.; Phillips, J.M. Pair-breaking effect of high current densities on the superconducting transition in $Y_1Ba_2Cu_3O_7$. *Phys. Rev. Lett.* **1994**, *72*, 752–75.

38. Liang, M.; Kunchur, M.N.; Fruchter, L.; Li, Z.Z. Depairing current density of infinite-layer $Sr_{1-x}La_xCuO_2$ superconducting films. *Physica C* **2013**, *492*, 178–180. [CrossRef]

39. Werthamer, N.R.; Helfand, E.; Hohenberg, P.C. Temperature and Purity Dependence of the Superconducting Critical Field, H_{c2}. III. Electron Spin and Spin-Orbit Effects. *Phys. Rev.* **1966**, *147*, 295–302. [CrossRef]

40. Gurevich, A. Enhancement of the upper critical field by nonmagnetic impurities in dirty two-gap superconductors. *Phys. Rev. B* **2003**, *67*, 184515–184527. [CrossRef]

41. He, Q.L.; Liu, H.; He, M.; Lai, Y. H.; He, H.; Wang, G.; Law, K.T.; Lortz, R.; Wang, J.; Sou, I.K. Two-dimensional superconductivity at the interface of a Bi_2Te_3/FeTe heterostructure. *Nat. Commun.* **2014**, *5*, 4247–4254. [CrossRef]

42. Campi, G.; Bianconi A.; Poccia, N.; Bianconi, G.; Barba, L.; Arrighetti, G.; Innocenti, D.; Karpinski, J.; Zhigadlo, N. D.; Kazakov, S. M.; et al. Inhomogeneity of charge-density-wave order and quenched disorder in a high-T_c superconductor. *Nature* **2015**, *525*, 359–362. [CrossRef]

43. Jarlborg, T.; Bianconi A. Multiple Electronic Components and Lifshitz Transitions by Oxygen Wires Formation in Layered Cuprates and Nickelates. *Condens. Matter* **2019**, *4*, 15. [CrossRef] [PubMed]

44. Campi, G.; Bianconi A. High Temperature superconductivity in a hyperbolic geometry of complex matter from nanoscale to mesoscopic scale. *J. Supercond. Nov. Magn.* **2016**, *29*, 627-631. [CrossRef]

45. Dean, C.L.; Kunchur, M.N.; He, Q.L.; Liu, H.; Wang, J.; Lortz, R.; Sou, I.K. Current driven vortex-antivortex pair breaking and vortex explosion in the Bi_2Te_3/FeTe interfacial superconductor. *Physica C* **2016**, *527*, 46–49. [CrossRef]

46. Dean, C.L.; Kunchur, M.N.; He, Q.L.; Liu, H.; Wang, J.; Lortz, R.; Sou, I.K. Current-induced depairing in the Bi_2Te_3/FeTe interfacial superconductor. *Phys. Rev. B* **2015**, *92*, 094502–094506. [CrossRef]

47. Pan, Z.-H.; Fedorov, A.V.; Gerdner, D.; Lee, Y.S.; Chu, S.; Valla, T. Measurement of an Exceptionally Weak Electron-Phonon Coupling on the Surface of the Topological Insulator Bi_2Se_3 Using Angle-Resolved Photoemission Spectroscopy. *Phys. Rev. Lett.* **2012**, *108*, 187001–187005. [CrossRef]

5

Mechanism of High-Temperature Superconductivity in Correlated-Electron Systems

Takashi Yanagisawa

National Institute of Advanced Industrial Science and Technology 1-1-1 Umezono, Tsukuba, Ibaraki 305-8568, Japan; t-yanagisawa@aist.go.jp

Abstract: It is very important to elucidate the mechanism of superconductivity for achieving room temperature superconductivity. In the first half of this paper, we give a brief review on mechanisms of superconductivity in many-electron systems. We believe that high-temperature superconductivity may occur in a system with interaction of large-energy scale. Empirically, this is true for superconductors that have been found so far. In the second half of this paper, we discuss cuprate high-temperature superconductors. We argue that superconductivity of high temperature cuprates is induced by the strong on-site Coulomb interaction, that is, the origin of high-temperature superconductivity is the strong electron correlation. We show the results on the ground state of electronic models for high temperature cuprates on the basis of the optimization variational Monte Carlo method. A high-temperature superconducting phase will exist in the strongly correlated region.

Keywords: strongly correlated electron systems; mechanisms of superconductivity; high-temperature superconductivity; kinetic driven superconductivity; optimization variational Monte Carlo method; Hubbard model; three-band d-p model

---

1. Introduction

It is a challenging research subject to clarify the mechanism of high temperature superconductivity, and indeed it has been studied intensively for more than 30 years [1–3]. For this purpose, it is important to clarify the ground state and phase diagram of electronic models with strong correlation because high temperature cuprates are strongly correlated electron systems.

Most superconductors induced by the electron–phonon interaction have s-wave pairing symmetry. We can understand physical properties of s-wave superconductivity based on the Bardeen–Cooper–Schrieffer (BCS) theory [4–6]. The critical temperature T_c of most of electron–phonon superconductors is very low except for exceptional compounds. Many unconventional superconductors that cannot be understood by the BCS theory have been discovered. They are, for example, heavy fermion superconductors, organic superconductors and cuprate superconductors for which the pairing mechanism is different from the electron–phonon interaction. In particular, cuprate superconductors exhibit relatively high T_c and have become of great interest. A common feature in both electron–phonon systems and correlated electron systems is that critical temperature may have a strong correlation with the energy scale of the interaction that induces electron pairing.

This paper has two parts. In the first part, we give a review on mechanisms of superconductivity in the electron–phonon system and in the correlated electron system. In the second part, we mainly discuss the mechanism of high-temperature cuprates.

The model for CuO_2 plane in cuprate superconductors is called the d-p model or the three-band Hubbard model [7–24]. It is certainly a very difficult task to elucidate the phase diagram of the d-p model. Simplified models are also used to investigate the mechanism of superconductivity, for example the two-dimensional (2D) single-band Hubbard model [25–50] and ladder model [51–56].

The Hubbard model was introduced to understand the metal–insulator transition [57] and was employed to understand various magnetic phenomena [58,59]. On the basis of the Hubbard model, it is possible to understand the appearance of inhomogeneous states reported for cuprates, such as stripes [60–67] and checkerboard-like density wave states [68–71]. It was also expected that the Hubbard model can account for high temperature superconductivity [72].

A variational Monte Carlo method is used to examine the ground state properties of strongly correlated electron systems, where we calculate the expectation values exactly using a numerical method [28–31,36–41]. We introduced the wave function of $\exp(-S)$-type in the study of superconductivity in the Hubbard model [73–75]. This wave function is very excellent in the sense that the energy expectation value is lower than that of any other wave functions [50].

The paper is organized as follows. In Section 2.1, we discuss the phonon mechanism of superconductivity. In Section 2.2, we discuss the electron mechanism of superconductivity. Section 2.3 is devoted to a discussion on superconductivity in correlated electron systems. In Section 3.1, we show the model for high temperature cuprates. We present the optimization variational Monte Carlo method (OVMC) in Section 3.2. We show the results on superconductivity based on the OVMC in Section 3.3. We discuss the stability of antiferromagnetic state in Section 3.4. We show the phase diagram when the hole doping rate is changed in Section 3.5. We give a summary in Section 4.

2. Part I. Superconductivity in Many-Electron Systems

2.1. Possibility of High-T_c Superconductivity

In the BCS theory, the electron–phonon interaction is assumed to induce attractive interaction between electrons and the pairing symmetry is s-wave [4–6]. There are many superconductors with s-wave pairing symmetry and most of them are due to the electron–phonon interaction. The BCS theory was successful to explain physical properties of these superconductors.

In the strong-coupling theory based on the Green function formulation [76,77], the critical temperature T_c was estimated as [78],

$$T_c = \frac{\theta_D}{1.45} \exp\left(-\frac{1.04(1+\lambda)}{\lambda - \mu^*(1+0.62\lambda)}\right), \tag{1}$$

where λ is the electron–phonon coupling constant, $\theta_D = \hbar\omega_D/k_B$ is the Debye temperature and μ^* is the renormalized Coulomb parameter defined by

$$\mu^* = \frac{\mu}{1 + \mu \ln(\epsilon_F/\omega_D)}, \tag{2}$$

for $\mu = U/\epsilon_F$ where U is the strength of the Coulomb interaction and ϵ_F is the Fermi energy. μ^* is the phenomenological parameter being approximately 0.1.

The electron–phonon coupling constant λ is expressed as

$$\lambda = 2\int_0^\infty d\omega \frac{\alpha(\omega)^2 F(\omega)}{\omega}, \tag{3}$$

where $\alpha(\omega)$ is the averaged electron–phonon coupling over the Fermi surface and $F(\omega)$ indicates the product of the spectral function of phonon and the density of states. This is approximately written as

$$\lambda \simeq \frac{\rho(\epsilon_F)\langle I\rangle^2}{M\omega_D^2}, \tag{4}$$

where $\rho(\epsilon_F)$ is the density of states at the Fermi surface and M is the mass of an atom. McMillan predicted that T_c would have a limit being of the order of 30 K from the analysis for this formula [78].

The McMillan formula was modified by replacing $\theta_D/1.45$ by logarithmic Debye frequency $\omega_{\ln}$ where [79]

$$\omega_{\ln} = \exp\left(\frac{2}{\lambda}\int_0^\infty d\omega\alpha(\omega)^2 F(\omega)\frac{\ln\omega}{\omega}\right). \tag{5}$$

It was predicted that high critical temperature would be possible for large λ since $T_c \propto \sqrt{\lambda}$ for $\lambda \gg 1$. If $\omega_{\ln}$ is large, λ is also large, and the crystal is stable, high T_c would be realized. It was predicted that high T_c would be realized in hydrogen solid with high Debye temperature [80]. In fact, high temperature superconductors with T_c above 200 K were discovered under extremely high pressure (160~200 GPa) in hydrogen compounds such as H_3S and LaH_{10} [81–83].

It is important to consider multi-band superconductors in the search for high temperature superconductors. In fact, MgB_2 and iron based superconductors are multi-band superconductors. An important role of Lifshitz transition in iron based superconductors and MgB_2 multi-band superconductors has been predicted [84]. An interesting point is that the possibility of high-T_c superconductivity in materials where tuning the chemical potential shows a quasi-1D Fermi surface topology as in organics and hydrides [85]. A layered superconductor such as cuprate superconductor can be regarded as a multiband superconductor due to interlayer couplings. A multi-band superconductivity has been investigated as a generalization of the BCS theory since early works on the two-band superconductivity [86–89]. There will appear many interesting properties in superconductors with multiple gaps such as time-reversal symmetry breaking [90–103], the existence of massless modes [104–109], unusual isotope effect [110–114] and fractional-flux quantum vortices [115–119]. When we have multiple order parameters, there appear multiple Nambu–Goldstone bosons and Higgs bosons [104,120–129]. This will result in significant excitation modes that are unique in multi-band superconductors.

It is important to include in a theoretical picture the presence of multiple electronic components with anomalous normal state properties in the charge and spin sector, e.g., the well known Fermi arcs and charge pseudogap phenomenology. The "shape resonance" scenario of multigap BCS-BEC crossover has been proposed [130,131]. The study of the electronic structure of the cuprates superconductors $Bi_2Sr_2CaCuO_{8+y}$ and La_2CuO_{4+y} doped by mobile oxygen interstitials using local probes has shown a scenario made of two electronic components: a strongly correlated Fermi liquid which coexists with stripes made of anisotropic polarons condensed into a generalized Wigner charge density wave [132–134].

2.2. Electron Correlation and Superconductivity

We discuss the electron correlation due to the Coulomb interaction between electrons. The on-site Coulomb interaction is important in the study of the metal insulator transition and magnetic properties of materials. The Hubbard model is written as [25]

$$H = \sum_{ij\sigma} t_{ij}c_{i\sigma}^\dagger c_{j\sigma} + U\sum_i n_{i\uparrow}n_{i\downarrow}, \tag{6}$$

where t_{ij} indicates the transfer integral and the second term denotes the Coulomb interaction with the strength U. t_{ij} are chosen as follows. $t_{ij} = -t$ when i and j are nearest-neighbor pairs $\langle ij\rangle$ and $t_{ij} = -t'$ when i and j are next-nearest neighbor pairs. In the following, N is the number of lattices, and N_e denotes the number of electrons.

When two electrons spin up and down at the same site, the energy becomes higher by U where U denotes the on-site Coulomb energy. In the case of half-filling, the Mott transition occurs when $U(> 0)$ is as large as the bandwidth and the ground state is an insulator. The effective Hamiltonian is derived in the limit of large U/t [135–137], based on the canonical transformation $H_{\text{eff}} = e^{iS} H e^{-iS}$. In the limit $U/t \to \infty$, the double occupancy is not allowed. The effective Hamiltonian is written as

$$H_{\text{eff}} = H + i[S, H] + \frac{i^2}{2}[S, [S, H]] + \cdots. \tag{7}$$

We write the Hamiltonian as $H = \tilde{H}_0 + H_1$ where

$$\tilde{H}_0 = \sum_{ij\sigma} t_{ij}(a_{i\sigma}^\dagger a_{j\sigma} + d_{i\sigma}^\dagger d_{j\sigma}) + U \sum_i n_{i\uparrow} n_{i\downarrow}, \tag{8}$$

$$H_1 = \sum_{ij\sigma} t_{ij}(a_{i\sigma}^\dagger d_{j\sigma} + d_{j\sigma}^\dagger a_{i\sigma}). \tag{9}$$

Here, we defined $a_{i\sigma} = c_{i\sigma}(1 - n_{i,-\sigma})$ and $d_{i\sigma} = c_{i\sigma} n_{i,-\sigma}$. $a_{i\sigma} = c_{i\sigma}(1 - n_{i,-\sigma})$ is the electron operator without double occupancy. We choose S to satisfy $i[S, \tilde{H}_0] + H_1 = 0$, so that H_{eff} reads in the subspace of no double occupancy,

$$H_{\text{eff}} = \sum_{ij\sigma} t_{ij} a_{i\sigma}^\dagger a_{j\sigma} + \frac{i}{2}[S, H_1] + \frac{i^2}{3}[S, [S, H_1]] + \cdots. \tag{10}$$

When we consider only the nearest-neighbor transfer $t_{ij} = -t$, the effective Hamiltonian reads

$$\begin{aligned}
H_{\text{eff}} &= -t \sum_{\langle ij \rangle \sigma} a_{i\sigma}^\dagger a_{j\sigma} - \frac{t^2}{U} \sum_{j\mu\mu'} [a_{j+\mu\uparrow}^\dagger a_{j\downarrow}^\dagger a_{j\downarrow} a_{j+\mu'\uparrow} \\
&\quad + a_{j\uparrow}^\dagger a_{j+\mu\downarrow}^\dagger a_{j+\mu'\downarrow} a_{j\uparrow} + a_{j+\mu\uparrow}^\dagger a_{j\downarrow}^\dagger a_{j+\mu'\downarrow} a_{j\uparrow} + a_{j\uparrow}^\dagger a_{j+\mu\downarrow}^\dagger a_{j\downarrow} a_{j+\mu\uparrow}],
\end{aligned} \tag{11}$$

where $j + \mu$ and $j + \mu'$ denote the nearest-neighbor sites in the μ and μ' directions, respectively. The second term being proportional to t^2/U contains the nearest-neighbor exchange interaction and also three-site terms when $\mu \neq \mu'$. The three-site terms are of the same order as the exchange interaction. When we neglect the three-site terms, the effective Hamiltonian reduces to the t-J model given by

$$H_{\text{eff}} = -\sum_{\langle ij \rangle \sigma} (a_{i\sigma}^\dagger a_{j\sigma} + \text{h.c.}) + J \sum_{\langle ij \rangle} \left(\mathbf{S}_i \cdot \mathbf{S}_j - \frac{1}{4} \tilde{n}_i \tilde{n}_j \right), \tag{12}$$

where $J = 4t^2/U$ and $\tilde{n}_i = \tilde{n}_{i\uparrow} + \tilde{n}_{i\downarrow}$ with $\tilde{n}_{i\sigma} = a_{i\sigma}^\dagger a_{i\sigma}$.

High-temperature cuprates and heavy fermion systems are typical correlated electron systems and many superconductors have been reported. Most of superconductors in these systems have nodes in the superconducting gap, namely, the Cooper pair is anisotropic. This indicates that superconductivity is unconventional and does not conform to the conventional BCS theory. The mechanism of superconductivity is certainly non-phonon mechanism. We show several characteristic properties of cuprate high-temperature superconductors:

1. The Cooper pair has d-wave symmetry.
2. The superconducting phase exists near the antiferromagnetic phase and parent materials are a Mott insulator.
3. The CuO_2 plane is commonly contained and the on-site Coulomb repulsive interaction works between d electrons.
4. The size of Cooper pair is very small being of order of 2 Å.
5. The CuO_2 plane is high anisotropic and there is a weak Josephson coupling between two layers.

The small size of Cooper pair also supports the non-phonon mechanism of cuprate superconductivity [138–141]. A plausible non-phonon mechanism is due to the Coulomb interaction on the same atom. Because the energy scale of the Coulomb interaction is very large, which is of the order of eV, we can expect superconductivity with high critical temperature T_c. The critical temperature of heavy fermion materials is, however, very low, although superconductivity occurs due to strong Coulomb interaction between f electrons. This is because the effective mass of f electrons is very large in heavy fermion systems owing to the large self-energy correction. The effective mass enhancement of heavy fermion materials becomes as large as 100–1000, which means that the effective cutoff becomes very small. As a result, the characteristic energy scale is reduced considerably and the critical temperature T_c becomes very low begin of the order of 1 K. In heavy fermion systems, the characteristic energy scale is given by the Kondo temperature T_K. The ratio of the effective mass m^* to the band mass m_0 is approximately given as $m^*/m_0 \simeq D/T_K$ for the bandwidth D and T_K. Thus, the effective bandwidth for heavy fermions is given by the Kondo temperature $T_K \simeq D/(m^*/m_0)$. Empirically, T_c is lowered as the effective mass increases. This is expressed as follows:

$$k_B T_c \simeq 0.1t/(m^*/m_0), \tag{13}$$

where t denotes the transfer integral proportional to the bandwidth. The estimated values of the transfer t, the ratio m^*/m_0 and T_c for several compounds are shown in Table 1. For cuprates, the transfer t is estimated as $t \sim 0.51$ eV. The bandwidth for iron pnictides is about five times smaller than that for cuprates. A list of typical superconductors in correlated electron systems is shown in Table 2.

Table 1. The transfer integral t, effective mass m^* and critical temperature T_c in correlated electron systems, where m_0 denotes the band mass. The orders of these quantities are shown in the table. For heavy fermions, $t/(m^*/m_0)$ corresponds to the Kondo temperature T_K. For Hydrides, the Debye frequency $\omega_{\ln}$ is shown.

	t or $\omega_{\ln}$	m^*/m_0	$t/(m^*/m_0)$	T_c	
Cuprate superconductors	5000 K	5	1000	100 K	$t \sim 0.51$ eV [142]
Iron pnictides	1000 K	~ 2	500	50 K	$t \sim 0.1$ eV [143]
Heavy fermion materials	10,000 K	100~1000	10~100	1~10 K	[144–146]
Organic superconductors	200~500 K	2~5	100	10 K	[147]
Hydrides H_3S	1000 K	~ 1	1000	100 K	$\omega_{\ln}$ [148]

Table 2. Superconducting materials.

Materials	T_c	Pair Symmetry	Crystal Structure	
$CeCu_2Si_2$	0.6 K	s or d	bc tetragonal	[149,150]
UPt_3	0.52 K	p or f	Hexagonal	[151]
UBe_{13}	0.86 K	p	Cubic	[152]
URu_2Si_2	1.2 K		bc tetragonal	[153–155]
$CeRu_2$	6.2 K	s	Laves Cubic	[156]
UPd_2Al_3	2 K	d	Hexagonal	[157–159]
UNi_2Al_3	1 K	p?	Hexagonal	[158,160]
$CeCoIn_5$	2.3 K	d	$HoCoGa_5$ type	[161,162]
$CeRhIn_5$	2.1 K	d	$HoCoGa_5$ type	[163]
	(16.3 kbar)		bc tetragonal	
$CeRh_2Si_2$	0.35 K		bc tetragonal	[164]
	(9 kbar)			
UGe_2	0.8 K	p?	Orthorhombic	[165]
	(13.5 kbar)			
$URhGe$	0.25 K	p?	Orthorhombic	[166]
Sr_2RuO_4	1.4 K	p or f	Perovskite	[167]
$PrOs_4Sb_{12}$	1.85 K	line nodes?	Skutterudite	[168]
$Na_xCoO_{2-y} \cdot H_2O$	5 K	p?	Triangular lattice	[169]
$Ba_{1-x}K_xBiO_3$	30 K	s	Perovskite	[170]
MgB_2	39 K	s	Hexagonal	[171]
$La_{2-x}Sr_xCuO_4$	36 K	d	Perovskite	
$YBa_2Cu_3O_{6+x}$	90 K	d	Perovskite	
$Tl_2Ba_2Ca_{n-1}Cu_nO_{2n+4}$	125 K	d	Perovskite	
$HgBa_2Ca_{n-1}Cu_nO_{2n+2+\delta}$	135 K	d	Perovskite	
$LaO_{1-x}F_xFeAs$	26 K		ZrCuSiAs type	[172]
$NdFeAsO_{1-y}$	54 K		ZrCuSiAs type	[173]
H_3S	203 K	s	(under pressure)	[81]
LaH_{10}	260 K	s	(under pressure)	[82,83,174,175]

2.3. Superconductivity in Strongly Correlated Electron Systems

The possibility of superconductivity in strongly correlated electron systems has been discussed intensively. The perturbative calculations such as the fluctuation-exchange approximation (FLEX) have been performed to investigate the superconducting ground state [176–178]. There were, however, the results by quantum Monte Carlo methods, which did not support the existence of high-temperature superconductivity in the two-dimensional Hubbard model [32,33,45]. In quantum Monte Carlo calculations, the strength of the Coulomb interaction U is not large enough because the range of accessible U is very restricted. It is now certain that there is a superconducting phase in the strongly correlated region [50]. The simplest wave function of superconducting state with strong electron correlation is the Gutzwiller-projected BCS wave function:

$$\psi_{BCS-G} = P_G \prod_{\mathbf{k}} (u_{\mathbf{k}} + v_{\mathbf{k}} c^{\dagger}_{\mathbf{k}\uparrow} c^{\dagger}_{-\mathbf{k}\downarrow}) |0\rangle, \tag{14}$$

where $u_{\mathbf{k}}$ and $v_{\mathbf{k}}$ are BCS parameters and P_G is the Gutzwiller operator to control the on-site electron correlation. P_G is written as

$$P_G = \prod_{j} (1 - (1-g) n_{j\uparrow} n_{j\downarrow}), \tag{15}$$

where g is a variational parameter in the range of $0 \leq g \leq 1$. The ratio of $u_{\mathbf{k}}$ and $v_{\mathbf{k}}$ is given as

$$\frac{v_{\mathbf{k}}}{u_{\mathbf{k}}} = \frac{\Delta_{\mathbf{k}}}{\zeta_{\mathbf{k}} + (\zeta_{\mathbf{k}}^2 + \Delta_{\mathbf{k}}^2)^{1/2}}, \tag{16}$$

where $\zeta_{\mathbf{k}}$ denotes the electron dispersion relation measured from the Fermi energy and $\Delta_{\mathbf{k}}$ is the gap function. We use the following form for the gap function in the two-dimensional case:

$$\Delta_{\mathbf{k}} = \Delta(\cos k_x - \cos k_y) \quad d-\text{wave}, \tag{17}$$

$$\Delta_{\mathbf{k}} = \Delta(\cos k_x + \cos k_y) \quad \text{anisotropic}-s-\text{wave}, \tag{18}$$

$$\Delta_{\mathbf{k}} = \Delta \quad \text{isotropic } s-\text{wave}. \tag{19}$$

Δ is a constant and is treated as a variational parameter. The wave function ψ_{BCS-G} is just the wave function that Anderson proposed as a wave function of the resonate-valence-bond (RVB) state [72].

It has been shown that the ground-state energy has a minimum at finite Δ for the BCS-Gutzwiller wave function with d-wave symmetry in the two-dimensional Hubbard model by using the variational Monte Carlo method [37]. The superconducting condensation energy E_{cond} per site in the limit of large system size was estimated as [37,38]

$$E_{cond}/N \simeq 0.2 \text{ meV}, \tag{20}$$

where the transfer integral t is set at 0.5 eV. The similar result was obtained for the three-band d-p model [18]. Thus, the condensation energy per atom is of the order of 10^{-4} eV.

The superconducting condensation energy E_{cond} for cuprate high-temperature superconductors was evaluated by using the result of specific heat measurement for YBCO as 0.17–0.26 meV per Cu atom [37,179]. The estimation of E_{cond} from the data of critical magnetic field gives the similar result [180]. The obtained results by theoretical calculations and experimental measurements are very close each other. This agreement is very remarkable. Thus, this value indicates the characteristic energy for cuprate high-temperature superconductors. This result may support that the superconductivity in cuprate high temperature superconductors is caused by the strong electron correlation and the 2D Hubbard model includes essential ingredients.

3. Part II. Mechanism of Superconductivity in Cuprates

We discuss the mechanism of superconductivity in this part. We show numerical results obtained by using the optimized wave functions.

3.1. Model for High-T_c Cuprates

The Hamiltonian of the d-p model for high-T_c cuprates is

$$
\begin{aligned}
H_{dp} &= \epsilon_d \sum_{i\sigma} d^\dagger_{i\sigma} d_{i\sigma} + \epsilon_p \sum_{i\sigma} (p^\dagger_{i+\hat{x}/2\sigma} p_{i+\hat{x}/2\sigma} + p^\dagger_{i+\hat{y}/2\sigma} p_{i+\hat{y}/2\sigma}) \\
&+ t_{dp} \sum_{i\sigma} [d^\dagger_{i\sigma} (p_{i+\hat{x}/2\sigma} + p_{i+\hat{y}/2\sigma} - p_{i-\hat{x}/2\sigma} - p_{i-\hat{y}/2\sigma}) + \text{h.c.}] \\
&+ t_{pp} \sum_{i\sigma} [p^\dagger_{i+\hat{y}/2\sigma} p_{i+\hat{x}/2\sigma} - p^\dagger_{i+\hat{y}/2\sigma} p_{i-\hat{x}/2\sigma} - p^\dagger_{i-\hat{y}/2\sigma} p_{i+\hat{x}/2\sigma} + p^\dagger_{i-\hat{y}/2\sigma} p_{i-\hat{x}/2\sigma} + \text{h.c.}] \\
&+ t'_d \sum_{\langle\langle ij \rangle\rangle\sigma} \epsilon_{ij} (d^\dagger_{i\sigma} d_{j\sigma} + \text{h.c.}) + U_d \sum_i d^\dagger_{i\uparrow} d_{i\uparrow} d^\dagger_{i\downarrow} d_{i\downarrow} \\
&+ U_p \sum_i (n^p_{i+\hat{x}/2\uparrow} n^p_{i+\hat{x}/2\downarrow} + n^p_{i+\hat{y}/2\uparrow} n^p_{i+\hat{y}/2\downarrow}).
\end{aligned}
\tag{21}
$$

Since we use the hole picture in this paper, $d_{i\sigma}$ and $d^\dagger_{i\sigma}$ represent the operators for the d hole. $p_{i\pm\hat{x}/2\sigma}$ and $p^\dagger_{i\pm\hat{x}/2\sigma}$ denote the operators for the p holes at the site $R_{i\pm\hat{x}/2}$, and in a similar way $p_{i\pm\hat{y}/2\sigma}$ and $p^\dagger_{i\pm\hat{y}/2\sigma}$ are defined. $n^p_{i\pm\hat{x}/2\sigma}$ and $n^p_{i\pm\hat{y}/2\sigma}$ are the number operators of p holes at $R_{i\pm\hat{x}/2}$ and $R_{i\pm\hat{y}/2}$, respectively. t_{dp} is the transfer integral between adjacent Cu and O orbitals and t_{pp} is that between nearest p orbitals. t'_d indicates that between d orbitals where $\langle\langle ij \rangle\rangle$ denotes a next nearest-neighbor pair of copper sites. ϵ_{ij} takes the values ± 1 (See Figure 1). This value is determined from the sign of the transfer integral between next nearest-neighbor d orbitals. U_d indicates the strength of the on-site Coulomb repulsion between d holes and U_p is that between p holes.

The values of band parameters were evaluated by several works [181–185]. We show an example: $U_d = 10.5$, $U_p = 4.0$ and $U_{dp} = 1.2$ eV [182]. Here, U_{dp} is the nearest-neighbor Coulomb interaction between holes on adjacent Cu and O orbitals and is small compared to U_d. U_{dp} is neglected in this paper. We write $\Delta_{dp} = \epsilon_p - \epsilon_d$. The number of sites is denoted as N, and the energy is measured in units of t_{dp}.

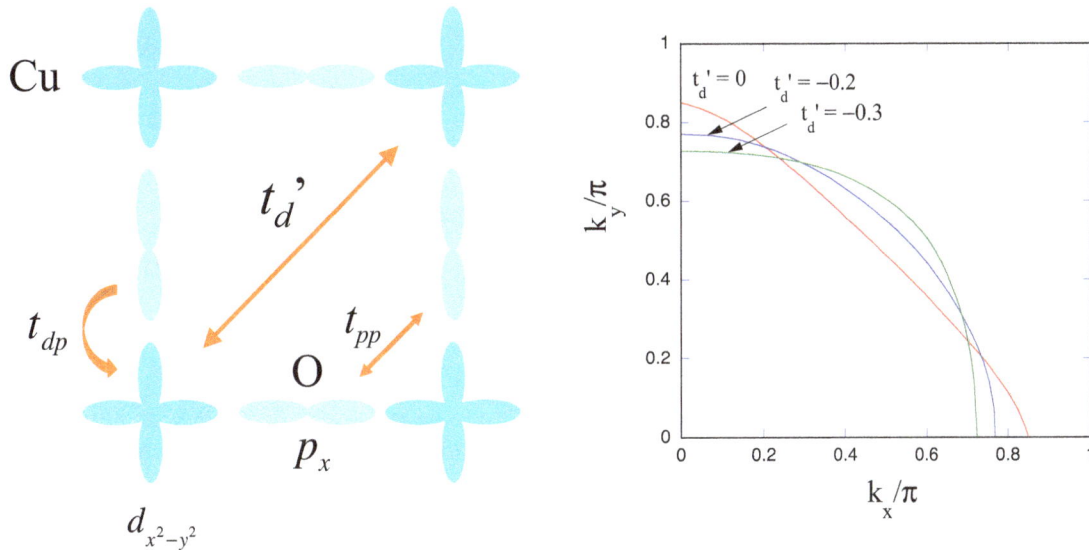

Figure 1. (**Left**) The transfer integral t'_d in the CuO_2 plane. t_{dp} and t_{pp} are conventionally defined. (**Right**) Fermi surface of the d-p model with the hole density 0.13 [74]. We put $t_{pp} = 0.4t_{dp}$ and $\epsilon_p - \epsilon_d = 2t_{dp}$ for $t'_d = 0$, $-0.2t_{dp}$ and $-0.3t_{dp}$.

3.2. Optimization Variational Monte Carlo Method

3.2.1. Off-Diagonal Wave Function

The Gutzwiller wave function is

$$\psi_G = P_G \psi_0, \tag{22}$$

where ψ_0 is a one-particle state. Our purpose is to improve the Gutzwiller function. We multiply the Gutzwiller function by an exponential-type operator. The wave function is given as as [50,73,186–190]

$$\psi_\lambda = \exp(-\lambda K)\psi_G, \tag{23}$$

where K denotes the kinetic part of the Hamiltonian. λ is a newly introduced real variational parameter [41,73,187,191]. There are other methods to improve the Gutzwiller function [43,192]. The following Jastrow operator is used [43],

$$P_{Jdh} = \prod_j \left(1 - (1-\eta)\prod_\tau \left[d_j(1-e_{j+\tau}) + e_j(1-d_{j+\tau}) \right] \right), \tag{24}$$

where d_j is the operator for the doubly-occupied site given as $d_j = n_{j\uparrow}n_{j\downarrow}$ and e_j is that for the empty site given by $e_j = (1-n_{j\uparrow})(1-n_{j\downarrow})$. η is the variational parameter in the range of $0 \leq \eta \leq 1$. The wave function is

$$\psi_\eta = P_{Jdh}\psi_G. \tag{25}$$

In this paper, we use the wave function of exponential type in Equation (23) because the energy is further lowered when we use this wave function [50]. The wave function for the d-p model is formulated similarly. An initial state ψ_0 contains many variational parameters ($\tilde{t}_{dp}, \tilde{t}_{pp}, \tilde{t}'_d$, and $\tilde{\epsilon}_p - \tilde{\epsilon}_d$):

$$\psi_0 = \psi_0(\tilde{t}_{dp}, \tilde{t}_{pp}, \tilde{t}'_d, \tilde{\epsilon}_p - \tilde{\epsilon}_d). \tag{26}$$

We use $\tilde{t}_{dp} = t_{dp}$ as the energy unit. We consider the following wave function that is improved from the Gutzwiller wave function [50,73,186–190]:

$$\psi_\lambda = \exp(-\lambda K)\psi_G. \tag{27}$$

The expectation values are evaluated by using the auxiliary field method [73,191]. The kinetic part K also contains the band parameters t_{pp}, t'_d and $\epsilon_p - \epsilon_d$ as variational parameters:

$$K = K(\hat{t}_{pp}, \hat{t}'_d, \hat{\epsilon}_p - \hat{\epsilon}_d). \tag{28}$$

We take $\hat{t}_{pp} = \tilde{t}_{pp}$, $\hat{t}'_d = \tilde{t}'_d$ and $\hat{\epsilon}_p - \hat{\epsilon}_d = \tilde{\epsilon}_p - \tilde{\epsilon}_d$, for simplicity. Thus, we have g, $\tilde{t}_{pp}$, $\tilde{t}'_d$, $\hat{\epsilon}_p - \hat{\epsilon}_d = \tilde{\epsilon}_p - \tilde{\epsilon}_d$, and λ as variational parameters. The expectation values for this type of wave function are calculated on the basis of the variational Monte Carlo method. One can evaluate the expectation value correctly within statistical errors.

3.2.2. Antiferromagnetic Wave Function

The AF one-particle state ψ_{AF} is formulated by the eigenfunction of the AF trial Hamiltonian:

$$H_{AF} = \sum_{ij\sigma} t_{ij}c^\dagger_{i\sigma}c_{j\sigma} - \Delta_{AF}\sum_{i\sigma}(-1)^{x_i+y_i}\sigma n_{i\sigma}, \tag{29}$$

where Δ_{AF} is the AF order parameter and (x_i, y_i) represents the coordinates of the site i. With ψ_{AF}, the wave function is given as

$$\psi_{\lambda,AF} = \exp(-\lambda K)P_G\psi_{AF}. \tag{30}$$

3.2.3. Superconducting Wave Function

We start from the BCS wave function

$$\psi_{BCS} = \prod_k (u_k + v_k c_{k\uparrow}^\dagger c_{-k\downarrow}^\dagger)|0\rangle, \tag{31}$$

with coefficients u_k and v_k satisfying $u_k^2 + |v_k|^2 = 1$. We choose $u_k/v_k = \Delta_k/(\xi_k + \sqrt{\xi_k^2 + \Delta_k^2})$ for the gap function Δ_k and $\xi_k = \epsilon_k - \mu$. We assume $\Delta_k = \Delta_{SC}(\cos k_x - \cos k_y)$. The Gutzwiller-projected BCS wave function is

$$\psi_{G-BCS} = P_{N_e} P_G \psi_{BCS}, \tag{32}$$

where P_{N_e} indicates the operator to extract the state with N_e electrons. The exponential-BCS wave function is given by

$$\psi_\lambda = e^{-\lambda K} P_G \psi_{BCS}. \tag{33}$$

In this wave function, we perform the electron–hole transformation for down-spin electrons:

$$d_k = c_{-k\downarrow}^\dagger, \quad d_k^\dagger = c_{-k\downarrow}; \tag{34}$$

and not for up-spin electrons: $c_k = c_{k\uparrow}$. The electron pair operator $c_{k\uparrow}^\dagger c_{-k\downarrow}^\dagger$ denotes the hybridization operator $c_k^\dagger d_k$ in this formulation.

3.3. Correlated Superconductivity

We first discuss the superconducting (SC) state in the two-dimensional Hubbard model. In the optimization Monte Carlo method, the SC state becomes indeed stable when the Coulomb interaction U is large to be of the order of the bandwidth. We show the ground-state energy as a function of the superconducting order parameter Δ in Figure 2 (left). The simple Gutzwiller-projected BCS wave function predicted the possibility of superconductivity in the Hubbard model, and the improved wave function also shows a stability of the SC state.

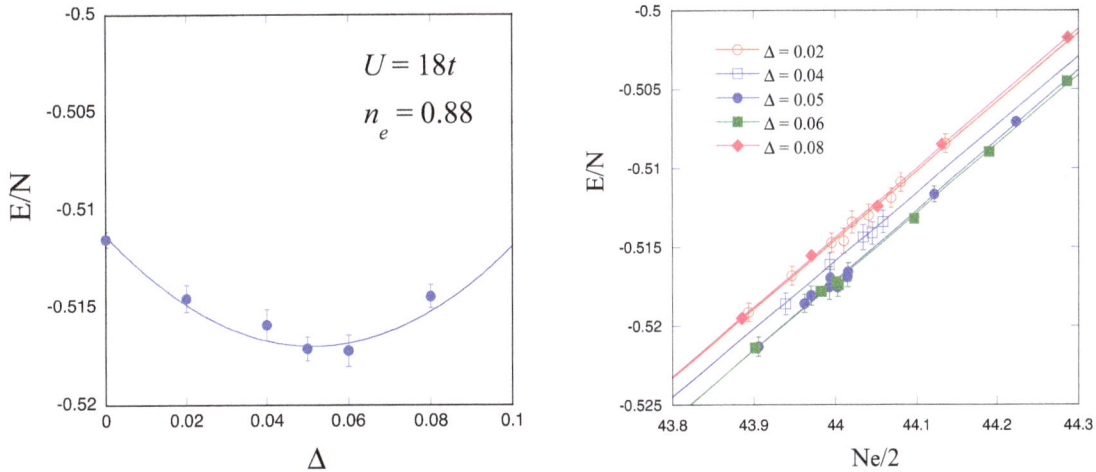

Figure 2. (**Left**) The ground-state energy as a function of the superconducting gap Δ for the optimized wave function ψ_λ for the Hubbard model on a 10×10 lattice with $U/t = 18$ and $t' = 0$. The electron density is $n_e = 0.88$. (**Right**) The ground-state energy as a function of the electron number where Δ is fixed for each line [50].

We show the SC and antiferromagnetic (AF) order parameters as a function of U in Figure 3. The AF order parameter has a peak when $U/t \sim 10$, which is of the order of the bandwidth, and the SC one also has a peak at U_c that is greater than the bandwidth. This indicates that there is the possibility of high-temperature superconductivity in the strongly correlated region.

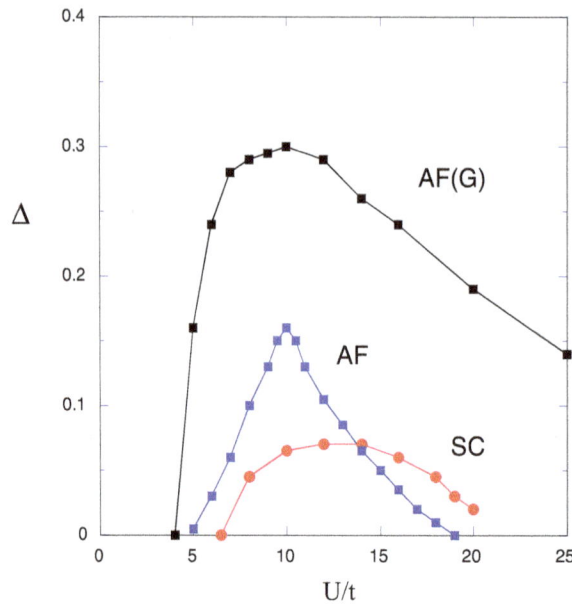

Figure 3. AF and SC order parameters as a function of U/t when $N_e = 88$ for the 2D Hubbard model on a 10×10 lattice. The periodic boundary conditions are periodic in one direction and antiperiodic in the other direction [50]. AF(G) indicates the result obtained for the simple Gutzwiller function.

The AF correlation is maximized at $U \sim U_c$ and decreases when U is larger than U_c. We show schematic pictures in Figure 4, where the SC condensation energy as a function of U is shown in the left panel, and the AF and SC gap functions are shown in the right panel. There is a crossover from weakly correlated region to the strongly correlated region. The superconducting state is most favorable when the AF correlation is gradually suppressed in the strongly correlated region. Thus, high temperature superconductivity is highly promising in the strongly correlated region where U is as large as the bandwidth D or larger than D.

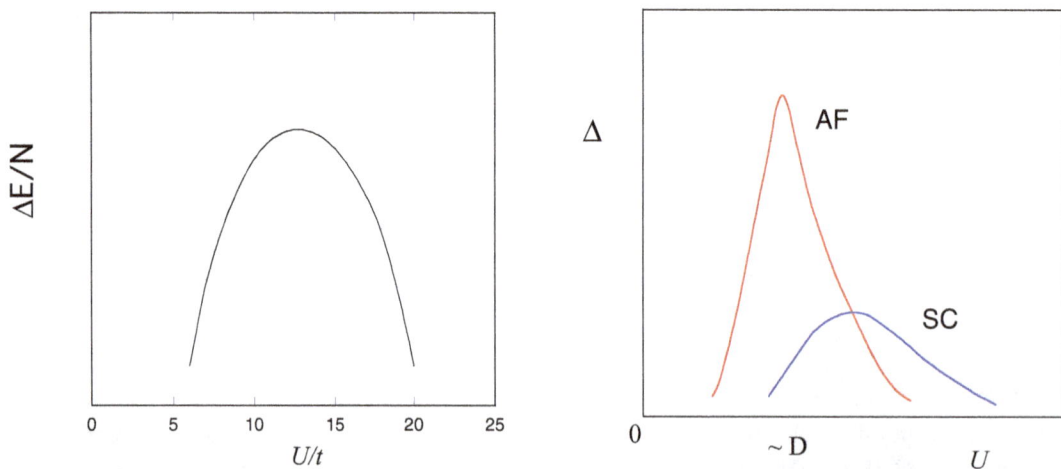

Figure 4. (**Left**) A schematic picture of the superconducting condensation energy as function of U for the 2D Hubbard model. (**Right**) A schematic picture of the gap function of AF and AC states as a function of U for the 2D Hubbard model.

3.4. Stability of Antiferromagnetic State

3.4.1. Hubbard Model

Let us examine the stability of AF state. There are two parameters U and t', and there is the AF region in the parameter space. High temperature superconductivity is expected near the boundary between the AF phase and the paramagnetic phase. We show the AF condensation energy ΔE_{AF} as a function of $1 - n_e$ in Figure 5a for $t' = 0$ and Figure 5b for $t' = -0.2t$. The AF region becomes larger as $|t'|$ increases. When $t' = -0.2t$, the AF region extends up to about 20% doping. From the competition between superconductivity and AF order, $t' = 0$ is most favorable for superconductivity.

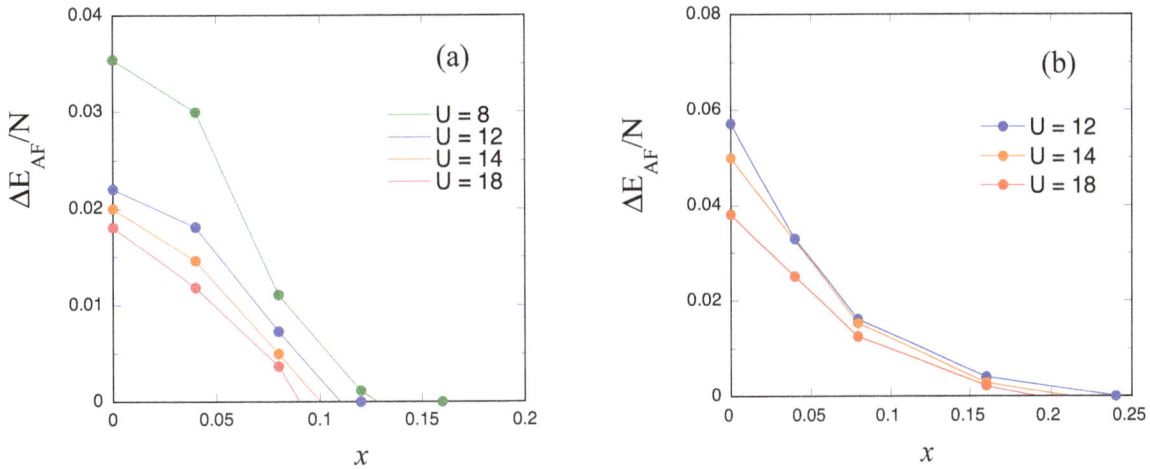

Figure 5. AF condensation energy ΔE_{AF} as a function of the hole doping rate $x = 1 - n_e$ on a 10×10 lattice for: $t' = 0$ (**a**); and $t' = -0.2t$ (**b**) [75]. We put $U/t = 12, 14$ and 18.

3.4.2. Three-Band d-p Model

In general, in the three-band d-p model, the AF correlation is very strong and the AF state is more stable than in the single-band Hubbard model. This is because d electrons are localized and easily form magnetic order [13]. To investigate the possibility of high temperature superconductivity in the d-p model, it is necessary to reveal regions with weak AF order. There are many parameters in the d-p model to control the strength of the AF correlation. Among them, the Coulomb repulsion between d electrons U_d, the level difference $\Delta_{dp} = \epsilon_p - \epsilon_d$, and the hole density x are important. The AF region is shown in Figure 6 where U_d and $\epsilon_p - \epsilon_d$ are varied, and the hole density is fixed at 0.1875. The AF region increases when the hole density decreases. We expect that high temperature superconductivity will occur near the boundary between AFM and PM regions. This boundary exists in the region when Δ_{dp} is small. High temperature superconductivity is likely occur when Δ_{dp} is small. There is a "on-site attractive region" when Δ_{dp} is large where two d electrons prefer to occupy the same site. In this region, a charge-density wave or an s-wave superconducting state will be realized.

We proposed to introduce the transfer integral t'_d to control the strength of AF correlation [74]. We show the AF region at half-filling in the $t_{pp} - t'_d$ plane in Figure 7. As $-t'_d$ increases, there is a phase transition from the AF insulator to the paramagnetic insulator (PMI). We expect that t'_d and t_{pp} will play an important role to suppress AF correlation when holes are doped in the d-p model.

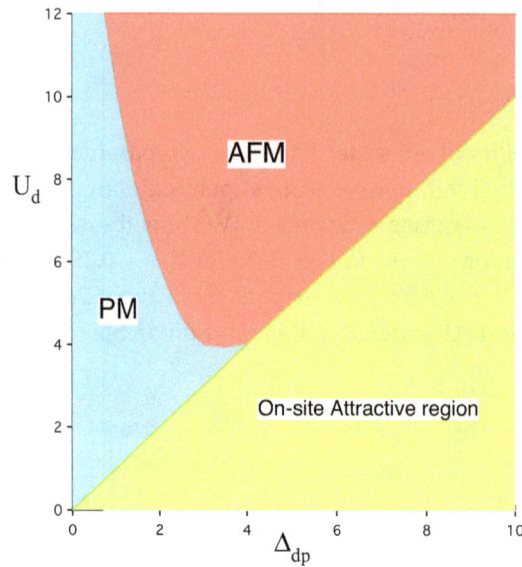

Figure 6. Antiferromagnetic and paramagnetic regions in the plane of U_d and $\Delta_{dp} = \epsilon_p - \epsilon_d$ for the d-p model. We put $t_{pp} = 0.4$ and $t'_d = 0$. There are 76 holes on a 8×8 lattice with 192 atoms in total. The energy unit is given by t_{dp}. AFM and PM denote the antiferromagnetic metal and paramagnetic metal, respectively. There is a "negative-U" region when the level difference is large where two d electrons prefer to occupy the same site. The ground state may be a charge-density wave state or an s-wave superconducting state. This is not clear yet.

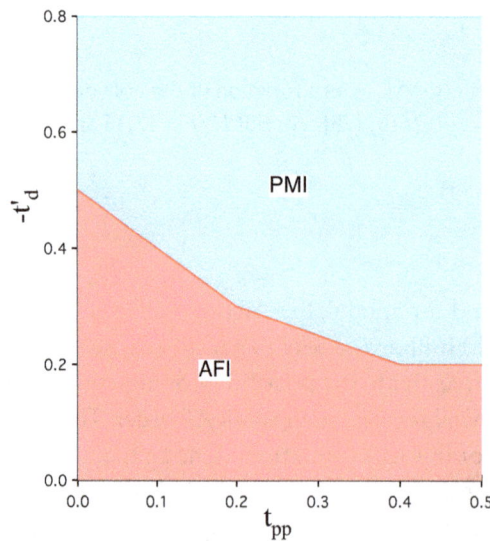

Figure 7. AF and paramagnetic insulator phases for the d-p model on a 6×6 lattice [193]. Parameters are $U_d = 8t_{dp}$, $U_p = 0$, $\epsilon_p - \epsilon_d = t_{dp}$.

3.5. Phase Diagram for the Hubbard Model

We discuss the phase diagram when carrier holes are doped in the CuO_2 plane. We evaluate the energy lowering when we include the order parameter Δ. We define

$$\Delta E = E(\Delta = 0) - E(\Delta_{min}), \tag{35}$$

where $E(\Delta)$ takes a minimum at $\Delta = \Delta_{min}$. We show ΔE as a function of the hole doping rate x in Figure 8 where we put $U/t = 18$ and $t' = 0$. This phase diagram contains several interesting features. There are three phases: antiferromagnetic insulator (AFI), coexistent state (AFSC) and superconducting phase (SC). When the hole doping rate x is large, e.g., $x > 0.09$, the pure d-wave stat is stable. There is the possibility of high (and room) temperature superconductivity in this phase. In the underdoped region, approximately $0.06 < x < x_{dSC}$ with $0.08 < x_{dSC} < 0.09$, there is the coexistent state of antiferromagnetism and superconductivity. This is the mixed phase of AF and SC. x_{dSC} could not be determined precisely. There is the possibility that both the AFSC and SC states are found for $x_{dSC} < x < 0.09$, but the SC solution will have lower energy. There is the AFSC-SC transition at $x = x_{dSC}$. The AFI state exists near half-filling for about $x < 0.06$, where doped holes form clusters and localize.

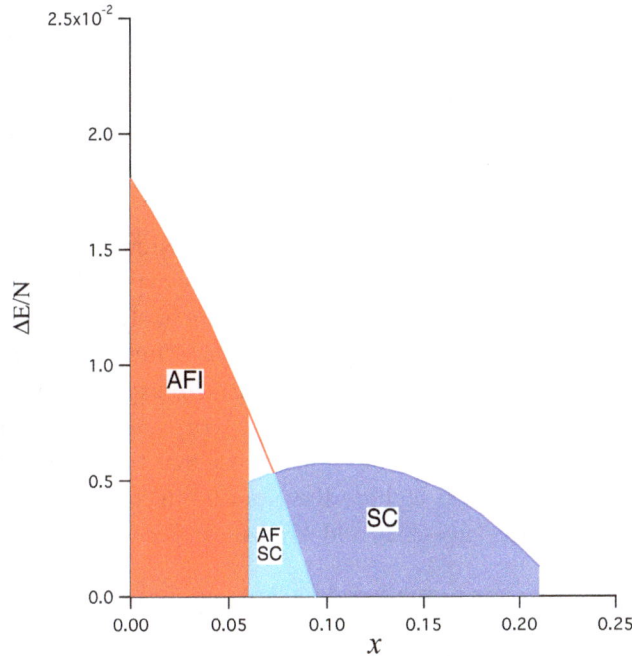

Figure 8. The condensation energy per site for the two-dimensional Hubbard model as a function of the hole doping rate x. The calculation was carried out on a 10×10 lattice. AFI indicates the antiferromagnetic insulating state and SC denotes the d-wave SC phase. There is the coexistent state indicated as AF-SC between these states. Parameters are $t' = 0$ and $U/t = 18$.

The existence of AFI phase is closely related to the phase separation [140,141] when the hole density is very small. In the phase-separated phase, the doped holes are localized and cannot be conductive. The existence of AFI phase is determined by the quantity

$$\delta^2 E(N_e) \equiv [E(N_e + \delta N_e) - 2E(N_e) + E(N_e - \delta N_e)]/(\delta N_e)^2, \tag{36}$$

where $E(N_e)$ is the ground-state energy with N_e electrons. $\delta^2 E(N_e)$ is approximately the second derivative of the energy $E(N_e)$ and is proportional to the charge susceptibility. When $\delta^2 E(N_e)$ is negative, the phase separation occurs. As shown in Figure 8, the phase separation occurs for $x < 0.06$. Concerning the phase separation, the parameter t' is important because the phase separation region decreases as $-t'$ increases. Thus, the AFI phase will decrease as $-t'$ increases. The phase separation disappears for $t' = -0.2t$.

4. Summary

We have discussed the possibility of high temperature superconductivity in many-electron systems. The critical temperature T_c may increase as the characteristic energy of the interaction increases. Empirically, T_c is proportional to the inverse of the effective mass of electrons. T_c is low when the effective mass is very heavy. A candidate of high (room) temperature superconductivity may be in materials with strong electron correlation and with small effective mass enhancement. From this view point, the repulsive Coulomb interaction can be a candidate of the origin of high temperature superconductivity.

We have shown phase diagrams for the 2D Hubbard model and the three-band d-p model. The diagram in Figure 8 exhibits the characteristic property of cuprate superconductors. This supports that the origin of high temperature superconductivity is the strong correlation between electrons. That is, the mechanism of high-T_c superconductivity is the electron-pair formation due to the strong on-site repulsive Coulomb interaction. The competition between antiferromagnetism and superconductivity is important in realizing high temperature superconductivity. High-T_c superconductivity is expected in the region near the boundary between AF phase and paramagnetic phase. In the phase diagram for the Hubbard model, the SC phase exists near the AF phase, and AF order and superconductivity coexist where the doping rate is approximately $0.05 \sim 0.06 < x < x_{dSC}$ and $0.08 < x_{dSC} < 0.09$. We expect that this coexistence may be related to anomalous metallic behavior in the underdoped region. The AF phase near half-filling is insulating, which is approximately for $x < 0.06$. There is the pure d-wave phase for $x > x_{dSC}$.

In the d-p model, the AF region exists in the multi-dimensional parameter space. The AF-PM boundary is a multi-dimensional region in this space. Since we expect that superconductivity occurs near the boundary, high temperature superconductivity is more likely to occur in the d-p model. There is the AF–PM boundary when the level difference Δ_{dp} is small. Thus, T_c of high temperature cuprates will be high when Δ_{dp} is small. This tendency is consistent with experimental T_c of cuprates.

We give a comment on the crossover between weakly correlated region and strongly correlated region. We expect that this crossover is universal in the sense that similar phenomena occur in nature. There may be a universal class. It will include the Kondo effect [194–196], QCD [197], BCS-BEC crossover [198], sine-Gordon model [199–202], and Gross–Neveu model [203].

Funding: This work was supported by a Grant-in-Aid for Scientific Research from the Ministry of Education, Culture, Sports, Science and Technology of Japan (Grant No. 17K05559).

Acknowledgments: A part of the computations was supported by the Supercomputer Center of the Institute for Solid State Physics, the University of Tokyo.

Conflicts of Interest: The author declares no conflict of interest.

Abbreviations

The following abbreviations are used in this manuscript:

OVMC   optimization variational Monte Carlo method
AF   antiferromagnetic
SC   superconductivity or superconducting
2D   two-dimensional
AFI   antiferromagnetic insulator
PI   paramagnetic insulator

References

1. Bednorz, J.B.; Müller, K.A. Possible high T_c superconductivity in the Ba-La-Cu-O system. *Z. Phys. B Condens. Matter* **1986**, *B64*, 189–193. [CrossRef]
2. Keimer, B.; Kivelson, S.A.; Norman, M.R.; Uchida, S.; Zaanen, J. From quantum matter to high-temperature superconductivity in copper oxides. *Nature* **2015**, *518*, 179. [CrossRef] [PubMed]
3. Rybicki, D.; Jurkutat, M.; Reichart, S.; Kapusta, C.; Haase, J. Perspective on the phase diagram of cuprate high-temperature superconductors. *Nat. Commun.* **2016**, *7*, 11413. [CrossRef] [PubMed]
4. Cooper, L.N. Bound Electron Paira in a Degenerate Fermi Gas. *Phys. Rev.* **1956**, *104*, 1189. [CrossRef]
5. Bardeen, J.; Cooper, L.N.; Schrieffer, J.R. Microscopic Theory of Superconductivity. *Phys. Rev.* **1957**, *106*, 162. [CrossRef]
6. Bardeen, J.; Cooper, L.N.; Schrieffer, J.R. Theory of Superconductivity. *Phys. Rev.* **1957**, *108*, 1175. [CrossRef]

7. Emery, V.J. Theory of high-T_c superconductivity in oxides. *Phys. Rev. Lett.* **1987**, *58*, 2794. [CrossRef] [PubMed]

8. Hirsch, J.E.; Loh, E.Y.; Scalapino, D.J.; Tang, S. Pairing interaction in CuO clusters. *Phys. Rev. B* **1989**, *39*, 243. [CrossRef]

9. Scalettar, R.T.; Scalapino, D.J.; Sugar, R.L.; White, S.R. Antiferromagnetic, charge-transfer, and pairing correlations in the three-band Hubbard model. *Phys. Rev. B* **1991**, *44*, 770. [CrossRef]

10. Unger, P.; Fulde, P. Spectral function of holes in the Emergy model. *Phys. Rev. B* **1993**, *48*, 16607. [CrossRef]

11. Oguri, A.; Asahatam, T.; Maekawa, S. Gutzwiller wave function in the three-band Hubbard model: A variational Monte Carlo study. *Phys. Rev. B* **1994**, *49*, 6880. [CrossRef] [PubMed]

12. Koikegami, S.; Yamada, K. Antiferromagnetic and superconducting correlations on the d-p model. *J. Phys. Soc. Jpn.* **2000**, *69*, 768. [CrossRef]

13. Yanagisawa, T.; Koike, S.; Yamaji, K. Ground state of the three-band Hubbard model. *Phys. Rev. B* **2001**, *64*, 184509. [CrossRef]

14. Koikegami, S.; Yanagisawa, T. Superconducting gap of the two-dimensional d-p model with small U_d. *J. Phys. Soc. Jpn.* **2001**, *70*, 3499–3502. [CrossRef]

15. Yanagisawa, T.; Koike, S.; Yamaji, K. Lattice distortions, incommensurability, and stripes in the electronic model for high-T_c cuprates. *Phys. Rev. B* **2003**, *67*, 132408. [CrossRef]

16. Koikegami, S.; Yanagisawa, T. Superconductivity in Sr_2RuO_4 mediated by Coulomb scattering. *Phys. Rev. B* **2003**, *67*, 134517. [CrossRef]

17. Koikegami, S.; Yanagisawa, T. Superconductivity in multilayer perovskite. *J. Phys. Soc. Jpn.* **2006**, *75*, 034715. [CrossRef]

18. Yanagisawa, T.; Miyazaki, M.; Yamaji, K. Incommensurate antiferromagnetism coexisting with superconductivity in two-dimensional d-p model. *J. Phys. Soc. Jpn.* **2009**, *78*, 031706. [CrossRef]

19. Weber, C.; Lauchi, A.; Mila, F.; Giamarchi, T. Orbital currents in extended Hubbard model of High-T_c cuprate superconductors. *Phys. Rev. Lett.* **2009**, *102*, 017005. [CrossRef]

20. Lau, B.; Berciu, M.; Sawatzky, G.A. High spin polaron in lightly doped CuO_2 planes. *Phys. Rev. Lett.* **2011**, *106*, 036401. [CrossRef]

21. Weber, C.; Giamarchi, T.; Varma, C.M. Phase diagram of a three-orbital model for high-T_c cuprate superconductors. *Phys. Rev. Lett.* **2014**, *112*, 117001. [CrossRef] [PubMed]

22. Avella, A.; Mancini, F.; Paolo, F.; Plekhanov, E. Emery vs. Hubbard model for cuprate superconductors: A composite operator method study. *Eur. Phys. J.* **2013**, *B86*, 265. [CrossRef]

23. Ebrahimnejad, H.; Sawatzky, G.A.; Berciu, M. Differences between the insulating limit quasiparticles of one-band and three-band cuprate models. *J. Phys. Condens. Matter* **2016**, *28*, 105603. [CrossRef] [PubMed]

24. Tamura, S.; Yokoyama, H. Variational study of magnetic ordered state in d-p model. *Phys. Procedia* **2016**, *81*, 5–8. [CrossRef]

25. Hubbard, J. Electron correlations in narrow energy bands. *Proc. R. Soc. Lond.* **1963**, *276*, 238–257.

26. Hubbard, J. Electron correlations in narrow energy bands III. *Proc. R. Soc. Lond.* **1964**, *281*, 401–419.

27. Gutzwiller, M.C. Effect of correlation on the ferromagnetism of transition metals. *Phys. Rev. Lett.* **1963**, *10*, 159. [CrossRef]

28. Ceperley, D.; Chester, G.V.; Kalos, K.H. Monte Carlo simulation of a many-fermion study. *Phys. Rev. B* **1977**, *16*, 3081. [CrossRef]

29. Gros, C.; Joynt, R.; Rice, T.M. Antiferromagnetic correlations in almost-localized Fermi liquids. *Phys. Rev. B* **1987**, *36*, 381. [CrossRef]

30. Yokoyama, H.; Shiba, H. Variational Monte Carlo studies of Hubbard model I. *J. Phys. Soc. Jpn.* **1987**, *56*, 1490–1506. [CrossRef]

31. Giamarchi, T.; Lhuillier, C. Phase diagrams of the two-dimensional Hubbard and t-J models by a variational Monte Carlo study. *Phys. Rev. B* **1991**, *43*, 12943. [CrossRef] [PubMed]

32. Zhang, S.; Carlson, J.; Gubernatis, J.E. Constrained path Monte Carlo method for fermion ground states. *Phys. Rev. B* **1997**, *55*, 7464. [CrossRef]

33. Zhang, S.; Carlson, J.; Gubernatis, J.E. Pairing correlation in the two-dimensional Hubbard model. *Phys. Rev. Lett.* **1997**, *78*, 4486. [CrossRef]

34. Yanagisawa, T.; Shimoi, Y. Exact results in strongly correlated electrons. *Int. J. Mod. Phys.* **1996**, *B10*, 3383–3450. [CrossRef]

35. Yanagisawa, T.; Shimoi, Y. Ground state of the Kondo-Hubbard model at half-filling. *Phys. Rev. Lett.* **1995**, *74*, 4939. [CrossRef] [PubMed]

36. Nakanishi, T.; Yamaji, K.; Yanagisawa, T. Variational Monte Carlo indications of d-wave superconductivity in the two-dimensional Hubbard model. *J. Phys. Soc. Jpn.* **1997**, *66*, 294–297. [CrossRef]

37. Yamaji, K.; Yanagisawa, T.; Nakanishi, T.; Koike, S. Variational Monte Carlo study on the superconductivity in the two-dimensional Hubbard model. *Physica C* **1998**, *304*, 225–238. [CrossRef]

38. Yamaji, K.; Yanagisawa,T.; Koike, S. Bulk limit of superconducting condensation energy in 2D Hubbard model. *Physica B* **2000**, *284–288*, 415–416. [CrossRef]

39. Yamaji, K.; Yanagisawa, T.; Miyazaki, M.; Kadono, R. Superconducting condensation energy of the two-dimensional Hubbard model in the large-negative-t' region. *J. Phys. Soc. Jpn.* **2011**, *80*, 083702. [CrossRef]

40. Hardy, T.M.; Hague, P.; Samson, J.H.; Alexandrov, A.S. Superconductivity in a Hubbard-Fröhlich model in cuprates. *Phys. Rev. B* **2009**, *79*, 212501. [CrossRef]

41. Yanagisawa, T.; Miyazaki, M.; Yamaji, K. Correlated-electron systems and high-temperature superconductivity. *J. Mod. Phys.* **2013**, *4*, 33. [CrossRef]

42. Bulut, N. $d_{x^2-y^2}$ superconductivity and the Hubbard model. *Adv. Phys.* **2002**, *51*, 1587–1667. [CrossRef]

43. Yokoyama, H.; Tanaka, Y.; Ogata, M.; Tsuchiura, H. Crossover of superconducting properties and kinetic-energy gain in two-dimensional Hubbard model. *J. Phys. Soc. Jpn.* **2004**, *73*, 1119–1122. [CrossRef]

44. Yokoyama, H.; Ogata, M.; Tanaka, Y. Mott transitions and *d*-wave superconductivity in half-filled-band Hubbard model on square lattice with geometric frustration. *J. Phys. Soc. Jpn.* **2006**, *75*, 114706. [CrossRef]

45. Aimi, T.; Imada, M. Does simple two-dimensional Hubbard model account for high-T_c superconductivity in copper oxides? *J. Phys. Soc. Jpn.* **2007**, *76*, 113708. [CrossRef]

46. Miyazaki, M.; Yanagisawa, T.; Yamaji, K. Diagonal stripe states in the light-doping region in the two-dimensional Hubbard model. *J. Phys. Soc. Jpn.* **2004**, *73*, 1643–1646. [CrossRef]

47. Yanagisawa, T. Phase diagram of the t-U^2 Hamiltonian of the weak coupling Hubbard model. *New J. Phys.* **2008**, *10*, 023014. [CrossRef]

48. Yanagisawa, T. Enhanced pair correlation functions in the two-dimensional Hubbard model. *New J. Phys.* **2013**, *15*, 033012. [CrossRef]

49. Yokoyama, H.; Ogata, M.; Tanaka, Y.; Kobayashi, K.; Tsuchiura, H. Crossover between BCS superconductor and doped Mott insulator of *d*-wave pairing state in two-dimensional Hubbard model. *J. Phys. Soc. Jpn.* **2013**, *82*, 014707. [CrossRef]

50. Yanagisawa, T. Crossover from wealy to strongly correlated regions in the two-dimensional Hubbard model—Off-diagonal Monte Carlo studies of Hubbard model II—. *J. Phys. Soc. Jpn.* **2016**, *85*, 114707. [CrossRef]

51. Noack, R.M.; White, S.R.; Scalapino, D.J. The doped two-chain Hubbard model. *EPL* **1995**, *30*, 163. [CrossRef]

52. Noack, R.M.; Bulut, N.; Scalapino, D.J.; Zacher, M.J. Enhanced $d_{x^2-y^2}$ pairing correlations in the two-leg Hubbard ladder. *Phys. Rev. B* **1997**, *56*, 7162. [CrossRef]

53. Yamaji, K.; Shimoi, Y.; Yanagisawa, T. Superconductivity indications of the two-chain Hubbard model due to the two-band effect. *Physica C* **1994**, *235*, 2221. [CrossRef]

54. Yanagisawa, T.; Shimoi, Y.; Yamaji, K. Superconducting phase of a two-chain Hubbard model. *Phys. Rev. B* **1995**, *52*, R3860. [CrossRef] [PubMed]

55. Nakano, T.; Kuroki, K.: Onari, S. Superconductivity due to spin fluctuations originating from multiple Fermi surfaces in the double chain superconductor $Pr_2Ba_4Cu_7O_{15-\delta}$. *Phys. Rev. B* **2007**, *76*, 014515. [CrossRef]

56. Koike, S.; Yamaji, K.; Yanagisawa, T. Effect of the medium-range transfer energies to the superconductivity in the two-chain Hubbard model. *J. Phys. Soc. Jpn.* **1999**, *68*, 1657–1663. [CrossRef]

57. Mott, N.F. *Metal-Insulator Transitions*; Taylor and Francis Ltd.: London, UK, 1974.

58. Moriya, T. *Spin Fluctuations in Itinerant Electron Magnetism*; Springer: Berlin, Germany, 1985.

59. Yosida, K. *Theory of Magnetism*; Springer: Berlin, Germany, 1996.

60. Tranquada, J.M.; Axe, J.D.; Ichikawa, N.; Nakamura, Y.; Uchida, S.; Nachumi, B. Neutron-scattering study of stripe-phase order of holes and spins in $La_{1.48}Nd_{0.4}Sr_{0.12}CuO_4$. *Phys. Rev. B* **1996**, *54*, 7489. [CrossRef] [PubMed]

61. Suzuki, T.; Goto, T.; Shinoda, T.; Fukase, T.; Kimura, H.; Yamada, K.: Ohashi, M.; Yamaguchi, Y. Observation of modulated magnetic long-range order in $La_{1.88}Sr_{0.12}CuO_4$. *Phys. Rev. B* **1998**, *57*, R3229. [CrossRef]

62. Yamada, K.; Lee, C.H.; Kurahashi, K.; Wada, J.; Wakimoto, S.; Ueki, S.; Kimura, H.; Endoh, Y.; Hosoya, S.; Shirane, G.; et al. Doping dependence of the spatially modulated dynamical spin correlations and the superconducting-transition temperature in $La_{2-x}Sr_xCuO_4$. *Phys. Rev. B* **1998**, *57*, 6165. [CrossRef]

63. Arai, M.; Nishijima, T.; Endoh, Y.; Egami, T.; Tajima, S.; Tomimoto, K.; Shiohara, Y.; Takahashi, M.; Garrett, A.; Bennington, S.M. Incommensurate spin dynamics of underdoped superconductor $YBa_2Cu_3Y_{6.7}$. *Phys. Rev. Lett.* **1999**, *83*, 608. [CrossRef]

64. Mook, H.A.; Pengcheng, D.; Dogan, F.; Hunt, R.D. One-dimensional nature of the magnetic fluctuations in $YBa_2Cu_3O_{6.6}$. *Nature* **2000**, *404*, 729. [CrossRef] [PubMed]

65. Wakimoto, S.; Birgeneau, R.J.; Kastner, M.A.; Lee, Y.S.; Erwin, R.; Gehring, P.M.; Lee, S.H.; Fujita, M.; Yamada, K.; Endoh, Y.; et al. Direct observation of a one-dimensional static spin modulation in insulating $La_{1.95}Sr_{0.05}CuO_4$. *Phys. Rev. B* **2000**, *61*, 3699. [CrossRef]

66. Bianconi, A.; Saini, N.L.; Lanzara, A.; Missori, M.; Rossetti, T.; Oyanagi, H.; Yamaguchi, H.; Oka, K.; Ito, T. Determination of the local lattice distortions in the CuO_2 plane of $La_{1.85}Sr_{0.15}CuO_4$. *Phys. Rev. Lett.* **1996**, *76*, 3412. [CrossRef] [PubMed]

67. Bianconi, A. Quantum materials: Shape resonances in superstripes. *Nat. Phys.* **2013**, *9*, 536. [CrossRef]

68. Hoffman, J.E.; McElroy, K.; Lee, D.-H.; Lang, K.M.; Eisaki, H.; Uchida, S.; Davis, J.C. Imaging quasiparticle interference in $Bi_2Sr_2CaCu_2O_{8+\delta}$. *Science* **2002**, *295*, 466. [CrossRef] [PubMed]

69. Wise, W.D.; Boyer, M.C.; Chatterjee, K.; Kondo, T.; Takeuchi, T.; Ikuta, J.; Wang, Y.; Hudson, E.W. Charge-density-wave origin of cuprate checkerboard visualized by scanning tunnelling microscopy. *Nat. Phys.* **2008**, *4*, 696. [CrossRef]

70. Hanaguri, T.; Lupien, C.; Kohsaka, Y.; Lee, D.-H.; Azuma, M.; Takano, M.; Takagi, H.; Davis, J. A checkerboard electronic crystal state in lightly hole-doped $Ca_{2-x}Na_xCuO_2Cl_2$. *Nature* **2004**, *430*, 1001. [CrossRef]

71. Miyazaki, M.; Yanagisawa, T.; Yamaji, K. Checkerboard states in the two-dimensional Hubbard model with the Bi2212-type band. *J. Phys. Soc. Jpn.* **2009**, *78*, 043706. [CrossRef]

72. Anderson, P.W. The resonating valence bond states in La_2CuO_4. *Science* **1987**, *253*, 1196. [CrossRef]

73. Yanagisawa, T.; Koike, S.; Yamaji, K. Off-diagonal wave function Monte Carlo Studies of Hubbard model I. *J. Phys. Soc. Jpn.* **1998**, *67*, 3867–3874. [CrossRef]

74. Yanagisawa, T.; Miyazaki, M. Mott transition in cuprate high-temperature superconductors. *EPL* **2014**, *107*, 27004. [CrossRef]

75. Yanagisawa, T. Antiferromagnetism, superconductivity and phase diagram in the two-dimensional Hubbard model—Off-diagonal wave function Monte Carlo studies of Hubbard model III—. *J. Phys. Soc. Jpn.* **2019**, *88*, 054702. [CrossRef]

76. Eliashberg, G.M. Interactions between electrons and lattice vibrations in a superconductor. *Sov. Phys. JETP* **1960**, *11*, 696–702.

77. Carbotte, J.P. Properties of boson-exchange superconductors. *Rev. Mod. Phys.* **1990**, *62*, 1027. [CrossRef]

78. McMillan, W.L. Transition temperature of strong-coupled superconductors. *Phys. Rev.* **1968**, *167*, 331. [CrossRef]

79. Allen, P.B.; Dynea, R.C. Transition temperature of strong-coupled superconductors reanalyzed. *Phys. Rev. B* **1968**, *12*, 905. [CrossRef]

80. Ashcroft, N.W. Metallic hydrogen: A high-temperature superconductor? *Phys. Rev. Lett.* **1968**, *21*, 1748. [CrossRef]

81. Drozdov, A.P.; Eremets, M.I.; Troyan, I.A.; Ksenofontov, V.; Shylin, S.I. Conventional superconductivity at 203 kelvin at high pressures in the sulfur hydride system. *Nature* **2015**, *525*, 73. [CrossRef] [PubMed]

82. Peng, F.; Sun, Y.; Pickard, C.J.; Needs, R.J.; Wu, Q.; Ma, Y. Hydrogen clathrate structures in rare earth hydrides at high pressures: Possible route to room-temperature superconductivity. *Phys. Rev. Lett.* **2017**, *119*, 107001. [CrossRef] [PubMed]

83. Liu, H.; Naumov, I.I.; Hoffmann, R.; Schcroft, N.W.; Hemley, R.J. Potential high-T_c superconducting lanthanum and yttrium hydrides at high pressure. *Proc. Natl. Acad. Sci. USA* **2017**, *114*, 6990–6995. [CrossRef] [PubMed]

84. Innocenti, D.; Poccia, N.; Ricci, A.; Valletta, A.; Caprara, S.; Perali, A.; Bianconi, A. Resonant and crossover phenomena in a multiband superconductor: Tuning the chemical potential near a band edge. *Phys. Rev. B* **2010**, *82*, 184528. [CrossRef]

85. Mazziotti, M.V.; Valletta, A.; Campi, G.; Innocenti, D.; Perali, A.; Bianconi, A. Possible Fano resonance for high-T_c multi-gap superconductivity in p-Terphenyl doped at the Lifshitz transition. *EPL* **2017**, *118*, 37003. [CrossRef]

86. Moskalenko, V.A. *Fiz. Metal Metallored.* **1959**, *8*, 2518.

87. Suhl, H.; Mattias, B.T.; Walker, L.R. Bardeen-Cooper-Schrieffer theory of superconductivity in the case of overlapping bands. *Phys. Rev. Lett.* **1959**, *3*, 552. [CrossRef]

88. Peretti, J. Superconductivity of transition elements. *Phys. Lett.* **1962**, *2*, 275. [CrossRef]

89. Kondo, J. Superconductivity in transition metals. *Prog. Theor. Phys.* **1963**, *29*, 1–9. [CrossRef]

90. Stanev, V.; Tesanovic, Z. Three-band superconductivity and the order parameter that breaks time-reversal symmetry. *Phys. Rev. B* **2010**, *81*, 134522. [CrossRef]

91. Tanaka, Y.; Yanagisawa, T. Chiral ground state in three-band superconductors. *J. Phys. Soc. Jpn.* **2010**, *79*, 114706. [CrossRef]

92. Tanaka, Y.; Yanagisawa, T. Chiral state in three-gap superconductors. *Solid State Commun.* **2010**, *150*, 1980–1982. [CrossRef]

93. Dias, R.G.; Marques, A.M. Frustrated multiband superconductivity. *Superconduct. Sci. Technol.* **2011**, *24*, 085009. [CrossRef]

94. Yanagisawa, T.; Tanaka, Y.; Hase, I.; Yamaji, K. Vortices and chirality in multi-band superconductors. *J. Phys. Soc. Jpn.* **2012**, *81*, 024712. [CrossRef]

95. Hu, X.; Wang, Z. Stability and Josephson effect of time-reversal-symmetry-broken multicomponent superconductivity induced by frustrated intercomponent coupling. *Phys. Rev. B* **2012**, *85*, 064516. [CrossRef]

96. Stanev, V. Model of collective modes in three-band superconductors with repulsive interband interactions. *Phys. Rev. B* **2012**, *85*, 174520. [CrossRef]

97. Platt, C.; Thomale, R.; Homerkamp, C.; Zhang, S.C.; Hanke, W. Mechanism for a pairing with time-reversal symmetry breaking in iron-based superconductors. *Phys. Rev. B* **2012**, *85*, 180502. [CrossRef]

98. Maiti, S.; Chubukov, A.V. $s + is$ state with broken time-reversal symmetry in Fe-based superconductors. *Phys. Rev. B* **2013**, *87*, 144511. [CrossRef]

99. Wilson, B.J.; Das, M.P. Time-reversal-symmetry-broken state in the BCS formalism for a multi-band superconductor. *J. Phys. Condens. Matter* **2013**, *25*, 425702. [CrossRef] [PubMed]

100. Ganesh, R.; Baskaran, G.; van den Brink, J.; Efremov, D.V. Theoretical prediction of a time-reversal broken chiral superconducting phase driven by electronic correlations in a single $TiSe_2$ layer. *Phys. Rev. Lett.* **2014**, *113*, 177001. [CrossRef]

101. Yerin, Y.S.; Omelyanchouk, A.N.; Il'ichev, E. Dc SQUID based on a three-band superocnductor with broken time-reversal symmetry. *Superconduct. Sci. Technol.* **2015**, *28*, 095006. [CrossRef]

102. Hillier, A.D.; Quintanilla, J.; Cywinskii, R. Evidence for time-reversal symmetry breaking in the noncentrosymmetric superconductor $LaNiC_2$. *Phys. Rev. Lett.* **2009**, *102*, 117007. [CrossRef]

103. Hase, I.; Yanagisawa, T. Electronic structure of $RNiC_2$ (R = La, Y, and Th). *J. Phys. Soc. Jpn.* **2009**, *78*, 084724. [CrossRef]

104. Yanagisawa, T.; Hase, I. Massless modes and abelian gauge fields in multi-band superconductors. *J. Phys. Soc. Jpn.* **2013**, *82*, 124704. [CrossRef]

105. Lin, S.Z.; Hu, X. Phase solitons in multi-band superconductors with and without time-reversal symmetry. *New J. Phys.* **2012**, *14*, 063021. [CrossRef]

106. Kobayashi, K.; Machida, M.; Ota, Y.; Nori, F. Massless collective excitations in frustrated multiband superconductors. *Phys. Rev. B* **2013**, *88*, 224516. [CrossRef]

107. Koyama, T. Collective modes in multiband superconductors: Rigorous study based on the Ward-Takahashi relations. *J. Phys. Soc. Jpn.* **2014**, *83*, 074715. [CrossRef]

108. Yanagisawa, T.; Tanaka, Y. Fluctuation-induced Nambu-Goldstone bosons in a Higs-Josephson model. *New J. Phys.* **2014**, *16*, 123014. [CrossRef]

109. Tanakai, Y.; Hase, I.; Yanagisawa, T.; Kato, G.; Nishio, T.; Arisawa, S. Current-induced massless mode of the interband phase difference in two-band superconductors. *Physica C* **2015**, *516*, 10. [CrossRef]

110. Valletta, A.; Bianconi, A.; Perali, A.; Saini, N.L. Electronic and superconducting properties of a superlattice of quantum stripes at the atomic limit. *Z. Physik B Condens. Matter* **1997**, *104*, 707. [CrossRef]

111. Choi, H.Y.; Yun, J.H.; Bang, Y.; Lee, H.C. Model for the inverse isotope effect of FeAs-based superconductors in the π-phase-shifted pairing state. *Phys. Rev. B* **2009**, *80*, 052505. [CrossRef]

112. Shirage, P.M.; Kihou, K.; Miyazawa, K.; Lee, C.-H.; Kito, H.; Eisaki, H.; Yanagisawa, T.; Tanaka, Y.; Iyo, A. Inverse iron isotope effect on the transition temperature of the (Ba,K)Fe$_2$As$_2$ superconductor. *Phys. Rev. Lett.* **2009**, *103*, 257003. [CrossRef]

113. Yanagisawa, T.; Odagiri, K.; Hase, I.; Yamaji, K.; Shirage, P.M.; Tanaka, Y.; Iyo, A.; Eisaki, H. Isotope effect in multi-band and multi-channel attractive systems and inverse isotope effect in iron-based superconductors. *J. Phys. Soc. Jpn.* **2009**, *78*, 094718. [CrossRef]

114. Perali, A.; Innocenti, D.; Valletta, A.; Bianconi, A. Anomalous isotope effect near a 2.5 Lifshitz transition in a multi-band multi-condensates superconductor made of a superlattice of stripes. *Superconduct. Sci. Technol.* **2012**, *25*, 124002. [CrossRef]

115. Izyumov, Y.A.; Laptev, V.M. Vortex structure in superconductors with a many-component order parameter. *Phase Transit.* **1990**, *20*, 95. [CrossRef]

116. Volovik, G.E. *The Universe in a Helium Droplet*; Oxford University Press: Oxford, UK, 2009.

117. Kuplevakhsky, S.V.; Omelyanchouk, A.N.; Yerin, Y.S. Soliton states in mesoscopic two-band superconducting cylinders. *J. Low Temp. Phys.* **2011**, *37*, 667. [CrossRef]

118. Tanaka, Y.; Yamamori, H.; Yanagisawa, T.; Nishio, T.; Arisawa, S. Experimental formation of a fractional vortex in a superconducting bi-layer. *Physica C* **2018**, *548*, 44. [CrossRef]

119. Yanagisawa, T.; Hase, I.; Tanaka, Y. Massless and quantized modes of kinks in the phase space of superconducting gaps. *Phys. Lett. A* **2018**, *382*, 3483. [CrossRef]

120. Littlewood, P.B.; Varma, C.M. Amplitude collective modes in superconductors and their coupling to charge-density wave. *Phys. Rev. B* **1982**, *26*, 4883. [CrossRef]

121. Cea, T.; Benfatto, L. Nature and Raman signature of the Higgs amplitude modes in the coexisting superconducting and charge-density-wave mode. *Phys. Rev. B* **2014**, *90*, 224515. [CrossRef]

122. Pekker, D.; Varma, C.M. Amplitude/Higgs modes in condensed matter physics. *Annu. Rev. Condens. Matter Phys.* **2015**, *6*, 269–297. [CrossRef]

123. Cea, T.; Castellani, C.; Seibold, G.; Benfatto, L. Nonrelativistic dynamics of the amplitude (Higgs) mode in superconductors. *Phys. Rev. Lett.* **2015**, *115*, 157002. [CrossRef]

124. Koyama, T. Perturbative approach to the collective modes in the TRSB phase of multiband superconductors. *J. Phys. Soc. Jpn.* **2016**, *85*, 064715. [CrossRef]

125. Yanagisawa, T. Fluctuation modes in multi-gap superconductors. In *Vortices and Nanostructured Superconductors*; Crisan, A., Ed.; Springer: Berlin, Germany, 2017.

126. Yanagisawa, T. Nambu-Goldstone bosons characterized by the order parameters in spontaneous symmetry breaking. *J. Phys. Soc. Jpn.* **2017**, *86*, 104711. [CrossRef]

127. Aitchison, I.J.R.; Ao, P.; Thouless, D.; Zhu, X.M. Effective Lagrangian for BCS superconductors at $T = 0$. *Phys. Rev. B* **1995**, *51*, 6531. [CrossRef] [PubMed]

128. Murotani, Y.; Tsuji, N.; Aoki, H. Theory of light-induced resonances with collective Higgs and Leggett modes in multiband superconductors. *Phys. Rev. B* **2017**, *95*, 104503. [CrossRef]

129. Yanagisawa, T. Theory of Green's functions of Nambu-Goldstone and Higgs modes in superconductors. *J. Superconduct. Novel Magn.* **2019**. [CrossRef]

130. Perali, A.; Bianconi, A.; Lanzara, A.; Saini, N.L. The gap amplification at a shape resonance in a superlattice of quantum stripes: A mechanism for high T_c. *Solid State Commun.* **1996**, *100*, 181–186. [CrossRef]

131. Bianconi, A.; Valletta, A.; Perali, A.; Saini, N.L. Superconductivity of s striped phase at the atomic limit. *Physica C* **1998**, *296*, 269–280. [CrossRef]

132. Kusmartsev, F.V.; Di Castro, D.; Bianconi, G.; Bianconi, A. Transformation of strings into an inhomogeneous phase of stripes and itinerant carriers. *Phys. Lett. A* **2000**, *275*, 118. [CrossRef]

133. Müller, K.A.; Zhao, G.M.; Conder, K.; Keller, H. The ratio of small polarons to free carriers in derived from susceptibility measurements. *J. Phys. Condens. Matter* **1998**, *10*, L291. [CrossRef]

134. Bianconi, A. On the Fermi liquid coupled with a generalized Wigner polaronic CDW giving high T_c superconductivity. *Solid State Commun.* **1994**, *91*, 1–5. [CrossRef]

135. Harris, A.B.; Lange, R.V. Single-particle excitations in narrow energy bands. *Phys. Rev.* **1967**, *157*, 295. [CrossRef]

136. Chao, K.A.; Spalek, J.; Oleś, A.M. Kinetic exchange interaction in a narrow s-band. *J. Phys. C* **1977**, *10*, L271. [CrossRef]

137. Chao, K.A.; Spalek, J.; Oleś, A.M. Canonical perturbation expansion of the Hubbard model. *Phys. Rev. B* **1978**, *18*, 3453. [CrossRef]

138. Micnas, R.; Ranninger, J.; Robaszkiewicz, S. Superconductivity in narrow-band systems with local nonretarded attractive interactions. *Rev. Mod. Phys.* **1990**, *62*, 113. [CrossRef]

139. Robaszkiewicz, S.; Pawlowski, G. Effects of finite pair binding energy in a model for a superconductor with local electron pairing. *Physica C* **1993**, *210*, 61–79. [CrossRef]

140. Arrigoni, E.; Strinati, G.C. Doping-induced incommensurate antiferromagnetism in a Mott-Hubbard insulator. *Phys. Rev. B* **1991**, *44*, 7455. [CrossRef] [PubMed]

141. Kapcia, K.; Robaszkiewicz, S.; Micnas, R. Phase separation in a lattice model of a superconductor with pair hopping. *J. Phys. Condens. Matter* **2012**, *24*, 215601. [CrossRef] [PubMed]

142. Feiner, L.F.; Jefferson, J.H.; Raimondi, R. Effective single-band models for high-T_c cuprates I Coulomb interactions. *Phys. Rev. B* **1996**, *53*, 8751. [CrossRef]

143. Kuroki, K.; Onari, S.; Arita, R.; Usui, H.; Tanaka, Y.; Kontani, H.; Aoki, H. Unconventional Pairing Originating from the Disconnected Fermi Surfaces of Superconducting LaFeAsO$_{1-x}$F$_x$. *Phys. Rev. Lett.* **2008**, *101*, 087004. [CrossRef]

144. Grewe, N.; Steglich, F. Heavy fermions. *Handb. Phys. Chem. Of Rare Earths* **1991**, *14*, 343.

145. Hewson, A.C. *The Kondo Problem to Heavy Fermions*; Cambridge University Press: Cambridge, UK, 1993.

146. Onuki, Y. *Physics of Heavy Fermions: Heavy Fermions and Strongly Correlated Electron Systems*; World Scientific Pub Co Inc.: Singapore, 2018.

147. Ishiguro, T.; Yamaji, K.; Saito, G. *Organic Superconductors*; Springer: Berlin, Germany, 2012.

148. Akashi, R.; Kawamura, M.; Tsuneyuki, S.; Nomura, Y.; Arita, R. First-principles study of the pressure and crystal-structure dependences of the superconducting transition temperature in compressed sulfur hydrides. *Phys. Rev. B* **2015**, *91*, 224513. [CrossRef]

149. Steglich, R.; Asrts, J.; Bredl, C.D.; Lieke, W.; Meschede, D.; Franz, W.; Schäfer, H. Superconductivity in the presence of strong Pauli paramagnetism: CeCu$_2$Si$_2$. *Phys. Rev. Lett.* **1979**, *43*, 1892. [CrossRef]

150. Kittaka, S.; Aoki, Y.; Shimura, Y.; Sakakibara, T.; Seiro, S.; Geibel, C.; Steglich, F.; Ikeda, H.; Machida, K. Multiband superconductivity with unexpected deficiency of nodal quasiparticles in CeCu$_2$Si$_2$. *Phys. Rev. Lett.* **2014**, *112*, 0607002. [CrossRef] [PubMed]

151. Stewart, G.R.; Fisk, Z.; Smith, J.L.; Willis, J.O.; Wire, M.S. New heavy-fermion system NpBe$_{13}$ with a comparison to UBe$_{13}$ and PuBe$_{13}$. *Phys. Rev. Lett.* **1984**, *52*, 679. [CrossRef]

152. Ott, H.R.; Rudingier, H.; Delsing, P.; Fisk, Z. Magnetic ground state of a heavy-electron system: U$_2$Zn$_{17}$. *Phys. Rev. Lett.* **1984**, *52*, 1551. [CrossRef]

153. Palstra, T.T.; Menovsky, A.A.; van den Berg, J. Superconducting and magnetic transitions in the heavy-fermion system URu$_{12}$. *Phys. Rev. Lett.* **1985**, *55*, 2727. [CrossRef] [PubMed]

154. Amitsuka, H.; Sato, M.; Metoki, N.; Yokoyama, M.; Kuwahara, K.; Sakakibara, T.; Morimoto, H.; Kawarazaki, S.; Miyako, Y.; Mydosh, J.A. Effect of pressure on tiny antiferromagnetic moment in the heavy-electron compound URu$_2$Si$_2$. *Phys. Rev. Lett.* **1999**, *83*, 5114. [CrossRef]

155. Ohkuni, H.; nada, Y.; Tokiwa, Y.; Sakurai, K.; Settai, R.; Honma, T.; Haga, Y.; Yamamoto, E.; Onuki, Y.; Yamagami, H.; et al. Fermi surface properties and de Haas-van Alphen oscillation in both the normal and superconducting mixed states of URu$_2$Si$_2$. *Philos. Mag.* **1999**, *B79*, 1045.

156. Hedo, M.; Inada, Y.; Yamamoto, E.; Haga, Y.; Onuki, Y.; Aoki, Y.; Matsuda, D.; Sato, H.; Takahashi, S. Superconducting properties of CeRu$_2$. *J. Phys. Soc. Jpn.* **1998**, *67*, 272–279. [CrossRef]

157. Geibel, G.; Schank, C.; Thies, S.; Kitazawa, H.; Bredl, C.D.; Bohm, A.; Rau, M.; Grauel, A.; Caspary, R.; Helfrich, R.; et al. Heavy-fermion superconductivity at $T=$2K in the antiferromagnet UPd$_2$Al$_3$. *Z. Phys. B* **1991**, *84*, 1–2. [CrossRef]

158. Kyogaku, M.; Kitaoka, Y.; Asayama, K.; Geibel, C.; Schank, C.; Steglich, F. NMR and NQR studies of magnetism and superconductivity in the antiferromagnetic heavy fermion superconductors UM$_2$Al$_3$ (M = Ni and Pd). *J. Phys. Soc. Jpn.* **1993**, *62*, 4016–4030. [CrossRef]

159. Inada, Y.; Yamagami, H.; Haga, Y.; Sakurai, K.; Tokiwa, Y.; Honma, T.; Yamamoto, E.; Onuki, Y.; Yanagisawa, T. Fermi surface and de Haas-van Alphen oscillation in both the normal and superconducting mixed states of UPd$_2$Al$_3$. *J. Phys. Soc. Jpn.* **1999**, *68*, 3643–3654. [CrossRef]

160. Ishida, K.; Mukuda, H.; Kitaoka, Y.; Asayama, K.; Mao, Z.Q.; Mori, Y.; Maeno, Y. Spin-triplet superconductivity in UNi_2Al_3 revealed by the ^{27}Al knight shift measurement. *Phys. Rev. Lett.* **2002**, *89*, 037002. [CrossRef] [PubMed]

161. Petrovic, C.; Pagliuso, P.G.; Hundley, M.F.; Movshovich, R.; Sarrao, J.L.; Thompson, J.D.; Fisk, Z.; Monthoux, P. heavy-fermion superconductivity in $CeCoIn_5$ at 2.3 K. *J. Phys. Condens. Matter* **2001**, *13*, L337. [CrossRef]

162. Izawa, K.; Yamaguchi, H.; Matsuda, Y.; Shishido, H.; Settai, R.; Onuki, Y. Angular position of nodes in the superconducting gap of quasi-2D heavy-fermion superconductor $CeCoIn_5$. *Phys. Rev. Lett.* **2001**, *87*, 057002. [CrossRef] [PubMed]

163. Hegger, H.; etrovic, C.; Moshopoulou, E.G.; Hundley, M.F.; Sarrao, J.L.; Fisk, Z.; Thompson, J.D. Pressure-induced superconductivity in quasi-2D $CeRhIn_5$. *Phys. Rev. Lett.* **2000**, *84*, 4986. [CrossRef] [PubMed]

164. Movshovich, R.; Graf, T.; Mandrus, D.; Thompson, J.D.; Smith, J.L.; Fisk, Z. Superconductivity in heavy-fermion $CeRh_2Si_2$. *Phys. Rev. B* **1996**, *53*, 8241. [CrossRef] [PubMed]

165. Saxena, S.S.; Agarwal, P.; Ahilan, K.; Grosche, F.M.; Haselwimmer, R.K.W.; Steiner, M.J.; Pugh, E.; Walker, I.R.; Julian, S.R.; Monthoux, P.; et al. Superconductivity on the border of itinerant-electron ferromagnetism in UGe_2. *Nature* **2000**, *406*, 587. [CrossRef] [PubMed]

166. Aoki, D.; Huxley, A.; Ressouche, E.; Braithwaite, D.; Flouquet, J.; Brison, J.-P.; Lhotel, E.; Paulsen, C. Coexistence of superconductivity and ferromagnetism in URhGe. *Nature* **2001**, *413*, 613. [CrossRef] [PubMed]

167. Maeno, Y.; Hashimoto, H.; Yoshida, K.; Nishizaki, S.; Fujita, T.; Bednorz, J.G.; Lichtenberg, F. Superconductivity in a layered perovskite without copper. *Nature* **1994**, *372*, 532. [CrossRef]

168. Bauer, E.D.; Frederick, N.A.; Ho, P.-C.; Zapf, V.S.; Maple, M.B. Superconductivity and heavy fermion behavior in $PrOs_4Sb_{12}$. *Phys. Rev. B* **2002**, *65*, R100506. [CrossRef]

169. Takada, K.; Sakurai, H.; Takayama-Muromachi, E.; Izumi, F.; Dilanian, F.A.; Sakai, T. Superconductivity in two-dimensional CoO_2 layers. *Nature* **2003**, *422*, 53. [CrossRef] [PubMed]

170. Chaillout, C.; Remeika, J.P.; Santoro, A.; Mareizo, M. The determination of the Bi valence state in $BaBiO_3$ by neutron powder diffraction data. *Solid State Commun.* **1985**, *56*, 829–831. [CrossRef]

171. Nagamatsu, J.; Nakagawa, N.; Muranaka, T.; Zenitani, Y.; Akimitsu, J. Superconductivity at 39 K in magnesium diboride. *Nature* **2001**, *410*, 63. [CrossRef] [PubMed]

172. Kamihara, Y.; Watanabe, T.; Hirano, M.; Hosono, H. Iron-based layered superconductor $La(O_{1-x}F_x)FeAs$ ($x = 0.05 - 0.12$) with $T_c = 26$ K. *J. Am. Chem. Soc.* **2000**, *130*, 3296. [CrossRef] [PubMed]

173. Kito, H.; Eisaki, H.; Iyo, A. Superconductivity at 54 K in F-free $NdFeAsO_{1-y}$. *J. Phys. Soc. Jpn.* **2008**, *77*, 063707. [CrossRef]

174. Drozdov, A.P.; Minkov, V.S.; Besedin, S.P.; Kong, P.P.; Kuzovnikov, M.A.; Knyazev, D.A.; Eremets, M.I. Superconductivity at 250 K in lanthanum hydride at high pressures. *Nature* **2019**, *569*, 528. [CrossRef] [PubMed]

175. Somayazulu, M.; Ahart, M.; Mishra, A.K.; Geballe, Z.M.; Baldini, M.; Meng, Y.; Struzhkin, V.V.; Hemley, R.J. Evidence for superconductivity above 260 K in lanthanum superhydride at megabar pressures. *Phys. Rev. Lett.* **2019**, *122*, 027001. [CrossRef] [PubMed]

176. Bickers, N.E.; Scalapino, D.J.; White, S.R. Convserving Approximations for Strongly Correlated Electron Systems: Bethe-Salpeter Equation and Dynamics for the Two-Dimensional Hubbard Model. *Phys. Rev. Lett.* **1989**, *62*, 961. [CrossRef]

177. Pao, C.-H.; Bickers, N.E. Anisotropic superconductivity in the 2D Hubbard model: Gap function and interaction weight. *Phys. Rev. B* **1994**, *49*, 1586. [CrossRef]

178. Monthoux, P.; Scalapino, D.J. Self-consistent $d_x^2 - d_y^2$ pairing in a two-dimensional Hubbard model. *Phys. Rev. Lett.* **1994**, *72*, 1874. [CrossRef]

179. Loram, J.W.; Mirza, K.A.; Cooper, J.R.; Liang, W.Y. Electronic specific heat of $YBa_2Cu_3O_{6+x}$ from 1.8 to 300 K. *Phys. Rev. Lett.* **1993**, *71*, 1740. [CrossRef]

180. Hao, Z.; Clem, J.R.; McElfresh, M.W.; Civale, L.; Malozemoff, A.P.; Holtzberg, F. Model for the reversible magnetization of high-κ type-II superconductors: Application to high-T_c superconductors. *Phys. Rev. B* **1991**, *43*, 2844. [CrossRef] [PubMed]

181. Weber, C.; Haule, K.; Kotliar, G. Critical weights and waterfalls in doped charge-transfer insulators. *Phys. Rev. B* **2008**, *78*, 134519. [CrossRef]

182. Hybertsen, M.S.; Schlüter, M.; Christensen, N.E. Calculation of Coulomb-interaction parameter for La_2CuO_4 using a constrained-density-functional approach. *Phys. Rev. B* **1989**, *39*, 9028. [CrossRef] [PubMed]

183. Eskes, H.; Sawatzky, G.A.; Feiner, L.F. Effective transfer for singlets formed by hole doping in the high-T_c superconductors. *Physica C* **1989**, *160*, 424–430. [CrossRef]

184. McMahan, A.K.; Annett, J.F.; Martin, R.M. Cuprate parameters from numerical Wannier functions. *Phys. Rev. B* **1990**, *42*, 6268. [CrossRef]

185. Eskes, H.; Sawatzky, G. Single-, triple , or multiplel-band Hubbard models. *Phys. Rev. B* **1991**, *43*, 119. [CrossRef]

186. Otsuka, H. Variational Monte Carlo studies of the Hubbard model in one- and two-dimensions. *J. Phys. Soc. Jpn.* **1992**, *61*, 1645–1656. [CrossRef]

187. Yanagisawa, T.; Koike, S.; Yamaji, K. *d*-wave state with multiplicative correlation factors for the Hubbard model. *J. Phys. Soc. Jpn.* **1999**, *68*, 3608–3614. [CrossRef]

188. Eichenberger, D.; Baeriswyl, D. Superconductivity and antiferromagnetism in the-dimensional Hubbard model: A variational study. *Phys. Rev. B* **2007**, *76*, 180504. [CrossRef]

189. Baeriswyl, D.; Eichenberger, D.; Menteshashvii, M. Variational ground states of the two-dimensional Hubbard model. *New J. Phys.* **2009**, *11*, 075010. [CrossRef]

190. Baeriswyl, D. Superconductivity in the repulsive Hubbard model. *J. Superconduct. Novel Magn.* **2011**, *24*, 1157–1159. [CrossRef]

191. Yanagisawa, T. Quantum Monte Carlo diagonalization for many-fermion systems. *Phys. Rev. B* **2007**, *75*, 224503. [CrossRef]

192. Misawa, T.; Imada, M. Origin of high-T_c superconductivity in doped Hubbard models and their extensions: Roles of uniform charge fluctuations. *Phys. Rev. B* **2014**, *90*, 115137. [CrossRef]

193. Yanagisawa, T.; Miyazaki, M.; Yamaji, K. Crossover induced electron pairing and superconductivity by kinetic renormalization in correlated electron systems. *Condens. Matter* **2018**, *3*, 26. [CrossRef]

194. Kondo, J. *The Physics of Dilute Magnetic Alloys*; Cambridge University Press: Cambridge, UK, 2012.

195. Yanagisawa, T. Kondo effect in the presence of spin-orbit coupling. *J. Phys. Soc. Jpn.* **2012**, *81*, 094713. [CrossRef]

196. Yanagisawa, T. Kondo effect in Dirac systems. *J. Phys. Soc. Jpn.* **2015**, *84*, 074705. [CrossRef]

197. Ellis, R.K.; Stirling, W.J.; Webber, B.R. *QCD and Collider Physics*; Cambridge University Press: Cambridge, UK, 1996.

198. Nozieres, P.; Schmitt-Rink, S. Bose condensation in an attractive fermi gas: From weak to strong coupling superconductivity. *J. Low Temp. Phys.* **1985**, *59*, 195–211. [CrossRef]

199. Rajaraman, R. *Solitons and Instantons*; North-Holland: Amsterdam, The Netherlands, 1989.

200. Solyom, J. The Fermi gas model of one-dimensional conductors. *Adv. Phys.* **1979**, *28*, 201–303. [CrossRef]

201. Yanagisawa, T. Chiral sine-Gordon model. *EPL* **2016**, *113*, 41001. [CrossRef]

202. Yanagisawa, T. Renormalization group analysis of the hyperbolic sine-Gordon model. *Prog. Theor. Exp. Phys.* **2019**, *2019*, 023A01. [CrossRef]

203. Gross, D.; Neveu, A. Dynamical Symmetry breaking in asymptotically free field theories. *Phys. Rev. D* **1974**, *10*, 3235. [CrossRef]

6

Impact of Surface Roughness on Ion-Surface Interactions Studied with Energetic Carbon Ions $^{13}C^+$ on Tungsten Surfaces

Maren Hellwig [1,*]🆔**, Martin Köppen** [1]🆔**, Albert Hiller** [2]**, Hans Rudolf Koslowski** [2]🆔**,
Andrey Litnovsky** [2]**, Klaus Schmid** [3]**, Christian Schwab** [4] **and Roger A. De Souza** [4]

[1] Independent researchers, Heinersdorfer Str. 52, 13086 Berlin, Germany; martin_koeppen@gmx.de
[2] Forschungszentrum Jülich GmbH, Institut für Energie—und Klimaforschung—Plasmaphysik, 52428 Jülich,
    Germany; a.hiller@fz-juelich.de (A.H.); h.r.koslowski@fz-juelich.de (H.R.K.); a.litnovsky@fz-juelich.de (A.L.)
[3] Max-Planck-Institut für Plasmaphysik, Boltzmannstraße 2, 85748 Garching, Germany;
    klaus.schmid@ipp.mpg.de
[4] Institut für Physikalische Chemie, RWTH Aachen University, Landoltweg 2, 52074 Aachen, Germany;
    schwab@pc.rwth-aachen.de (C.S.); desouza@pc.rwth-aachen.de (R.A.D.S.)
* Correspondence: maren-hellwig@t-online.de

Abstract: The effect of surface roughness on angular distributions of reflected and physically sputtered particles is investigated by ultra-high vacuum (UHV) ion-surface interaction experiments. For this purpose, a smooth ($R_a = 5.9\,\text{nm}$) and a rough ($R_a = 20.5\,\text{nm}$) tungsten (W) surface were bombarded with carbon ions $^{13}C^+$ under incidence angles of 30° and 80°. Reflected and sputtered particles were collected on foils to measure the resulting angular distribution as a function of surface morphology. For the qualitative and quantitative analysis, secondary ion mass spectrometry (SIMS) and nuclear reaction analysis (NRA) were performed. Simulations of ion-surface interactions were carried out with the SDTrimSP (Static Dynamic Transport of Ions in Matter Sputtering) code. For rough surfaces, a special routine was derived and implemented. Experimental as well as calculated results prove a significant impact of surface roughness on the angular distribution of reflected and sputtered particles. It is demonstrated that the effective sticking of C on W is a function of the angle of incidence and surface morphology. It is found that the predominant ion-surface interaction process changes with fluence.

Keywords: roughness; ion-surface interaction; angular distribution; reflection; physical sputtering; deposition; sticking; plasma-wall interaction; secondary ion mass spectrometry; nuclear reaction analysis; SDTrimSP

1. Introduction

Ion-surface interactions such as reflection, physical sputtering, and chemical erosion are of key importance in future nuclear-fusion devices. Nuclear fusion aims for the production of electrical power by using the fusion reaction between the two hydrogen isotopes deuterium and tritium. When the hydrogen isotopes D and T fuse, a He ion and a fast neutron are generated. The He ions as well as other impurities in the plasma are magnetically directed to the armored targets in the divertor region and undergo ion-surface interactions. The armored targets are called plasma-facing components, which will be made of tungsten in future fusion devices due to the beneficial physical and chemical properties of W. The surface of these armored targets will not be polished so that the impact of surface roughness, which will be most likely in a range of R_a ~20 nm due to manufacturing processes, needs to be quantified. Since these components need to withstand extreme heat loads, they are segmented into small tiles, also termed "castellated". Due to their castellated design the crack propagation is limited, and the formation of eddy currents is reduced. However, the segmentation of tiles leads to an enlargement of the total surface area [1]. Since the total amount of retained radioactive T needs to be limited, hydrogen retention in gaps between adjacent tiles is of concern. In present tokamaks, hydrogen is co-deposited with carbon which is eroded from other components and transported along magnetic field lines. Mixed amorphous hydrocarbon layers grow dependent on exposure conditions and period. Typical particle-surface interaction processes which are of key importance for the layer growth are particle reflection, particle sticking, physical sputtering, and chemical erosion. Experiments in the tokamaks TEXTOR and DIII-D [2] have demonstrated that mixed layers of hydrogen and carbon are located at the plasma-closest gap edges within the first mm. Since the used tungsten components have technically finished, but non-polished surfaces [2], the surface roughness is one potential reason for the observed behavior.

The interaction between a projectile and a target atom depends on the species itself, the projectile energy, and angle of incidence. For reflected and sputtered particles, the angular distribution of particles exiting the solid after the interaction is of key importance for successive interaction processes which are responsible for the particle transport into the narrow gaps. The kinematics of the interaction between a projectile and target atoms is described by successive binary collisions within the so-called binary collision approximation (BCA) [3]. The interaction in a collision is determined by the interaction potential which is typically chosen as a screened Coulomb potential [4]. Codes such as the Monte Carlo code SDTrimSP (Static Dynamic Transport of Ions in Matter Sputtering) [5] and MARLOWE [6,7] use the BCA to calculate particle and energy reflection coefficients as well as sputtering yields. Since the chemical composition of the surface and the bulk changes during particle bombardment, the fraction of reflected and sputtered particles is derived fluence dependent in SDTrimSP.

The influence of the surface morphology on the reflection of D on smooth and rough graphite was previously investigated experimentally and modeled with TRIM.SP [8]. It was found that the surface roughness has a minor influence on the reflection coefficient as a function of the angle of incidence. The influence of the surface roughness on the sputtering yields showed a larger re-deposition fraction for unpolished samples [9,10], experimentally as well as in modeling. Furthermore, it was found that the surface roughness modifies the angular dependence of the sputtering yield. Reflection coefficients and sputtering yields were determined for fusion relevant species combinations, but the angular distribution of reflected and sputtered particles of rough surfaces was not investigated so far.

To study the impact of the surface morphology on reflection and physical sputtering, dedicated ion-surface experiments are performed in a specialized ultra-high vacuum (UHV) apparatus. Furthermore, modeling of angular distributions of reflected and sputtered particles is performed with the Monte Carlo code SDTrimSP. Since SDTrimSP does not take surface morphology into account a dedicated set of pre- and post-processing routines for the given experimental surface morphology were developed [11]. The numerical and experimental methods and results are shown, compared, and the discrepancies are discussed in the following.

2. Materials and Methods

2.1. SDTrimSP Simulations

Simulations are performed with the SDTrimSP code for smooth surfaces in three dimensions (3D). To enhance the SDTrimSP results to consider rough surfaces as well, a pre- and post-processing of input and output data is performed as follows: The individual surface topology of rough surfaces induces shadowed regions where projectiles with a fixed angle of incidence cannot directly impinge. Consecutively after the interaction with the solid, the topology limits the number of possible trajectories of particles leaving the solid. Exiting particles can directly interact successively with a hilly surface morphology as depicted in Figure 1. For the simulation procedure, measured surface roughness profiles were taken to apply realistic hilly structures. The multi-step indirect implementation of the surface roughness effect on particle-surface interaction processes is performed via following steps:

1. Pre-processing of SDTrimSP input parameters by computing a normalized angular distribution of incidence angles $\sigma_{in}(\rho_{in}(\alpha_{in}))$ with respect to the smooth surface. (α_{in} = fixed incidence angle to smooth surface, ρ_{in} = incidence angles to rough plane, as depicted in Figure 2);

2. SDTrimSP simulations of the particle bombardment;

3. Post-processing of calculated angular distributions of particles exiting the solid by rotation of the reference system;

4. Test for shadowing: The angular distribution derived in step 3 are convoluted with a reflection probability matrix. A second interaction is treated as deposition on the surface in this approximation.

The transformation of the input data determines a normalized angular distribution of angles of incidence with respect to a global (constant) incidence angle to a smooth sample surface similar to that performed in [9,10]. As depicted in Figure 2, two reference systems are needed for the calculations. SDTrimSP uses the reference system labeled as smooth (grey lines). The second reference system relates to the rough surface (black lines) on the microscopic scale. The rough plane displayed has been determined by an atomic force microscope (AFM) [12] scan of a W sample. The incoming projectile is highlighted with a bold arrow, and two possible trajectories of reflected particles are shown by dashed-dotted lines. AFM scans of each sample surface are performed and respective angular distributions of incoming particles are computed. The surface roughness measurements were performed in three dimensions (3D). Several two-dimensional (2D) line scans were used for the calculation of the normalized angular distribution of incidence angles. It was seen that the measured surface was similar in such a way that the normalized angular distributions of incidence angles of different line scans were very similar. Thus, the surface roughness can only be approximated with SDTrimSP by considering an angular distribution of incoming particles in contrast to one incidence angle for smooth surfaces. The present procedure takes a 3D problem, approximates it in 2D and calculates a 3D output.

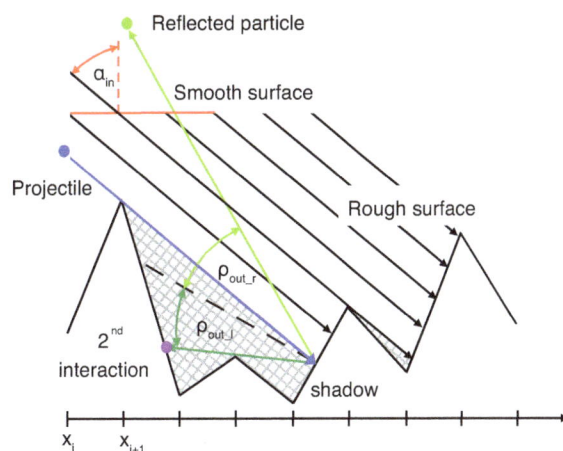

Figure 1. Illustration of interaction angles on rough surfaces showing a range of reflection angles which result in a second interaction.

To simulate the different experimental conditions, SDTrimSP simulations are performed with particle fluences in the range of 1.7×10^{20} m^{-2} and 10×10^{20} m^{-2}, 1×10^7 test particles and a projectile energy of 950 eV. The Krypton-Carbon interaction potential [4] was used. This interaction potential defines the scattering angle of the projectile and target atom in the surface. This interaction potential approximates best the screening length in the given projectile-target constellation. Furthermore, an inelastic energy loss model which combines the Oen-Robinson and Lindhard-Scharff models [8] was used. For smooth surfaces, the incidence angle α_{in} is constant, whereas for rough surfaces the normalized angular distribution of incidence angles $\sigma_{in}(\rho_{in}(\alpha_{in}))$ is used. The computed azimuthal and polar angles are taken as SDTrimSP output and transformed into spherical coordinates. The spherical coordinates are binned into intervals which corresponds to an $(2 \times 2)°$ area. This corresponds to the measurement area of the SIMS measurements as described in Section 2.3.2. Hence, computed particles are sorted according to their coordinates.

After the SDTrimSP simulations for rough surfaces, the results of the angular distributions of particles leaving the solid need to be transformed due to the angle γ_{AFM} between the smooth surface normal and the rough surface normal by rotating the reference systems (like depicted in Figure 2).

In an additional step, the surface topology on the macroscopic scale is accounted for by calculating all possible particle trajectories in 3D for each measured slope from the AFM scan. Trajectories which do not intersect with another surface of a hilly structure, which was taken from a 2D-scan, further away from the initial interaction point are counted as being reflected with a probability of 1. Those trajectories which intersect are considered as deposited particles and are labeled with a reflection probability of 0. Thus, a reflection probability matrix of the surface is created which is used for the convolution of the modeled pre- and post-processed angular distribution of particles leaving the solid. Furthermore, particles with an azimuthal angle of $\pm 1°$ were selected of the 3D-simulation results to be comparable to the experimentally analyzable area of the secondary ion mass spectrometry (SIMS) measurements. In the following, the conduced ion-surface experiments will be described.

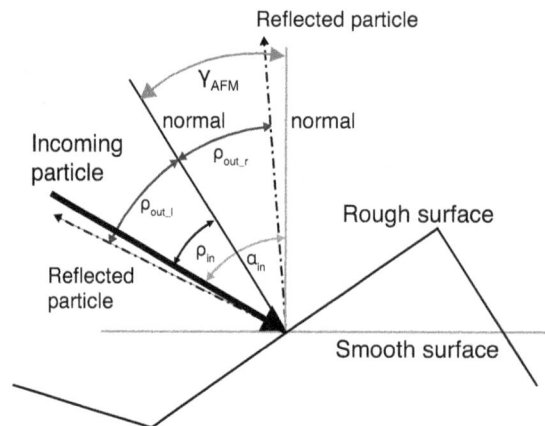

Figure 2. Definition of angles and reference systems for the pre- and post-processing: α_{in} represents the fixed incidence angle to the smooth surface, ρ_{in} the angles of incidence to the rough plane and Γ_{AFM} the angle which is measured with an atomic force microscope (AFM). The angles ρ_{out_l} and ρ_{out_r} represent possible exit angles of one reflected particle. In this picture $|\rho_{out\_l}| = |\rho_{out\_r}|$. For a cosine angular distribution with its center around the surface normal, both angles have the same likelihood.

2.2. Experiments

Investigations of the influence of the surface morphology on the angular distribution of reflected and sputtered particles are performed with C^+ ions impinging W surfaces at 950 eV. Projectiles at the given energy will uniformly bombard the sample surface in contrast to projectiles with energies close to the surface binding energy. It the latter case, the surface morphology with its slopes and sinks could affect the interaction process more strongly, which needs to be investigated in a different study.

C is chosen, since it is one of the main plasma impurities in present-day tokamaks and stellarators. To distinguish C atoms resulting from the individual ion-surface interaction processes from typical surface contamination, $^{13}C^+$ is used due to its low natural isotope abundance of 1.11%. Experiments are performed by bombarding smooth and rough W samples with 950 eV $^{13}C^+$ ions at incidence angles of 30° and 80° to the surface normal. Reflected and sputtered ^{13}C is captured on a cylindrically bent Ti foil for each of the four experiments individually. Exposed Ti foils are analyzed by means of SIMS.

In addition, the areal density of deposited ^{13}C on the W surface is quantified by nuclear reaction analysis (NRA).

2.2.1. Experimental Setup

Experiments were performed in a self-built UHV apparatus ALI [13], schematically depicted in Figure 3. The ion source ionizes the gas by electron impact ionization. After the extraction, an ion lens (1) collimates the ion beam to a spot with a diameter <1 mm. Consecutive deflector plates enable the alignment of the beam in two dimensions. The ions are mass selected in a magnetic sector field. A dedicated system of electrostatic lenses (2) and deflection plates (3) subsequently focuses the ions onto the target. Samples are mounted on a rotatable manipulator system from VG Scienta which is inserted into the vacuum recipient from the top flange. To compute the accumulated fluence during experimental operation, the sample current is recorded.

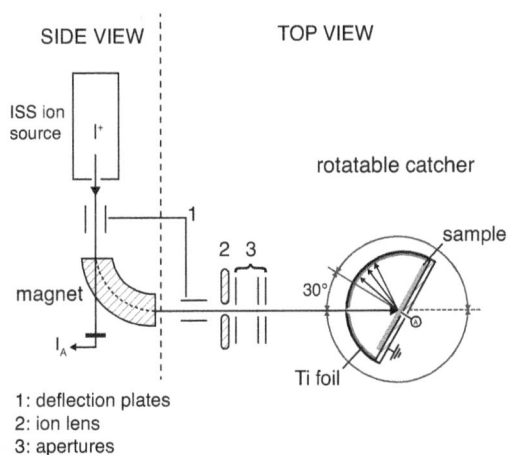

Figure 3. Illustration of the ultra-high vacuum ion beam experiment based on [13].

2.2.2. Sample Preparation

The surfaces of the W samples of 10 mm × 18 mm × 1 mm were prepared by chemical etching. This process yields the so-called *rough* samples. HNO_3 and HF were used as the etching medium. The *smooth* samples were additionally polished mechanically in several steps by using *FEPA P* silicon carbide paper (FEPA, Paris, France) *P800* (21.8 μm ± 1 μm), *P1200* (15.3 μm ± 1 μm), *P2500* (8.5 μm ± 1 μm), *P4000* (6 μm) followed by a diamond suspension with grains of 3 μm size, and a diamond suspension with 1 μm grains. Finally, the surface was polished with OP-S solution, a colloidal silica suspension with grains of 0.04 μm size. Microscopic images of both sample types are shown in Figure 4. The surface topology is measured with an Agilent 5400 AFM (Agilent Technologies, Santa Clara, CA, USA) on 100 μm × 100 μm areas, which was the maximum possible scanning area of the used AFM. The resolution of the measurements is 195 nm which equals 512 × 512 measured points. The scanned areas represent the sample morphology. For the simulations, AFM measurement results of lines scans are used and are depicted in Figure 5. The arithmetic (R_a), ten-point (R_{10}), and root-mean square (R_{rms}) roughness were calculated for both topology types [14] and are summarized in Table 1.

The Ti catcher foils of 125 μm thickness were prepared by cleaning the surface with a radio-frequency Ar plasma in the PADOS device [15]. Thus, surface impurities descending from the manufacturing process and due to air exposure are eliminated. After cleaning, the samples were stored under vacuum to reduce new deposition due to air exposure. It was shown by performing NRA measurements of cleaned and non-cleaned Ti foils that the cleaning process led to a reduction of ^{13}C by 10 times. Titanium was chosen due to a preferable combination of an easy bending of the foil, the interacting species, and the used surface analysis methods.

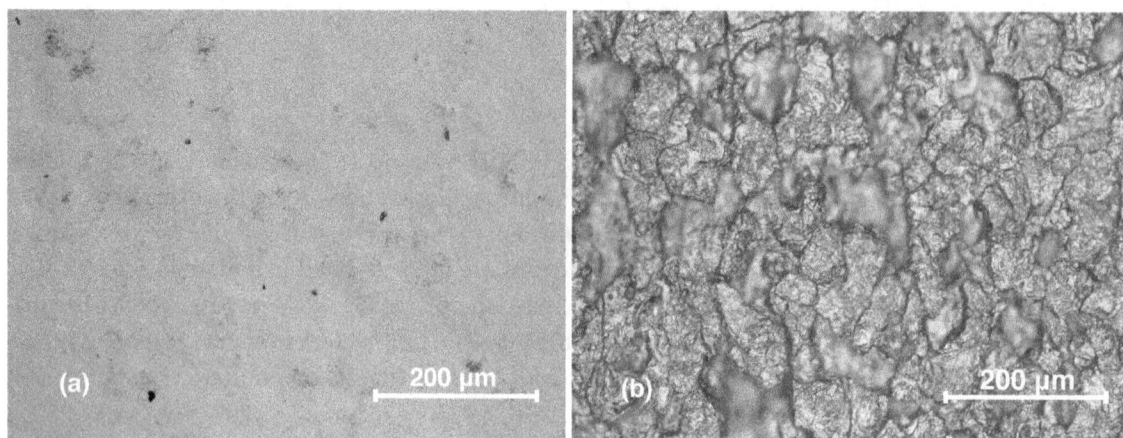

Figure 4. Images of the smooth (**a**) and rough (**b**) W sample taken with an optical microscope.

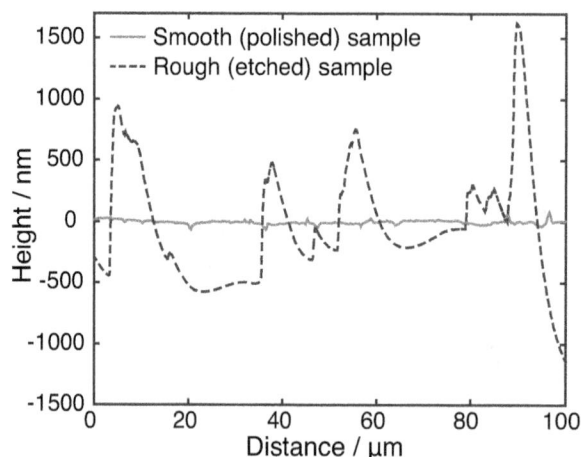

Figure 5. The AFM profiles of the surface topology of a smooth/polished (solid line) and a rough/etched (dashed line) W sample.

Table 1. Roughness parameters of the polished and etched W samples: The arithmetic (R_a), ten-point (R_{10}), and root-mean square (R_{rms}) roughness were calculated for both topology types with formulas from [14].

Surface Type	R_a [nm]	R_{10} [nm]	R_{rms} [nm]
Smooth	5.93	348.52	43.70
Rough	20.53	333.13	569.82

2.2.3. Target and Catcher Design

The target and catcher (Figure 6) are designed to allow an interaction of the projectile beam with the W surface and to collect the reflected and sputtered ^{13}C atoms on the Ti catcher foil in an azimuthal plane of $0°$. The catcher as well as the mounted Ti catcher foil are on ground potential and have a radius of 6 mm. Possible incidence angles of the projectile beam are defined by entry apertures in the catcher and start from normal incidence with $0°$, $30°$, $45°$, $60°$, and $80°$ with respect to the surface normal. For each experiment, a Ti foil with a hole aligned at one of the entry aperture positions is prepared. The W sample is insulated from the catcher with a Kapton foil (hatched) to allow monitoring of the ion beam current.

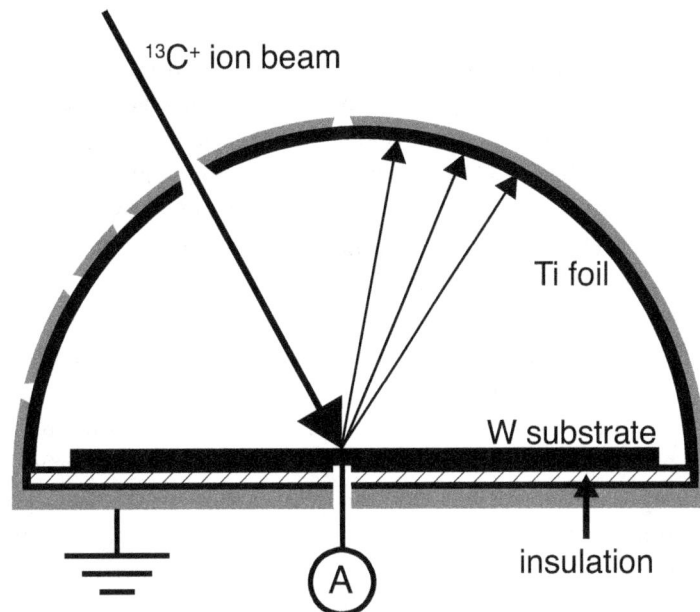

Figure 6. Illustration of the rotatable catcher of the ion-surface interaction experiment. Apertures at $0°$, $30°$, $45°$, $60°$, and $80°$ with respect to the surface normal are visible in the catcher setup (grey). For the experiments individual Ti foils are taken with a hole aligned at one of the entry aperture positions as shown for $30°$.

2.2.4. Experimental Parameters

For each experiment ^{13}C ions were produced by dissociative electron impact ionization of ^{13}CO. The W surface was bombarded up to 450 h to accumulate the ^{13}C deposition necessary for the post-mortem analysis of the Ti foil. The steady-state beam bombardment was controlled by using a high precision thermomechanical valve from Pfeiffer vacuum with thermomechanical control of the gas flux into the source. A base pressure below 5×10^{-8} hPa is achieved in the recipient leading to a mean free path for carbon ions of the order of 10^6 m. The accumulated particle fluence was determined by measuring the sample current as a function of time in combination with a determination of the spot area after the exposure.

2.3. Analyses

The bombarded W surfaces as well as the Ti foils are analyzed with respect to the deposited C. An absolute, quantitative analysis of the deposited C on the W surface as well as the reflected and sputtered C collected on the Ti foils is crucial to understand the underlying ion-surface interaction processes. By combining a NRA of W surfaces with an analyses with SIMS of the Ti foils, it is possible to compare absolute values of the conducted experiments and thus to isolate important mechanisms in the following.

2.3.1. NRA

The ^{13}C deposition on the W samples is analyzed with NRA [16] to determine quantitatively the areal density of ^{13}C. The measurements were performed with the reaction ^{13}C(^{3}He,p_0)^{15}N. A ^{3}He beam with 3.15 MeV and a spot diameter of 1 mm is used to trigger nuclear reactions with the ^{13}C deposition on the W samples. RBS (Rutherford backscattering spectrometry) and NRA spectra are measured under a scattering angle of $165°$ for two samples and two spots on each sample. The areal density of ^{13}C is determined with the SIMNRA program [17] by fitting experimentally obtained spectra to simulated spectra. A normalized particle balance for each experiment is estimated via the ratio of the deposited ^{13}C and the corresponding accumulated fluence on the sample.

2.3.2. SIMS

The angular distribution of reflected and sputtered ^{13}C is quantified by multiple SIMS [18] measurements on the Ti catcher foils. SIMS measurements are performed in fashion of a line scan along the median of the Ti foil at an azimuthal angle of 0° with a *Time-of-Flight SIMS IV* machine (IONTOF GmbH, Münster, Germany) at the RWTH Aachen University. Single measurement areas are scanned with a 2 keV Cs$^+$ sputter beam over an area of 200 µm $\times$ 200 µm and a 25 keV Ga$^+$ analysis beam which scanned 81.3 µm $\times$ 81.3 µm inside the sputtered area to eliminate edge effects. The analysis beam focus is adjusted for each measurement since the Ti foil remained slightly bent due to the experimental setup, although the bending of the Ti foil is minimized by the holder setup.

As mentioned above, absolute values are needed to compare the four conducted experiments. Thus, in the following the evaluation applied is described in detail since it deviates significantly from the standard evaluation. In the SIMS analysis the ^{13}C to ^{12}C ratio is estimated. Hydrocarbons are found and need to be considered additionally since ^{13}CH shows a non-proportional behavior to ^{12}CH into the depth of the sample. This is attributed to the foil exposure in ALI. Special care is taken during the measurement to optimize the separation of the ^{13}C and ^{12}CH as well as the ^{12}CH$_2$ and ^{13}CH mass peaks. To separate the isobaric interferences correctly, a non-standard analysis method is applied. The analysis procedure consists of the following steps, which will be described in the following:

1.   Reconstruction of mass spectra for recorded time steps.
2.   Dead time correction of mass peaks ^{12}C, ^{12}CH, ^{13}C, ^{12}CH$_2$, and ^{13}CH.
3.   Peak fitting of ^{12}C, ^{12}CH, ^{13}C, ^{12}CH$_2$, and ^{13}CH.
4.   Determination of the depth profiles of individual species.
5.   Background correction of the ^{13}C isotope measurement.

The data reconstruction and dead time correction is performed with the software Surface Lab 6.5 (ION-TOF GmbH, Münster, Germany) [19]. For the deconvolution of isobaric interferences of mass 13 and 14 the software CasaXPS 2.3.16 PR 1.6 (Casa Software Ltd., Teignmouth, UK) [20] is used. Especially for the first few analysis cycles the peak fitting results demonstrated severe differences to the standard SIMS method. This is caused by a strong peak asymmetry with a pronounced tail on the right side as depicted in Figure 7a. The peak shape originates from the measurement setup which causes an angular aberration and from calibration factors such as the high reflector voltage of U = −50 V which causes a spread in the arrival time of secondary ions in the detector [21]. Each side of the mass peak is thus fitted with a Lorentz distribution [20] with a full-width-half-maximum (FWHM) f and the peak position e with $x \leq e$ for the left side and $x > e$ for the right side:

$$L_{asym}\left(x : \alpha, \beta, f, e\right) = \begin{cases} [L\left(x : f, e\right)]^{\alpha} & x \leq e \\ [L\left(x : f, e\right)]^{\beta} & x > e \end{cases} \tag{1}$$

so that L_{asym} describes the left-to-right peak asymmetry. For continuity both Lorentz distributions are convoluted with a Gaussian distribution. In the present case $\alpha > \beta$ so that the asymmetric tail to the right side is reproduced. The exact peak shape is determined by fitting the non-disturbed ^{12}C peak. Shape shifting [22,23] to other peaks is applied in the fitting process to each time step for a single measurement crater. The integration of the fitted curves yields the depth profiles of the individual masses.

The background correction is required due to the natural ^{13}C abundance. Figure 7c shows that the Ti foil is contaminated with ^{12}C and ^{13}C throughout the analyzed layers. The ^{13}C deposition originating from the experiment is present in roughly the first 50 sputter seconds which corresponds to the depth. The intensity is corrected in a multi-step procedure. The minimum of the summed intensity counts of ^{13}C and ^{13}CH is computed (k) and serves as the start depth to calculate the mean ratio of background counts $\langle b \rangle$ for time steps i and the maximum number of single measurements N with

$$\langle b \rangle = \frac{1}{N} \sum_{i=k}^{N} \frac{I_{^{13}C,i} + I_{^{13}CH,i}}{I_{^{13}C,i} + I_{^{13}CH,i} + I_{^{12}C,i} + I_{^{12}CH,i} + I_{^{12}CH_2,i}} \tag{2}$$

$$\sigma(b) = \frac{1}{N} \sum_{i=k}^{N} (b_i - \langle b \rangle)^2 \tag{3}$$

at each measurement position individually. The natural abundance level of ^{13}C amounts to 1.11%. The experimentally determined abundance level of ^{13}C has a maximum error interval of (0.7–2.2)% which is a result of a small deviation in the analysis beam focus. The background intensity counts $(I_{^{13}C} + I_{^{13}CH})_{Bkg}$ are computed for each sputter time step i by

$$\left[(I_{^{13}C} + I_{^{13}CH})_{Bkg} \right]_i = \frac{\langle b \rangle}{1 - \langle b \rangle} \cdot \left[I_{^{12}C,i} + I_{^{12}CH,i} + I_{^{12}CH_2,i} \right] \tag{4}$$

as a fraction of the intensity counts of ^{12}C and its hydrocarbons. The background $(I_{^{13}C} + I_{^{13}CH})_{Bkg}$ is subtracted from the signal $(I_{^{13}C} + I_{^{13}CH})$ in order to obtain the ^{13}C deposition $(I_{^{13}C} + I_{^{13}CH})_{Exp}$ resulting from reflection and self-sputtering during the ion-surface interaction experiments for each time step i with

$$\left[(I_{^{13}C} + I_{^{13}CH})_{Exp} \right]_i = \left[I_{^{13}C,i} + I_{^{13}CH,i} \right] - \left[(I_{^{13}C} + I_{^{13}CH})_{Bkg} \right]_i . \tag{5}$$

An example of the background corrected ^{13}C depth profile is shown in red in Figure 7c.

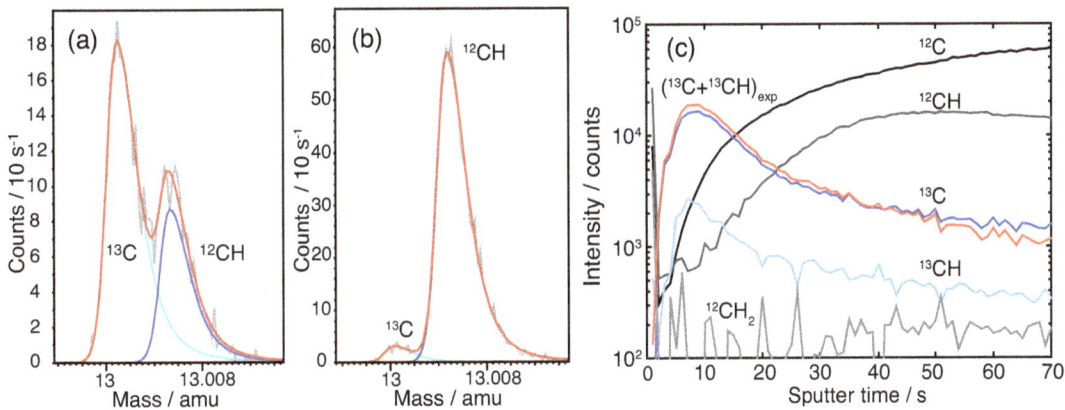

Figure 7. Extraction of the C intensities via peak fitting with CasaXPS [20] to reconstruct the corresponding depth profile for each SIMS crater. As an example the deconvolution of two mass spectra is shown. Mass spectrum (**a**) is taken after the 9th sputtering cycle and mass spectrum (**b**) after the 69th sputtering step. The mass spectrum illustrates the convolution of the ^{13}C and CH mass peaks in different sample depths. The light and dark blue curves show the ^{13}C and ^{12}CH fit, respectively. The red curve represents the envelope curve. Graph (**c**) depicts the results of the SIMS analysis for one crater. Each curve shows the contribution of the respective component to the mass spectrum as a function of sputter time.

3. Results

The results of the SDTrimSP simulations regarding effective reflection coefficients and sputtered fractions as a function of the fluence are shown and discussed. Furthermore, the experimentally obtained angular distributions are compared with corresponding simulations.

3.1. Incidence Angle of 30°

$^{13}C^+$ bombardment of a W surface with 950 eV leads to a composition change of the surface which is fluence dependent. For a smooth surface, the calculated fraction of reflected ^{13}C as well as the fraction of sputtered W is reduced by 0.03 as depicted in Figure 8a. A small increase of the self-sputtered fraction of ^{13}C up to 0.01 due to the deposition of ^{13}C during the exposure is observed. For the rough surface (Figure 8b) a significant different behavior is seen in comparison to the smooth surface. A greater amount of ^{13}C is deposited on the surface and the effective reflection decreases by 0.20. The sputtering of W decreases by 0.20 as well. The dominating process changes as a function of the fluence. In the last quarter of the experimental time which corresponds to particle fluence above $5.5 \times 10^{20} m^{-2}$ in this plot, self-sputtering of ^{13}C gets dominant compared to the decreasing reflected fraction of ^{13}C and sputtered fraction of W. At a fluence of roughly $7 \times 10^{20} m^{-2}$ self-sputtering occurs in 43% of the ^{13}C interactions.

Figure 9a displays the angular distributions obtained for the smooth surface. The angle of incidence corresponds to $-30°$. All particles which are reflected or sputtered with a negative velocity component v_y are shown at negative exit angles. Thus, all particles with a positive velocity component are depicted with positive exit angles since they are directed in forward direction. The experimental data which is derived via the SIMS analysis routine is depicted in red squares. A mainly specular reflection at $+30°$ is observed experimentally. The corresponding SDTrimSP simulation is shown in black. In comparison to the grey curve, which represents a cosine distribution, the simulation results show a similar broad, almost cosine angular distribution. Considering only interactions with collision cascades with less than (i) 20 collision (green) or (ii) less than 10 (blue) collisions in a cascade, a shift of the distribution maximum is observed. Thus, the location of the maximum depends on number of collisions in a cascade.

Figure 9b shows the angular distributions obtained when a rough surface topology is bombarded. The experimentally obtained distribution (red squares) is a symmetric distribution due to the dominant self-sputtering process. The calculated angular distribution which considers all collisions shows a broad, almost cosine distribution in black. Selecting the maximum number of collisions (in green and blue) does not show a change of the location of the maximum with the number of collisions in a collision cascade.

Furthermore, the NRA analysis of the smooth and rough W samples reveal a similar deposited ^{13}C areal density D_{NRA} with $D_{NRA, rough} / D_{NRA, smooth} \sim 0.92 \pm 0.08$. This result is compared to the SDTrimSP results of the integrated sputtered and reflected fractions: Although 34% less ^{13}C is reflected from the rough surfaces in comparison to the smooth surface over the experimental duration, a roughly equal amount of ejected ^{13}C is calculated for the rough surface. The integrated ^{13}C amount which is ejected by self-sputtering is a factor of 4.7 higher for the rough surface and thus compensates for the integrated smaller reflected fraction of ^{13}C of the rough surface. Therefore, an equivalent amount of ^{13}C is reflected ($A_{SDTrimSP}$) as well as deposited on the rough and smooth W sample for an incidence angle of 30° with $A_{SDTrimSP, rough} / A_{SDTrimSP, smooth} \sim 0.98$. The SDTrimSP results agree with the NRA analysis results for the ion-surface interaction experiment at an incidence angle of 30°.

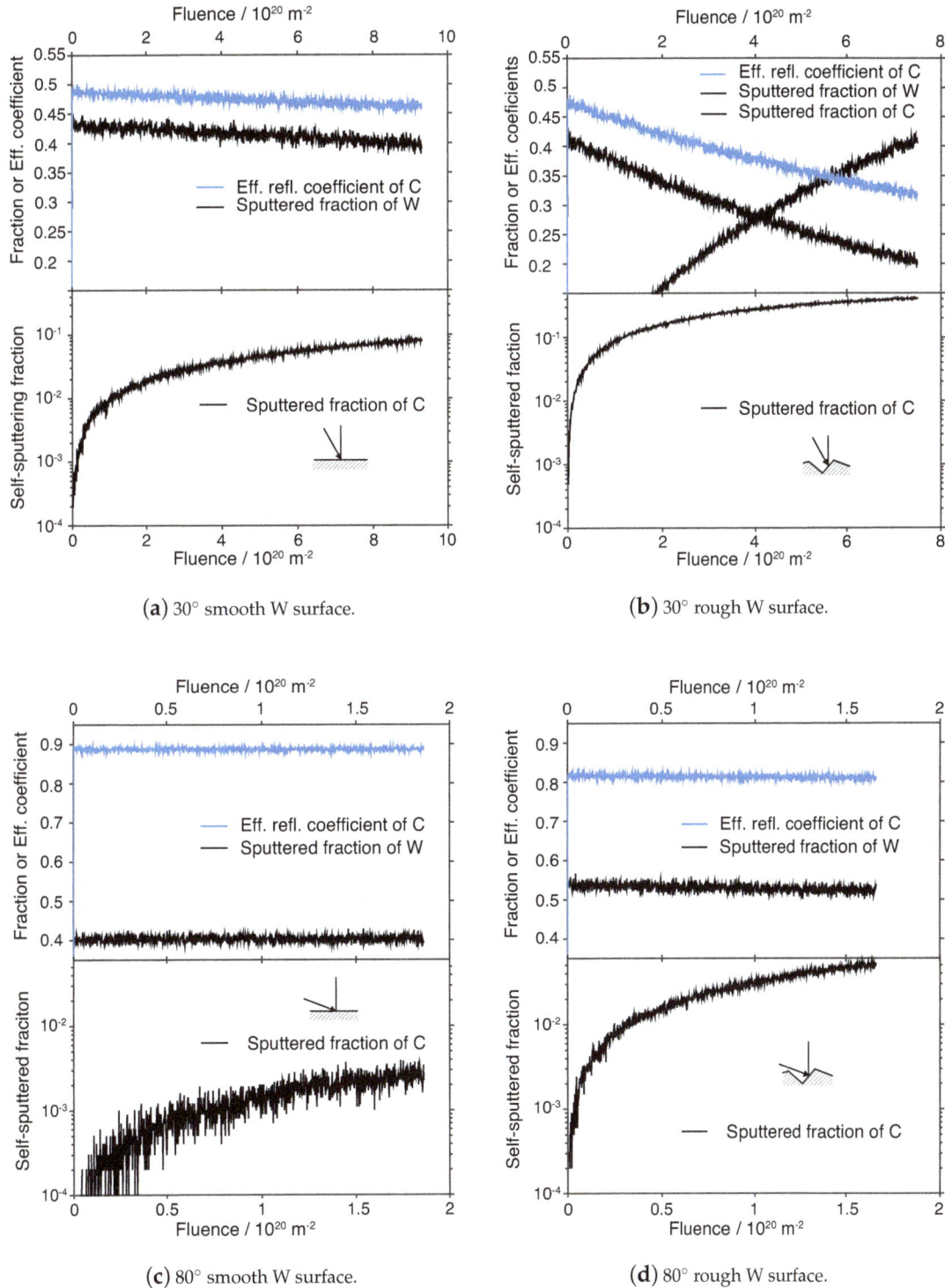

Figure 8. The fluence dependence of sputtered fractions and effective reflection coefficients computed by SDTrimSP is shown for different angles of incidence and surface morphology. The upper graph of each subplot illustrates the effective ^{13}C reflection coefficient and the sputtered fraction of W. The respective lower graph shows the self-sputtered fraction of ^{13}C which is plotted on a logarithmic scale for better visibility (In case (**b**) the self-sputtered fraction of ^{13}C is also drawn in the upper graph). (**a,b**) show the bombardment of ^{13}C under an incidence angle of 30° on a smooth and rough W surface, respectively. (**c,d**) show the bombardment of ^{13}C under an incidence angle of 80° on a smooth and rough W surface, respectively.

(a) 30° smooth W surface.

(b) 30° rough W surface.

(c) 80° smooth W surface.

(d) 80° rough W surface.

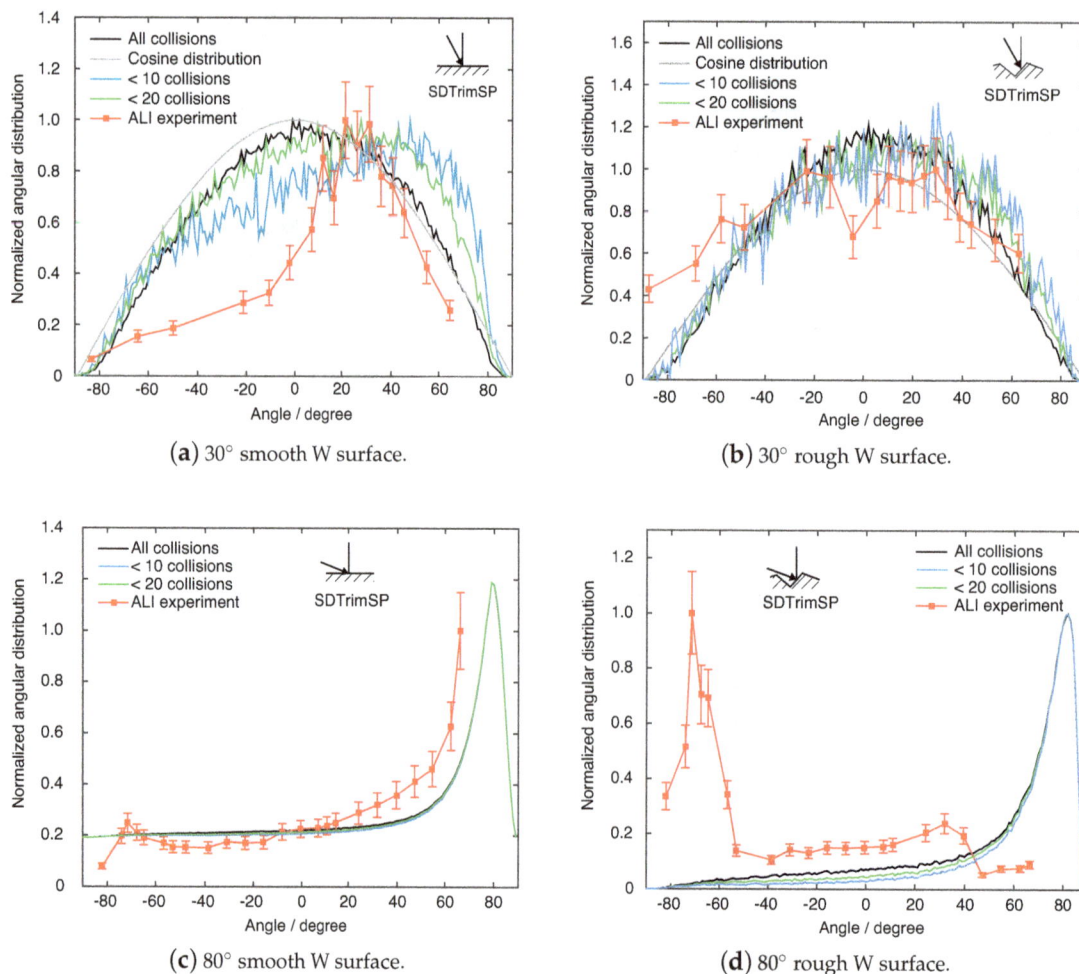

Figure 9. Comparison of experimentally and modeled angular distributions of ^{13}C ions impinging on W surfaces. The experimental data is depicted with red squares. The modeled data is illustrated in black, grey, green, and blue lines. Graph (**a**) shows results for the incidence angle of 30° on a smooth W sample. Graph (**b**) depicts the results for an incidence angle of 30° on a rough W sample. Graph (**c**) displays the results at an incidence angle of 80° on a smooth W sample and graph (**d**) the results of the bombardment at an incidence angle of 80° on a rough W sample.

3.2. Incidence Angle of 80°

Ions which bombard a smooth and rough surface under an incidence angle of 80° show a completely different behavior of reflected and sputtered fractions as well as angular distributions. The fraction of reflected ^{13}C from a smooth W surface (Figure 9c) is about 0.90 and remains steady. In addition, W is sputtered with a fraction of 0.40 at a constant level. Thus, the surface composition does not change as drastically as observed at an incidence angle of 30° under similar conditions, so that self-sputtering of ^{13}C has a sputtering yield of 0.003 for a smooth surface. For rough surfaces still a fraction of 0.82 of the ^{13}C is reflected and fraction of 0.53 of W is sputtered (Figure 8d). Both fractions decrease only slightly about 0.02 up to maximum fluence. The self-sputtered fraction of ^{13}C does increase more rapidly up to a fraction of 0.05 compared to the smooth case.

Figure 9c depicts the angular distribution of reflected and sputtered ^{13}C. The experimentally obtained distribution (red squares) shows a mainly specular reflection in forward direction. Due to technical limitations it was not possible to access data points beyond 70°. The simulation exhibits a similar distribution which shows specular reflection with a maximum at 78° (black curve). Filtered simulation results with a selected maximum number of collisions in a collision cascade are plotted for

(i) less than 20 (green) and (ii) less than 10 (blue) collisions. It can be seen that the location of maximum is not correlated with the number of collisions in a cascade for the given case.

Figure 9d shows the angular distributions which are obtained by the bombardment of a rough surface. The experimentally obtained distribution (red squares) demonstrates mainly backward reflected ^{13}C with a maximum at $-70°$. Calculated angular distributions however show a specular reflection in forward direction with a maximum at $+78°$. A limitation of the maximum number of collisions in a collision cascade does not lead to significant differences in the distribution. Again, no correlation of the maximum of the distribution with the number of collisions in a cascade is found.

The NRA analysis of the W samples exhibit that for particles bombarding the rough and smooth sample at incidence angle of $80°$ a roughly five times greater areal density of ^{13}C ($D_{NRA,rough} / D_{NRA,smooth} \sim 4.81 \pm 0.08$) is found on the rough surface. The SDTrimSP results of the integrated sputtered and reflected fractions indicate a similar trend: A 18% greater amount of ^{13}C is reflected from the smooth surface in comparison to the rough surface. In addition, the self-sputtered fraction of ^{13}C from the smooth surface is 5% of the self-sputtered fraction ^{13}C from the rough surface. The simulation results suggest in total a ratio of the ejected ^{13}C of $A_{SDTrimSP,rough} / A_{SDTrimSP,smooth} \sim 0.84$ which leads to a 1.19 greater deposited ^{13}C amount on the rough sample. Although the simulation results are not in agreement with the numbers of the NRA analysis results, a trend towards a greater deposition on the rough surface is demonstrated.

4. Discussion

It is observed that the number of collisions in the collision cascade is important with respect to the agreement between experimental and calculated angular distributions. Figure 10a illustrates the energy distribution of reflected and sputtered particles leaving the surface. Figure 10b shows the distribution of the number of collision of particles leaving the surface. The numerical values of these figures are compiled in Table 2. For an incidence angle of $30°$ the energy distribution is broad with a maximum at roughly 700 eV and an average of 500 eV for a smooth and rough surface. The average numbers of collisions in a cascade is calculated to amount to roughly 19 collisions with an average pathlength of roughly 45 nm before the particle exists the surface.

At an incidence angle of $80°$ the energy distribution shows that particles leave the solid with an average of 758 eV for the rough surface and 815 eV for the smooth surface. The maximum of the energy distribution is at roughly 900 eV. As a consequence, particles which interact with the solid collide on average about 7 times with target atoms with an average pathlength of 14–20 nm. The corresponding distribution of the number of collisions has its maximum around 4 collisions.

Additionally, to the SDTrimSP simulations, calculations were performed with the MARLOWE code [6,7,24] which also uses the BCA to compute particle-surface interactions. Differently to SDTrimSP, MARLOWE takes a crystalline solid into account. SDTrimSP results for the length of the collision cascades as well as the distributions are confirmed with MARLOWE simulations for the smooth surface (Table 3). The artifact, shown as a step in the energy distributions at 700 eV of the SDTrimSP simulations in Figure 10a is found with MARLOWE calculations as well. The reason for the artifact needs to be investigated further and is most likely generated by the implemented energy loss approximations in both codes. This investigation is beyond the scope of the presented study but must be pursuit further. In general, a larger angle of incidence leads to smaller number of collisions in a collision cascade (for $30°$: 19 collisions; for $80°$: 7 collisions) and higher exit energies of reflected particles.

The SDTrimSP results show a correlation between the number of collisions in a collision cascade and the corresponding shape of the angular distribution of reflected particles. Particles which are reflected after less than 6 collisions show a more specular angular distribution. Thus, the momentum components parallel to the surface is not significantly changed. Particles which are reflected after more than 6 collisions in a collision cascade are more likely to be ejected with a cosine distribution.

(**a**) Distribution as a function of the energy.

(**b**) Distribution as a function of the number of collisions.

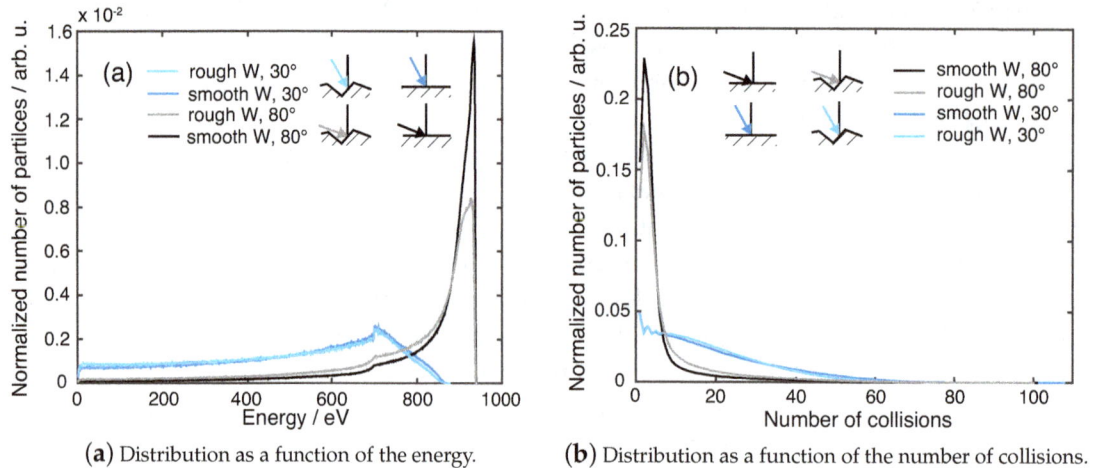

Figure 10. Normalized distributions of (**a**) the energy of the reflected and sputtered particles and (**b**) the number of collisions in the collision cascade for incidence angles of 30° and 80° calculated for smooth and rough W surfaces.

Table 2. Summary of the energy, number of collisions, and collision cascade length from reflected and sputtered ^{13}C projectiles which are calculated via SDTrimSP. Each simulation is performed with 10^7 particles, the number of particles which exit the target are shown as well.

Surface Type	Incidence Angle/°	Average Energy/eV	Max. # of Collisions	Average # of Collisions	Average Path-Length/nm	Max. Path-Length/nm	Particles Exited Solid
Smooth	30	500	108	19	47.6	266.6	4.7×10^6
	80	815	101	6	13.5	249.5	8.8×10^6
Rough	30	476	92	18	43.2	224.7	3.9×10^6
	80	758	100	8	19.6	240.1	8.1×10^6

Table 3. Summary of the energy and number of collisions from reflected ^{13}C projectiles which are computed with MARLOWE for a smooth W surface. For the simulation of the 30° case 5×10^6 particles and for the 80° case 2×10^6 particles are considered.

Incidence Angle/°	Average Energy/eV	Max. Energy/eV	Average # of Collisions	Max. # of Collisions	Particles Exited Solid
30	517	875	16	128	1.6×10^6
80	809	913	5	178	1.8×10^6

Furthermore, it is demonstrated that challenges arise from the comparison between experiment and simulation. One possible contribution to the found discrepancies is the approximation of a 3D surface morphology to a 2D problem for the input of SDTrimSP. Thus, 3D simulations with the 3D morphology should be performed in future studies. While particles, in the present study, impinging at an incidence angle of 30° on a rough W surface and 80° on a smooth surface, experimental and numerical angular distributions are in rather good agreement. For the case of particles impinging at an incidence angle of 30° on a smooth W surface and at an incidence angle of 80° on a rough W surface, discrepancies are observed and discussed in the following.

For particles impinging with an incidence angle of 30° on a smooth W surface, the experiment shows a higher specular reflection than observed in modeling. Particles in the simulation have a greater number of collisions and are ejected with a cosine distribution. Thus, by comparing experimental and numerical data, the location of the maximum of the angular distribution is correlated with the number of collisions in the collision cascade.

For ^{13}C ions which impinge a rough W surface under 80° a huge discrepancy is demonstrated since the experimentally found backward reflection is not reproduced in the modeling. Both cases which do not show a sufficient agreement between modeling and experiment show a common relation in the comparison of the average pathlengths. Comparing the pathlength of one angle of incidence and different surface roughness, the 30° smooth and 80° rough cases have a greater pathlength and more collisions in a collision cascade compared to the other topology.

5. Conclusions

The effect of the W surface roughness on the angular distribution of reflected and sputtered $^{13}C^+$ is investigated experimentally and numerically. By choosing $^{13}C^+$ projectiles, the natural abundance of 1.11% compared to ^{12}C is used in these experiments. The cleaning process of the Ti foil eliminates the impurities on the surface descending from the manufacturing process and from air exposure. The amount of ^{13}C is reduced by a factor of 10.

It is confirmed that sputtering leads to a cosine angular distribution, whereas reflection with projectile energies of 950 eV is demonstrated to be mainly specular. The surface roughness induces a change in the effective particle and energy reflection coefficients and sputtered fractions due to the fluence dependent change of the elemental composition of the surface (e.g., as seen in Figure 8b). In this study, it is shown that a higher surface roughness leads to an increase in absolute values of the effective reflection coefficient and sputtered fractions as the angle of incidence with respect to the surface normal is increased. Furthermore, an enhancement of ^{13}C sticking for large angles of incidence is demonstrated experimentally: A five times higher ^{13}C deposition on W was measured after the ion bombardment of the rough W surface under an angle of incidence of 80°. In the experiment we found differences of the deposited ^{13}C amount on W and the angular distribution of reflected and sputtered ^{13}C are likely to be of a geometrical origin. Particles impinging at a shallow angle to the surface almost directly get caught in the morphology surrounding the impingement point. Particles which are directly reflected correspond to a bombardment at 10° to the surface normal of a smooth surface which explains the experimentally found angular distribution in the 80° rough case. Performed simulations demonstrate the experimental trend of a higher deposition on the rough surface compared to the smooth surface by predicting a 1.19 higher deposition on the rough sample but show an angular distribution of projectiles which corresponds to forward scattering. The codes based on the BCA neither predict the experimental angular distributions in the given energy range and projectile-target combination nor reproduce the higher effective sticking probability for an angle of incidence of 80°. More complex simulations in 3D with a 3D surface topology are required here.

It is found that the angular distribution of reflected particles is correlated with the number of collisions in a collision cascade. For specular reflection, the collision cascade has a significantly lower number of collisions in a cascade. In addition, the momentum of the incoming particle parallel to the surface is better conserved when the collision cascade has a smaller number of collisions.

It is demonstrated that the surface roughness significantly alters the ion-surface interaction processes. The findings of this study not also play a role for fusion devices, but can also be used for vapor deposition methods. For future fusion devices such as ITER and DEMO, these findings are essential for codes which treat plasma-wall interaction processes. The codes need to consider the roughness induced effects. Otherwise consecutive reflection processes could lead to energy distributions of reflected particles which are shifted to too low energies compared to experimental energy distributions, since less collisions in a collision cascade result in higher particle energies of reflected particles. Thus, iterative erosion and deposition calculations could obtain a significant systematic error.

These implications lead to the need to further characterize surface roughness effects on ion-surface interaction processes with additional dedicated experiments. Future experiments should be accompanied by simulations with a complexity beyond the BCA and with a 3D surface morphology e.g., with molecular dynamics simulations. These studies should address the correlation of different roughness parameters with the effective sticking probability with respect to the angle of incidence of the projectile. It should be investigated if a critical threshold of the roughness is existent and the significance of its influence on surface interaction processes. Furthermore, the influence of different roughness parameters on the reflection and its angular distribution should be studied, such as the skewness, which gives a measure of the asymmetry of the surface topology or the kurtosis which is a measure of the sharpness of the surface topology. Additionally, it is interesting to investigate if the surface morphology changes over the bombardment time and particle fluence, respectively. Since the present study shows a proof of principle of the effect for one projectile-target combination, a huge parameter space yet needs to be investigated. The impact of the mass ratio between the projectile and target atoms (reduced mass) in correlation with the angle of incidence as well as charge effects also should be carefully investigated to elucidate the underlying physics.

Author Contributions: Conceptualization: M.H., M.K., H.R.K. and A.L.; Methodology: M.H.; Software: M.H. and K.S.; Validation: M.H., M.K., H.R.K., K.S., C.S. and R.A.D.S.; Formal analysis: M.H.; Investigation: M.H., M.K., A.H., H.R.K. and C.S.; Resources: H.R.K. and R.A.D.S.; Data curation: M.H., M.K.; Writing—original draft preparation: M.H.; Writing—review and editing: M.H., M.K., H.R.K., A.L., K.S. and R.A.D.S.; Visualization: M.H.; Supervision: H.R.K. and A.L.; Project administration: M.H. and H.R.K.; Funding acquisition: A.L.

Funding: This research received no external funding.

Acknowledgments: The ion-surface interaction experiments (ALI) were performed at the Institute for Energy and Climate Research: Plasma Physics at the Forschungszentrum Jülich GmbH. The NRA measurements were performed at the Peter-Grünberg-Institute - Institute of Semiconductor Nanoelectronics at the Forschungszentrum Jülich GmbH. The SIMS measurements were performed at the Institute for Physical Chemistry, RWTH Aachen University. This study is an excerpt of the Phd thesis "Influence of 3D geometry and surface roughness on plasma-wall interaction processes on tungsten surfaces" by Maren Hellwig at the Ruhr University Bochum and at the Institute for Energy and Climate Research: Plasma Physics at the Forschungszentrum Jülich GmbH. She gratefully thanks the referees of her Phd thesis Christian A. E. Linsmeier and Achim von Keudell.

Conflicts of Interest: The authors declare no conflict of interest.

Abbreviations

The following abbreviations are used in this manuscript:

AFM	Atomic Force Microscopy
BCA	Binary collision approximation
NRA	Nuclear reaction analysis
SIMS	Secondary ion mass spectrometry
UHV	Ultra-high vacuum

References

1.   Krieger, K.; Jacob, W.; Rudakov, D.; Bastasz, R.; Federici, G.; Litnovsky, A.; Maier, H.; Rohde, V.; Strohmayer, G.; West, W.; et al. Formation of deuterium-carbon inventories in gaps of plasma facing components. *J. Nucl. Mater.* **2007**, *363*, 870–876. [CrossRef]

2.   Litnovsky, A.; Hellwig, M.; Matveev, D.; Komm, M.; van den Berg, M.; Temmerman, G.D.; Rudakov, D.; Ding, F.; Luo, G.N.; Krieger, K.; et al. Optimization of tungsten castellated structures for the ITER divertor. *J. Nucl. Mater.* **2015**, *463*, 174–179. [CrossRef]

3.   Eckstein, W. *Computer Simulations of Ion-Solid Interactions*; Springer Series in Material Science; Springer: Berlin/Heidelberg, Germany, 1991; Volume 10.

4.   Eckstein, W.; Hackel, S.; Heinemann, D.; Fricke, B. Influence of the interaction potential on simulated sputtering and reflection data. *Z. Phys. D* **1992**, *24*, 171–176. [CrossRef]

5.   Eckstein, W.; Dohmen, R.; Mutzke, A.; Schneider, R. *SDTrimSP: A Monte-Carlo Code for Calculating Collision Phenomena in Randomized Targets*; Technical Report IPP 12/3; Max-Planck-Institut für Plasmaphysik: München, Germany 2007.

6.   Robinson, M.T.; Torrens, I.M. Computer simulation of atomic-displacement cascades in solids in the binary-collision approximation. *Phys. Rev. B* **1974**, *9*, 5008–5024. [CrossRef]

7.   Robinson, M.T. The temporal development of collision cascades in the binary-collision approximation. *Nucl. Instrum. Methods Phys. Res. Sect. B* **1990**, *48*, 408–413. [CrossRef]

8.   Mayer, M.; Eckstein, W.; Scherzer, B. Reflection of low energy hydrogen from carbon at oblique incidence. *J. Appl. Phys.* **1995**, *77*, 6609–6615. [CrossRef]

9.   Küstner, M.; Eckstein, W.; Dose, V.; Roth, J. The influence of surface roughness on the angular dependence of the sputter yield. *Nucl. Instrum. Methods Phys. Res. B* **1998**, *145*, 320–331. [CrossRef]

10.   Küstner, M.; Eckstein, W.; Hechtl, E.; Roth, J. Angular dependence of the sputtering yield of rough beryllium surfaces. *J. Nucl. Mater.* **1999**, *265*, 22–27. [CrossRef]

11.   Hellwig, M. Influence of 3D Geometry and Surface Roughness on Particle-Surface Interaction Processes with Tungsten Surfaces. Ph.D. Thesis, Ruhr Universität Bochum, Bochum, Germany, 2016.

12.   Meyer, E. Atomic force microscopy. *Prog. Surf. Sci.* **1992**, *41*, 3–49. [CrossRef]

13.   Baretzky, B. Investigation of the Surface Composition of Collision and of Segregation Deominated Systems for the Bombardment of Solids with Low-Energy Ions. Ph.D. Thesis, Ludwig-Maximilians-Universität München, Bochum, Germany, 1990.

14.   Gadelmawla, E.; Koura, M.; Maksoud, T.; Elwa, I.; Soliman, H.H. Roughness parameters. *J. Mater. Process. Technol.* **2002**, *123*, 133–145. [CrossRef]

15.   Möller, S.; Alegre, D.; Kreter, A.; Petersson, P.; Esser, H.G.; Samm, U. Thermo-chemical fuel removal from porous materials by oxygen and nitrogen dioxide. *Phys. Scr.* **2014**, *2014*. [CrossRef]

16.   Walls, J.M.; Smith, R. (Eds.) *Surface Science Techniques*; Elsevier: Pergamon, Turkey, 1994.

17.   Mayer, M. *SIMNRA User's Guide*; Max-Planck-Institut für Plasmaphysik: Garching, Germany, 2002.

18.   Van der Heide, P. *Secondary Ion Mass Spectroscopy—An Introduction to Principles and Practice*; John Wiley & Sons: Hoboken, NJ, USA, 2014.

19.   ION-TOF GmbH. *Surface Lab 6.5*; ION-TOF GmbH: Münster, Germany, 2014.

20.   Fairley, N. *CasaXPS Manual 2.3.15 Rev 1.0*; Casa Software Ltd.: Teignmouth, UK, 2009.

21.   Green, F.M.; Gilmore, I.S.; Seah, M.P. TOF-SIMS: Accurate Mass Scale Calibration. *J. Am. Soc. Mass. Spectrom.* **2006**, *17*, 514–523. [CrossRef] [PubMed]

22.   Cliff, J.; Gaspar, D.; Bottomley, P.; Myrold, D. Peak fitting to resolve CN-isotope ratios in biological and environmental samples using TOF-SIMS. *Appl. Surf. Sci.* **2004**, *231–232*, 912–916. [CrossRef]

23.   Fahey, A.; Messenger, S. Isotopic ratio measurements by time-of-flight secondary ion mass spectrometry. *Int. J. Mass Spectrom.* **2001**, *208*, 227–242. [CrossRef]

24.   Nuclear Energy Agency. Available online: https://www.oecd-nea.org/tools/abstract/detail/psr-0137 (accessed on 10 January 2019).

7

Dynamical Thermalization of Interacting Fermionic Atoms in a Sinai Oscillator Trap

Klaus M. Frahm [1]📵, **Leonardo Ermann** [2]📵 **and Dima L. Shepelyansky** [1,*]📵

[1] Laboratoire de Physique Théorique, IRSAMC, Université de Toulouse, CNRS, UPS, 31062 Toulouse, France
[2] Departamento de Física Teórica, GIyA, Comisión Nacional de Energía Atómica,
   CP1650 Buenos Aires, Argentina
*   Correspondence: dima@irsamc.ups-tlse.fr

Abstract: We study numerically the problem of dynamical thermalization of interacting cold fermionic atoms placed in an isolated Sinai oscillator trap. This system is characterized by a quantum chaos regime for one-particle dynamics. We show that, for a many-body system of cold atoms, the interactions, with a strength above a certain quantum chaos border given by the Åberg criterion, lead to the Fermi–Dirac distribution and relaxation of many-body initial states to the thermalized state in the absence of any contact with a thermostat. We discuss the properties of this dynamical thermalization and its links with the Loschmidt–Boltzmann dispute.

Keywords: quantum chaos; cold atoms; interacting fermions; thermalization; dynamical chaos; Sinai oscillator

---

1. Introduction

The problem of emergence of thermalization in dynamical systems started from the Loschmidt–Boltzmann dispute about time reversibility and thermalization in an isolated system of moving and colliding classical atoms [1,2] (see the modern overview in [3,4]). The modern resolution of this dispute is related to the phenomenon of dynamical chaos where an exponential instability of motion breaks the time reversibility at infinitely small perturbation (see e.g., [5–8]). The well known example of such a chaotic system is the Sinai billiard in which a particle moves inside a square box with an internal circle colliding elastically with all boundaries [9].

The properties of one-particle quantum systems, which are chaotic in the classical limit, have been extensively studied in the field of quantum chaos during the last few decades and their properties have been mainly understood (see, e.g., [10–12]). Thus, it was shown that the level spacing statistics in the regime of quantum chaos [13] is the same as for Random Matrix Theory (RMT) invented by Wigner for a description of spectra of complex nuclei [14,15]. This result became known as the Bohigas–Giannoni–Schmit conjecture [13,16]. Thus, classically chaotic systems (e.g., Sinai billiard) are usually characterized by Wigner–Dyson (RMT) statistics with level repulsion [13–15] while the classically integrable systems usually show Poisson statistics for level spacing distribution [11,12,16]. In this way, the level spacing statistics gives a direct indication for ergodicity (Wigner–Dyson statistics) or non-ergodicity (Poisson statistics) of quantum eigenstates. It was also established that the classical chaotic diffusion can be suppressed by quantum interference effects leading to an exponential localization of eigenstates [17–20] being similar to the Anderson localization in disordered solid-state systems [21]. The localized phase is characterized by Poisson statistics and the delocalized or metallic phase has RMT statistics. For billiard systems, the localized (nonergodic) and delocalized (ergodic) regimes appear in the case of rough billiards as described in [22,23].

It was also shown that, in the regime of quantum chaos, the Bohr correspondence principle [24] and the fully correct semiclassical description of quantum evolution remain valid only for a logarithmically short Ehrenfest time scale $t_E \sim \ln(1/\hbar)/h$ [17,19]. Here, $\hbar$ is an effective dimensionless Planck constant and h in the Kolmogorov–Sinai entropy characterizing the exponential divergence of classical trajectories. This result is in agreement with the Ehrenfest theorem, which states that the classical-quantum correspondence works on a time scale during which the wave packet remains compact [25]. However, for the classically chaotic systems, the Ehrenfest time scale is rather short due to an exponential instability of classical trajectories. After the Ehrenfest time scale t_E, the quantum out-of-time correlations (or OTOC as it is used to say now) stop decaying exponentially in contrast to exponentially decaying classical correlators [26,27]. For $t > t_E$, the decay of quantum correlations stops and they remain on the level of quantum fluctuations being proportional to $\hbar$ [26–28]. Since the level of quantum fluctuations is proportional to $\hbar$, the classical diffusive spreading over the momentum is affected by quantum corrections only on a significantly larger diffusive time scale $t_D \propto 1/\hbar^2 \gg t_E \propto \ln(1/\hbar)$ [17,19,26–28].

The problem of the emergence of RMT statistics and quantum ergodicity in many-body quantum systems is more complex and intricate as compared to one-particle quantum chaos. Indeed, it is well known that, in many-body quantum systems, the level spacing between nearest energy levels drops exponentially with the increase of number of particles or with energy excitation δE above the Fermi level in finite size Fermi systems, e.g., in nuclei [29]. Thus, at first glance, it seems that an exponentially small interaction between fermions should mix many-body quantum levels leading to RMT level spacing statistics (see, e.g., [30]).

Furthermore, the size of the Hamiltonian matrix of a many-body system grows exponentially with the number of particles, but, since all interactions have a two-body nature, the number of nonzero interaction elements in this matrix does not grow faster than the number of particles in the fourth power. Thus, we have a very sparse matrix being rather far from the RMT type. A two-body random interaction model (TBRIM) was proposed in [31,32] to consider the case of generic random two-body interactions of fermions in the limiting case of strong interactions when one-particle orbital energies are neglected. Even if the TBRIM matrix is very sparse, it was shown that the level spacing statistics $p(s)$ is described by the Wigner–Dyson or RMT distribution [33,34].

However, it is also important to analyze another limiting case when the two-body interaction matrix elements of strength U are weak or comparable with one-particle energies with an average level spacing Δ_1. In metallic quantum dots, this case with $U/\Delta_1 \approx 1/g$ corresponds to a large conductance of a dot $g = E_{Th}/\Delta_1 \gg 1$, where $E_{Th} = \hbar/t_D$ is the Thouless energy with t_D being a diffusion spread time over the dot [35–37]. In this case, the main question is about critical interaction strength U or excitation energy δE above the Fermi level of the dot at which the RMT statistics becomes valid. First, numerical results and simple estimates for a critical interaction strength in a model similar to TBRIM were obtained by Sven Åberg in [38,39]. The estimate of a critical interaction U_c, called the Åberg criterion [40], compares the typical two-body matrix elements with the energy level spacing Δ_c between quantum states *directly coupled by two-body interactions*. Thus, the Åberg criterion tells that the Poisson statistics is valid for many-body energy levels for $U < U_c \sim \Delta_c$ and the RMT statistics sets in for $U > U_c \sim \Delta_c$. In [41], this criterion, proposed independently of [38,39], was applied to the TBRIM of weakly interacting fermions in a metallic quantum dot being confirmed by extensive numerical simulations. It was also argued that the dynamical thermalization sets in an isolated finite fermionic system for energy excitations δE above the critical border δE_{ch} determined from the above criterion [41]:

$$\delta E > \delta E_{ch} \approx g^{2/3}\Delta_1 , \ \ g = \Delta_1/U. \tag{1}$$

The emergence of thermalization in an isolated many-body system induced by interactions between particles without any contact with an external thermostat represents the Dynamical Thermalization Conjecture (DTC) proposed in [41]. The validity of the Åberg criterion was numerically confirmed for various physical models (see [40] and references therein). An additional confirmation was given by the analytical derivation presented in [42] showing that, for three interacting particles, in a metallic dot the RMT sets in when the two-body matrix elements U become larger than the two-particle level spacing $\Delta_2 \sim \Delta_c$ being parametrically larger than the three-particle level spacing $\Delta_3 \ll \Delta_2$. The advanced theoretical arguments developed in [43,44] confirm the relation (1) for interacting fermions in a metallic quantum dot.

The test for the transition from Poisson to RMT statistics is rather direct and needs only the knowledge of energies' eigenvalues. However, the verification of DTC is much more involved since it requires the computation of system eigenstates. Thus, it is much more difficult to check numerically the relation (1) for DTC. However, it is possible to show that there is a transition from non-thermalized eigenstates at weak interactions (presumably for $\delta E < \delta E_{ch}$) to dynamically thermalized individual eigenstates at relatively strong interactions (presumably for $\delta E > \delta E_{ch}$). Thus, for the TBRIM with fermions, the validity of DTC for individual eigenstates at $U > U_c \sim \Delta_c$ has been demonstrated in [45,46] by the computation of energy E and entropy S of each eigenstate and its comparison with the theoretical result given by the Fermi–Dirac thermal distribution [47].

Even if the TBRIM represents a useful system for DTC tests, it is not so easy to realize it in real experiments. Thus, in this work, we investigate the DTC features in a system of cold fermionic atoms placed in the Sinai oscillator trap created by a harmonic two-dimensional potential with a repulsive circular potential created by a laser beam in a vicinity of the trap center. In such a case, the repulsive potential in the center is modeled as an elastic circle as in the case of Sinai billiard [9]. For one particle, it has been shown in [48] that the Sinai oscillator has an almost fully chaotic phase space and that, in the quantum case, the level spacing statistics is described by the RMT distribution. Due to one-particle quantum chaos in the Sinai oscillator, we expect that this system will have properties similar to the TBRIM. On the other side, the Sinai oscillator trap has been already experimentally realized with Bose–Einstein condensate of cold bosonic atoms [49–51]. At present, cold atom techniques allow for investigating various properties of cold interacting fermionic atoms [52,53] and we argue that the investigation of dynamical thermalization of such fermionic atoms, e.g., ^6Li, in a Sinai oscillator trap is now experimentally possible. Thus, in this work, we study properties of DTC of interacting fermionic atoms in a Sinai oscillator trap. Here, we consider the two-dimensional (2D) case of such a system assuming that the trap frequency in the third direction is small and that the 2D dynamics are not significantly affected by the adiabatically slow motion in the third dimension.

Finally, we note that, at present, the TBRIM model in the limit of strong interactions attracts a high interest in the context of field theory since, in this limit, it can be mapped on a black hole model of quantum gravity in $1 + 1$ dimensions known as the Sachdev–Ye–Kitaev (SYK) model linked also to a strange metal [54–59]. In fact, the SYK model, in its fermionic formulation [56], corresponds to the TBRIM considered with a conductance close to zero $g \to 0$. In these lines, the dynamical thermalization in TBRIM and SYK systems has been discussed in [45,46]. Furthermore, there is also a growing interest in dynamical thermalization for various many-body systems known also as the eigenstate thermalization hypothesis (ETH) and many-body localization (MBL) (see, e.g., [60–63]). We think that the system of interacting fermionic atoms in a Sinai oscillator trap captures certain features of TBRIM and SYK models and thus represents an interesting test ground to investigate nontrivial physics of these systems in real cold atom experiments.

This paper is composed as follows: in Section 2, we describe the properties of the one-particle dynamics in a Sinai oscillator; numerical results for dynamical thermalization on interacting atoms in this oscillator are presented in Section 3; the conditions of thermalization for fermionic cold atoms in realistic experiments are given in Section 4; the discussion of the results is presented in Section 5.

2. Quantum Chaos in Sinai Oscillator

The model of one particle in the 2D Sinai oscillator is described in detail in [48] with the Hamiltonian:

$$H_1 = \frac{1}{2m}(p_x^2 + p_y^2) + \frac{m}{2}(\omega_x^2 x^2 + \omega_y^2 y^2) + V_d(x,y). \tag{2}$$

Here, the first two terms describe the 2D oscillator with frequencies ω_x, ω_y and the last term gives the potential wall of elastic disk of radius r_d. We choose the dimensionless units with mass $m = 1$, frequencies $\omega_x = 1$, $\omega_y = \sqrt{2}$ and disk radius $r_d = 1$. The disk center is located at $(x_d, y_d) = (-1/2, -1/2)$ so that the disk bungs a hole in the center as it was the case in the experiments [49]. The Poincare sections at different energies are presented in [48] showing that the phase space is almost fully chaotic (see Figure 1 there). The quantum evolution is described by the Schrödinger equation with the quantized Hamiltonian (2), where the conjugate momentum and coordinate variables become operators with the commutation relation $[x, p_x] = [y, p_y] = i\hbar$ [48]. For the quantum problem, we use the value of the dimensionless Planck constant $\hbar = 1$ so that the ground state energy is $E_g = 1.685$. In the following, the energies are expressed in atomic like units of energy $E_u = \hbar \omega_x$ (for our choice of Sinai oscillator parameters, we also have $E_u = \hbar \omega_x = \hbar^2/(m r_d^2)$) [48] with the typical size of oscillator ground state being equal to the disk radius: $a_0 = \Delta x_{osc} = (\hbar/m\omega_x)^{1/2} = r_d$.

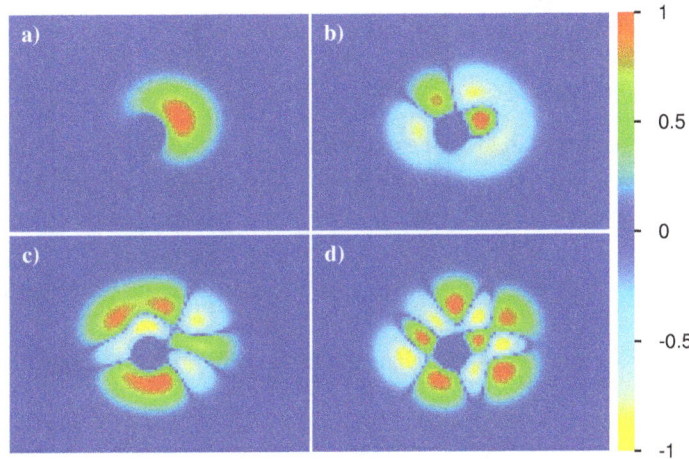

Figure 1. Color plot of one-particle eigenstates $\varphi_k(x, y)$ of the Sinai Hamiltonian in coordinate plane (x, y) with $-7.6 \leq x \leq 7.6$ and $-5.4 \leq y \leq 5.4$ for orbital numbers $k = 1$ (ground state) (**a**), $k = 6$ (**b**), $k = 11$ (**c**) and $k = 16$ (**d**). The numerical values of the color bar apply to the signed and nonlinearly rescaled wave function amplitude: $\mathrm{sgn}[\varphi_k(x, y)] |\varphi_k(x, y)/\varphi_{max}|^{1/2}$, where φ_{max} is the maximum of $|\varphi_k(x, y)|$ and the exponent $1/2$ provides amplification of regions of small amplitude.

In [48], it is shown that the classical dynamics of this system are almost fully chaotic. In the quantum case, the level spacing statistics is well described by the RMT distribution. The average dependence of energy level number k is well described by the theoretical dependence $k(\varepsilon) = \varepsilon^2/(2\sqrt{2}) - \varepsilon/2$ [48]. Thus, the one-particle density of states $\rho_1(\varepsilon)$ and corresponding level spacing Δ_1 are:

$$\rho_1(\varepsilon) = \frac{dk}{d\varepsilon} = \frac{\varepsilon}{\sqrt{2}} - \frac{1}{2} \quad , \quad \Delta_1 = \frac{1}{\rho} \approx \frac{\sqrt{2}}{\varepsilon} \approx \frac{0.84}{\sqrt{k}}. \tag{3}$$

Examples of several eigenstates, computed on a numerical grid of 28,341 spatial points, are shown in Figure 1. More details on the numerical diagonalization of (2) and other example eigenstates can be found in [48].

3. Sinai Oscillator with Interacting Fermionic Atoms

3.1. Two-Body Interactions of Fermionic Atoms

The two-body interaction of atoms appears usually due to van der Waals forces which drop rapidly with the distance between two atoms and the short ranged interaction can be described in the framework of the scattering length approach (see, e.g., [64,65]). Therefore, we assume that the finite effective interaction range r_c is significantly smaller than the disk radius r_d and the typical size of the wave function, i.e., $r_c \ll r_d$. Such a short range interaction is indeed used to modelize atomic interactions in harmonic traps (see, e.g., [66]). For example, in a typical experimental situation, the disk radius is of the order of micron $r_d \sim 1\,\mu m = 10^{-4}$ cm, while, for Li and other alkali atoms, we have $r_c \sim 3 \times 10^{-7}$ cm [64,65]. Of course, in the limit of small $r_c \ll r_d$, the interaction between two atoms takes place mainly in the s-wave scattering, so effectively the interaction operates between fermions with different quantum numbers of the Sinai oscillator. This feature is of course taken into account in our numerical simulations.

In the following, we use a simple interaction function having a constant amplitude U for $r \leq r_c$ and being zero for $r > r_c$, where we simply choose $r_c = 0.2r_d$, which corresponds well to the short range interaction regime. The precise value of r_c is not very important since a slight modification $r_c \to \bar{r}_c$ can be absorbed in a modified amplitude according to $U \to \bar{U} = U(\bar{r}_c/r_c)^2$, a relation we verified numerically for various values of $\bar{r}_c < r_d$. We numerically checked that, at a used $r_c = 0.2r_d$ value, we are in the regime when the interaction matrix elements are proportional to r_c^2 so that we are in the regime of short-range interactions. We mention that, in experiments, the strength of the interaction amplitude can be changed by a variation of the magnetic field via the Feshbach resonance [67].

3.2. Reduction to TBRIM Like Case and Its Analysis

Using the methods described in [48], we numerically compute a certain number of one-particle or orbital energy eigenvalues ε_k and corresponding eigenstates $\varphi_k(\mathbf{r})$ of the Sinai oscillator (2). As repulsive interaction potential $v(\mathbf{r})$, we choose the short ranged box function $v(\mathbf{r}) = U$ if $|\mathbf{r}| \leq r_c = 0.2$ (since $r_d = 1$) and $v(\mathbf{r}) = 0$, otherwise. Here, the parameter $U > 0$ gives the overall scale of the interaction strength depending on the charge of the particles and eventually other physical parameters.

Therefore, the corresponding many-body Hamiltonian with M one-particle orbitals and $0 \leq L \leq M$ spinless fermions takes the form:

$$H = \sum_{k=1}^{M} \varepsilon_k\, c_k^\dagger c_k + \sum_{i<j,k<l} V_{ij,kl}\, c_i^\dagger c_j^\dagger c_l c_k, \tag{4}$$

where, for $i < j$ and $k < l$, we have the interaction matrix elements:

$$V_{ij,kl} = \bar{V}_{ij,kl} - \bar{V}_{ij,lk} \quad,\quad \bar{V}_{ij,kl} = \int d\mathbf{r}_1 \int d\mathbf{r}_2\, \varphi_i^*(\mathbf{r}_1)\varphi_j^*(\mathbf{r}_2)\, v(\mathbf{r}_1 - \mathbf{r}_2)\varphi_k(\mathbf{r}_1)\varphi_l(\mathbf{r}_2) \tag{5}$$

and $c_k^\dagger$, c_k are fermion operators for the M orbitals satisfying the usual anticommutation relations. We note that, in the literature, when expressing a two-body interaction potential in second quantization, one usually uses the raw matrix elements $\bar{V}_{ij,kl}$ with an additional prefactor of $1/2$ and full independent sums for the four indices i, j, k and l. Using the particle exchange symmetry: $\bar{V}_{ij,kl} = \bar{V}_{ji,lk}$, one can reduce the i,j sums to $i < j$, which removes the prefactor $1/2$ (after exchanging the index names l and k for the $i > j$ contributions and exploiting that contributions at $i = j$ or $l = k$ obviously vanish). The definition of the anti-symmetrized interaction matrix elements $V_{ij,kl}$ according to (5) allows for also reducing the k,l sums to $k < l$. Furthermore, the ordering of the two fermion operators $c_l c_k$ in (4) is also important and necessary to obtain positive expectation values if the interaction is repulsive. The anti-symmetrized matrix elements $V_{ij,kl}$ correspond to a $M_2 \times M_2$ matrix with $M_2 = M(M-1)/2$. In order to avoid a global shift of the non-interacting eigenvalue spectrum due to the interaction, we also apply a diagonal shift $V_{ij,ij} \to V_{ij,ij} - (1/M_2)\sum_{k<l} V_{kl,kl}$ to ensure that this matrix has a vanishing trace (One can easily show that the trace of the $M_2 \times M_2$ anti-symmetrized interaction matrix is proportional to the trace of the interaction operator in the many-body Hilbert space with a factor depending on M and L). Of course, such a global energy shift does not affect the issues of thermalization, interaction induced eigenfunction mixing or the quantum time evolution with respect to the Hamiltonian H, etc.

We note that the transition from the classical Hamiltonian (2) to the quantum one (4) is done by the standard procedure of second quantization (see, e.g., [68]).

3.3. Aberg Parameter

In absence of interaction, the energy eigenvalues of (4) are given as the sum of occupied orbital energies:

$$E(\{n_k\}) = \sum_{k=1}^{M} \varepsilon_k \, n_k, \tag{6}$$

where $\{n_k\}$ represents a configuration such that $n_k \in \{0,1\}$ and $\sum_k n_k = L$. The associated eigenstates are the basis states where each orbital is either occupied (if $n_k = 1$) or unoccupied (if $n_k = 0$) and, in this work, we will denote these states in the usual occupation number representation: $|n_M \cdots n_2 \, n_1 \rangle$, where, for convenience, we write the lower index orbitals starting from the right side.

The distribution of the total one-particle energies (6) is numerically rather close to a Gaussian (since n_k act as quasi-random numbers) with mean and variance (see also Equation (A.4) of Ref. [46]):

$$E_{\text{mean}} = L\bar{\varepsilon} \quad , \quad \sigma_0^2 = \frac{L(M-L)}{M-1} \left(\overline{\varepsilon^2} - \bar{\varepsilon}^2 \right) \quad , \quad \overline{\varepsilon^n} = \frac{1}{M} \sum_{k=1}^{M} \varepsilon_k^n \quad , \quad n = 1, 2. \tag{7}$$

Therefore, the many-body level spacing Δ_{MB} or inverse Heisenberg time at the band center $E = E_{\text{mean}}$ is given by $\Delta_{\text{MB}} = 1/t_{\text{H}} = \sqrt{2\pi}(\sigma_0/d)$, where $d = M!/(L!(M-L)!)$ is the dimension of the fermion Hilbert space in the sector of M orbitals and L particles. In our numerical computations, we simply evaluated the quantities $\overline{\varepsilon^n}$ of (7) using the exact one-particle energy eigenvalues obtained from the numerical diagonalization of the one-particle Sinai Hamiltonian H_1 given in (2). However, to get some analytical simplification for large M, one may use the one-particle density of states (3), which gives, after replacing the sums by integrals and neglecting the constant term, $\overline{\varepsilon^n} \approx 2\,\varepsilon_M^n/(n+2)$ and $\overline{\varepsilon^2} - \bar{\varepsilon}^2 \approx \varepsilon_M^2/18 \approx \sqrt{2}\,M/9$.

For the question of whether the interaction strength is sufficiently strong to mix the non-interacting basis states, the important quantity is the effective level spacing of states coupled directly by the interaction $\Delta_c = \sqrt{2\pi}\,[\sigma_0(L=2)/K]$, where $K = 1 + L(M-L) + L(L-1)(M-L)(M-L-1)/4$ is the number of nonzero elements for a column (or row) of H [41,69] and we need to use the variance for only two particles:

$$\sigma_0^2(L=2) = \frac{2(M-2)}{M-1} \left(\overline{\varepsilon^2} - \bar{\varepsilon}^2 \right) \quad \Rightarrow \quad \frac{\sigma_0^2(L=2)}{\sigma_0^2} = \frac{2(M-2)}{L(M-L)} \tag{8}$$

because the interaction only couples states where (at least) $L-2$ particles are on the same orbital such that (at most) only the partial sum of two one-particle energies is different between two coupled states. Even though for two particles the hypothesis of a Gaussian distribution is theoretically not justified, the distribution is still sufficiently similar to a Gaussian and it turns out that the value of $1/\Delta_c = K/[\sqrt{2\pi}\,\sigma_0(L=2)]$ as the coupled two-particle density of states in the band center is numerically quite accurate with an error below 10% (for $M = 16$ and our choice of ε_k values).

According to the Åberg criterion [38,39,41], the onset of chaotic mixing happens for typical interaction matrix elements U comparable to Δ_c. Therefore, we compute the quantity $V_{\text{mean}} = \sqrt{\langle |V_{ij,kl}|^2 \rangle}$ (which is proportional to the interaction amplitude U) where the average is done with respect to all M_2^2 matrix elements of the interaction matrix. This quantity might be problematic and not correspond to a typical interaction matrix element in the case of a long tail distribution. However, in our case, it turns out that $V_{\text{mean}} \approx 2\exp(\langle \ln |V_{ij,kl}| \rangle)$, which excludes this scenario. Using this quantity, we introduce the dimensionless Åberg parameter and the critical interaction amplitude U_c by $A = V_{\text{mean}}/\Delta_c = U/U_c$ such that $A = 1$ if $U = U_c$. We expect [38,39,41] to be the onset of strong/chaotic mixing at $A \gg 1$ and a perturbative regime for $A \ll 1$, while, at $A = 1$, we have the critical interaction strength $U = U_c$. The value of U_c depends on the parameters L, M, σ_0 and the overlap of the one-particle eigenstates according to (5). To obtain some useful analytical expression of U_c, we note that the quantity V_{mean}, numerically computed for $4 \leq M \leq 30$, can be quite accurately fitted by $V_{\text{mean}} \approx 3 \times 10^{-4} U / \varepsilon_M$. Furthermore, we remind readers of the expression $\Delta_c = (1/K) \sqrt{4\pi(M-2)(\overline{\varepsilon^2} - \overline{\varepsilon}^2)/(M-1)}$, which can be simplified in the limit $M \gg 1$ and $L \gg 1$, such that $K \approx (M-L)^2 L^2 / 4$, resulting in: $\Delta_c = 4/3 \sqrt{2\pi} \, \varepsilon_M / [(M-L)^2 L^2]$. Here, we also used the found expression above $\overline{\varepsilon^2} - \overline{\varepsilon}^2 \approx \varepsilon_M^2 / 18$. From this, we find that $U_c = \Delta_c U / V_{\text{mean}} \approx C M / [(M-L)^2 L^2]$ with a numerical constant $C \approx 16 \times 10^4 \sqrt{\pi}/9 \approx 3.15 \times 10^4$, where we also used $\varepsilon_M^2 \approx 2\sqrt{2} M$ according to (3). Below, we will give more accurate numerical values of V_{mean}, Δ_c and U_c for the parameter choice of M and L numerically relevant in this work.

We note that this estimate for $A = U/U_c$ applies to energies close to the many body band center of H and that, for energies away from the band center, the value of Δ_c is enhanced, thus reducing the effective value of A. Furthermore, we computed V_{mean} by a simplified average over *all* interacting matrix elements not taking into account an eventual energy dependence according to the index values of i, j, k, l in (5).

3.4. Density of States

In this work, we present numerical results for the case of $M = 16$ orbitals and $L = 7$ particles corresponding to a many-body Hilbert space of dimension $d = M! / (L!(M-L)!) = 11440$ and the number $K = 820$ of directly coupled states of a given initial state by non-vanishing interaction matrix elements in (4). Thus, in our studies, the whole Hilbert space is built only on these $M = 16$ orbitals. We diagonalize numerically the many-body Hamiltonian (4) for various values of A in the range $0.025 \leq A \leq 200$. We have also verified that the results and their physical interpretation are similar for smaller cases such as $M = 12$, $L = 5$ (with $d = 792$, $K = 246$) or $M = 14$, $L = 6$ ($d = 3003$, $K = 469$).

We mention that, for $M = 16$ and $L = 7$, we find numerically that $V_{\text{mean}} = 3.865 \times 10^{-5} U$ and, from (8), that $\Delta_c = \sqrt{2\pi} [\sigma_0 (L-2)/K] = 6.1706 \times 10^{-3}$, where the quantities $\overline{\varepsilon^n}$ were exactly computed from the numerical orbitals energies ε_k. From this, we find that $U_c = \Delta_c U / V_{\text{mean}} \approx 159.65$. This expression is more accurate than the more general analytical estimate for arbitrary $M \gg 1$ and $L \gg 1$ given in the last section (which would provide $U_c \approx 127$ for $M = 16$ and $L = 7$).

Our first observation is that, even in the presence of interactions, the density of states has approximately a Gaussian form with the same center E_{mean} given in (7) for the case $A = 0$. This is simply due the fact that the interaction matrix has, by choice, a vanishing trace and does not provide a global shift of the spectrum. We determine the variance $\sigma^2(A)$ of the Gaussian density of states by a fit of the integrated density of states $P(E)$ using

$$P(E) = (1 + \text{erf}[q(E)])/2 \quad , \quad q(E) = (E - E_{\text{mean}})/[\sqrt{2}\sigma(A)], \tag{9}$$

such that $P'(E)$ is a Gaussian of width $\sigma(A)$ and center E_{mean} (see Appendix A of Ref. [46] for more details). From this, we find the behavior:

$$\sigma^2(A) = \sigma_0^2 (1 + \alpha A^2), \tag{10}$$

where α is a constant depending on M and L; for $M = 16$, $L = 7$ the fit values of σ_0 and α are $\sigma_0 = 3.013 \pm 0.009$ and $\alpha = 0.00877 \pm 0.00010$. It is also possible to determine $\sigma(A)$ using the expression $\sigma^2(A) = \text{Tr}_{\text{Fock}} [(H - E_{\text{mean}}\mathbf{1})^2] / d$ where the trace in Fock space can be evaluated either by using the matrix H before diagonalizing it or using its exact energy eigenvalues E_m. This provides the same behavior as (10) with the very similar numerical values $\sigma_0 = 3.013 \pm 0.007$ and $\alpha = 0.00858 \pm 0.00008$ (for $M = 16$, $L = 7$). We mention that the integrated Gaussian density of states (9) is not absolutely exact but quite accurate for values $A \leq 10$. For larger values of A, the deviations increase, but the overall form is still correct. As described in [46], the quality of the fit can be considerably improved if we replace in (9) the linear function $q(E)$ by a polynomial of degree 5. In this case, the precision of the fit is highly accurate for the full range of A values we consider. In particular, we use this improved fit to perform the spectral unfolding when computing the nearest level spacing distribution (shown below).

To obtain some theoretical understanding of (10), one can consider a model where the initial interaction matrix elements (5) are replaced by independent Gaussian variables with identical variance V_{mean}^2. In this case, one can show theoretically [46] that $\sigma^2(A) = \sigma_0^2 + K_2 V_{mean}^2$, where $K_2 = L(L-1)[1 + M - L + (M-L)(M-L-1)/4]$ is a number somewhat larger than K taking into account that certain non-vanishing interaction matrix elements in Fock space are given as a sum of *several* initial interaction matrix elements (5) (see Appendix A of [46] for details). The parameter K_2 takes for $M = 16$, $L = 7$ ($M = 14$, $L = 6$ or $M = 12$, $L = 5$) the value $K_2 = 1176$ ($K_2 = 690$ or $K_2 = 370$, respectively). Since $V_{mean} = A\Delta_c = A\sqrt{2\pi}\,\sigma_0(L = 2)/K$, we indeed obtain (10) with $\alpha = \alpha_{th} = 4\pi(M-2)K_2/[K^2(L(M-L))]$. For $M = 16$, $L = 7$, we find $\sigma_0 = 3.0279$ (see (7)) and $\alpha_{th} = 0.00488$. The latter is roughly by a factor of 2 smaller than the numerical value. We attribute this to the fact that the real initial interaction matrix elements (5) are quite correlated, and not independent uniform Gaussian variables, leading therefore to an effective increase of the number K_2 due to hidden correlations. The important point is that theoretically at very large values values of M and L, e.g., $M \approx 2L \gg 1$, we have $K_2 \approx K \approx L^4/4$ and $\alpha_{th} \approx 32\pi/L^5$, which is parametrically small for very large L. Therefore, there is a considerable range of values $1 < A < 1/\sqrt{\alpha}$ where the interaction strongly mixes the non-interacting many-body eigenstates but where the density of states is only weakly affected by the interaction. This regime is also known as the Breit–Wigner regime (see, e.g., [40], for the case of interacting Fermi systems).

3.5. Thermalization and Entropy of Eigenstates

In the following, we mostly concentrate on values $A \leq 10$ such that the effect of the increase of the spectral width $\sigma(A)$ is still small or at least quite moderate. The question arises if a given many-body state, either an exact eigenstate of H or a state obtained from a time evolution with respect to H, is thermalized according to the Fermi–Dirac distribution [47]. As in [45,46], we determine the occupation numbers $n_k = \langle c_k^\dagger c_k \rangle$ for such a state, as well as the corresponding fermion entropy S [47] and the effective total one-particle energy E_{1p} by :

$$S = -\sum_{k=1}^{M} \left(n_k \ln n_k + (1 - n_k) \ln(1 - n_k) \right) \quad , \quad E_{1p} = \sum_{k=1}^{M} \varepsilon_k n_k \tag{11}$$

based on the assumption of weakly interacting fermions. In the regime of modest interaction $A \lesssim 5$ (for $M = 16$, $L = 7$), corresponding to a constant spectral width $\sigma(A) \approx \sigma_0$, we have typically $E_{1p} \approx E_{ex}$ (for exact eigenstates of H) or $E_{1p} \approx \langle H \rangle$ (for other states). If the given state is thermalized, its occupation numbers n_k should be close to the theoretical Fermi–Dirac filling factor $n(\varepsilon_k)$ with $n(\varepsilon) = 1/(1 + \exp[\beta(\varepsilon - \mu)])$, where inverse temperature $\beta = 1/T$ and chemical potential μ are determined by the conditions:

$$L = \sum_{k=1}^{M} n(\varepsilon_k) \quad , \quad E = \sum_{k=1}^{M} \varepsilon_k n(\varepsilon_k). \tag{12}$$

Here, E is normally given by E_{1p}, but one may also consider the value E_{ex} (or $\langle H \rangle$) provided the latter is in the energy interval where the conditions (12) allow for a unique solution. Furthermore, for a given energy E, we can also determine the theoretical (or thermalized) entropy $S_{th}(E)$ using (11) with n_k being replaced by $n(\varepsilon_k)$ (where β, μ are determined from (12) for the energy E).

The many-body states with energies above E_{mean} are artificial since they correspond to negative temperatures due to the finite number of orbitals considered in our model. Therefore, we limit our studies to the lower half of the energy spectrum $29 \leq E \leq 39 \approx E_{mean}$ (for $M = 16$, $L = 7$). In Figure 2, we compare the thermalized Fermi–Dirac occupation number $n(\varepsilon)$ with the occupation numbers n_k for two eigenstates at level numbers $m = 123$ (1354, with $m = 1$ corresponding to the ground state) with approximate energy eigenvalue $E \approx 32$ ($E \approx 35$) for three different Åberg parameters $A = 0.35$, $A = 3.5$ and $A = 10$. These states are not too close to the ground state but still quite far below the band center.

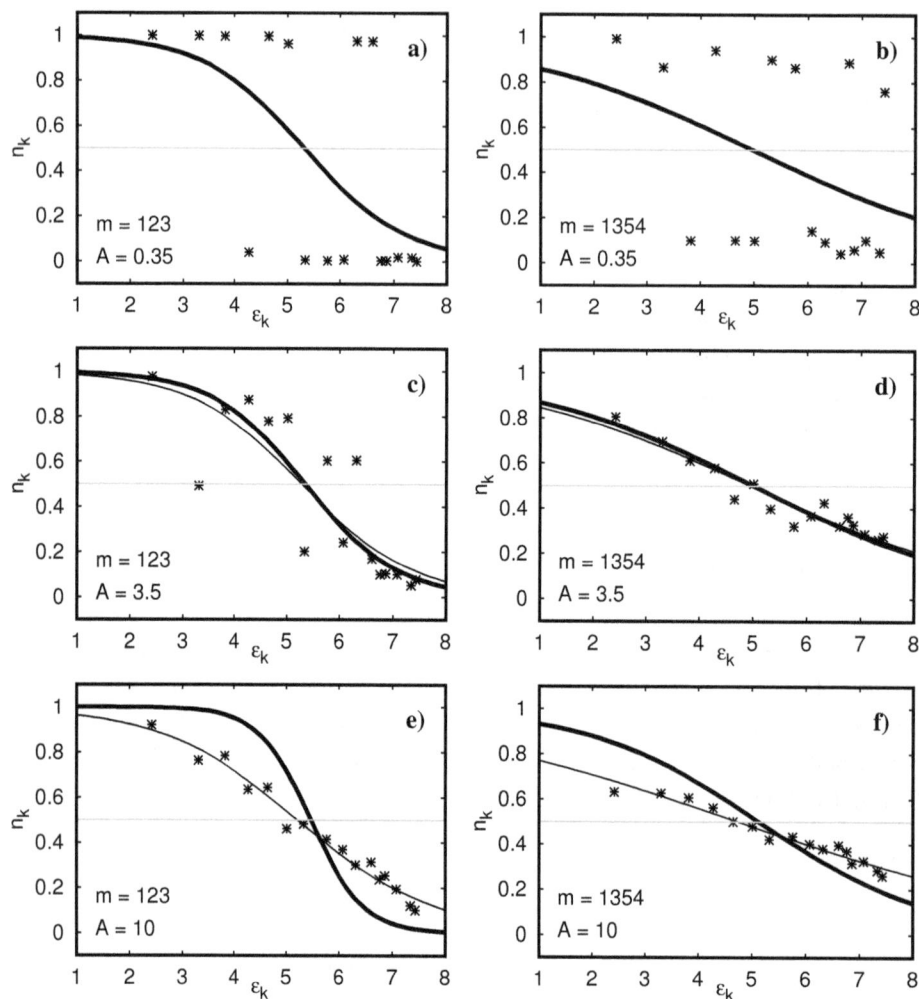

Figure 2. Orbital occupation number n_k versus orbital energies ε_k (black stars) of individual eigenstates at level numbers $m = 123$ (**a,c,e**), 1354 (**b,d,f**), and Åberg parameter $A = 0.35$ (**a,b**), $A = 3.5$ (**c,d**), $A = 10$ (**e,f**), (with $m = 1$ corresponding to the ground state). The thin blue (thick red) curves show the theoretical Fermi–Dirac occupation number $n(\varepsilon) = 1/(1 + \exp[\beta(\varepsilon - \mu)])$, where inverse temperature β and chemical potential μ are determined from (12) with $E = E_{1p}$ ($E = E_{ex}$). The horizontal green lines correspond to the constant value 0.5 whose intersections with the red or blue curves provide the positions of the chemical potential. In this and all subsequent figures, the orbital number is $M = 16$, the number of particles is $L = 7$ and the corresponding dimension of the many body Hilbert space is $d = 11440$. Table 1 gives for each of these levels the values of E_{ex}, E_{1p}, S, S_{th}, β, μ and for both energies for the latter three parameters.

We note that the regime of negative temperatures is natural for the TBRIM where the energy spectrum is inside a finite energy band (this regime has been discussed in [45,46]). However, for the Sinai oscillator, the energy spectrum is unbounded and, due to that, the regime of negative temperatures, appearing in the numerical simulations due to a finite number of one-particle orbital, is artificial.

Table 1. Parameters of the eigenstates corresponding to Figure 2. S is the entropy, E_{1p} the effective total one-particle energy, both given by (11), and E_{ex} is the exact energy eigenvalue. Inverse temperature β, chemical potential μ, theoretical entropy S_{th} are determined by (12) or (11) (with n_k replaced by $n(\varepsilon_k)$) for both energies E_{1p}, E_{ex}.

A	m	S	$S_{th}(E_{1p})$	E_{1p}	$\mu(E_{1p})$	$\beta(E_{1p})$	$S_{th}(E_{ex})$	E_{ex}	$\mu(E_{ex})$	$\beta(E_{ex})$
0.35	122	0.95	7.91	32.15	5.31	1.05	7.89	32.13	5.31	1.05
0.35	1353	4.91	10.16	35.29	4.98	0.45	10.16	35.30	4.98	0.45
3.5	122	6.99	8.28	32.52	5.28	0.95	7.54	31.81	5.34	1.15
3.5	1353	10.16	10.23	35.45	4.95	0.43	10.10	35.15	5.00	0.47
10	122	8.91	8.98	33.33	5.22	0.77	4.96	30.10	5.46	2.02
10	1353	10.52	10.54	36.28	4.75	0.32	9.53	34.12	5.14	0.63

At weak interaction, $A = 0.35$, both states are not at all thermalized with occupation numbers being either close to 1 or 0. Apparently, these states result from weak perturbations of the non-interacting eigenstates $|0000011000110111>$ or $|1000100011001011>$, where the n_k values are rounded to 1 (or 0) if $n_k > 0.5$ ($n_k < 0.5$). For $m = 1354$, the values of n_k are a little bit farther away from the ideal values 0 or 1 as compared to $m = 123$ but still sufficiently close to be considered as perturbative. Apparently, the state $m = 123$, which is lower in the spectrum (with larger effective two-body level spacing), is less affected by the interaction than the state 1354. In both cases, the entropy S is quite below the thermalized entropy S_{th} (see Table 1 for numerical values of entropies, energies, inverse temperature and chemical potential for the states shown in Figure 2).

At intermediate interaction, $A = 3.5$, the occupation numbers are closer to the theoretical Fermi–Dirac values but still with considerable deviations. Here, both entropy values S are rather close to S_{th}. The state 1354 seems to be better thermalized than the state $m = 123$, the latter having a slightly larger deviation between both entropy values. At stronger interaction, $A = 10$, both states are very well thermalized with a good matching of both entropy values (again with the state 1354 being a bit better thermalized than the state $m = 123$) provided we use E_{1p} as reference energy to compute temperature and chemical potential. The temperature obtained from E_{ex} is too small because here the increase of $\sigma(A)$ is already quite strong and E_{ex} rather strongly deviates from E_{1p}. In addition, the value of S_{th} using E_{ex} does not match S. Obviously, at stronger interaction values, it is necessary to use E_{1p} to test the thermalization hypothesis of a given state.

Figure 3 shows the mutual dependence between the three parameters β, μ on E when solving the conditions (12). The chemical potential as a function of $\beta = 1/T$ is rather constant except for smallest values of β where $\mu \sim 1/\beta$ with a negative prefactor. One can actually easily show from (12) that in the limit $\beta \to 0$ the chemical potential does not depend on ε_k and is given by $\mu = - \ln[1 + (M - 2L)/L]/\beta$ providing a singularity if $L \neq M/2$ with negative (positive) prefactor for $L < M/2$ ($L > M/2$) and $\mu = 0$ for $L = M/2$. The temperature (β^{-1}) vanishes for E close to the lower energy border and diverges for E close to the band center E_{mean}.

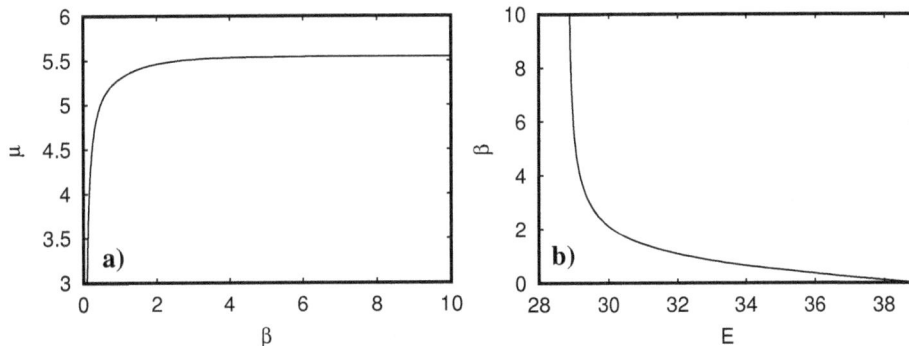

Figure 3. Dependence of chemical potential μ on inverse temperature $\beta = 1/T$ (**a**) and of $\beta = 1/T$ on energy E (**b**) where β and μ are determined from (12) for a given energy E.

In Figure 4, we present the nearest level spacing distribution $p(s)$ for different values of the Åberg parameter. To compute $p(s)$, we have used only the "physical" levels in the lower half of the energy spectrum and the unfolding has been done with the integrated density of states (9), where $q(E)$ is replaced by a fit polynomial of degree 5. For the smallest value $A = 0.2$, the distribution $p(s)$ is very close to the Poisson distribution with some residual level repulsion at very small spacings. This is a quite well known effect because typically the transition from Wigner–Dyson to Poisson statistics (when tuning some suitable parameter such as the Åberg parameter from strong to weak coupling) is non-uniform in energy and happens first at larger spacings (energy differences) and then at smaller spacings. The reason is simply that two levels which by chance are initially very close are easily repelled by a small residual coupling matrix element (when slightly changing a disorder realization or similar). For $A = 0.5$, there is somewhat more level repulsion at small spacings, but the distribution is still rather close to the Poisson distribution with some modest deviations for $s \leq 1.2$. For the larger Åberg values $A = 3.5$ and $A = 10$, we clearly obtain Wigner–Dyson statistics (taking into account the quite limited number of only $d/2 - 1 = 5719$ level spacing values for the histograms). These results clearly confirm that the transition from $A < 1$ to $A > 1$ corresponds indeed to a transition from a perturbative regime to a regime of chaotic mixing with Wigner–Dyson level statistics [14].

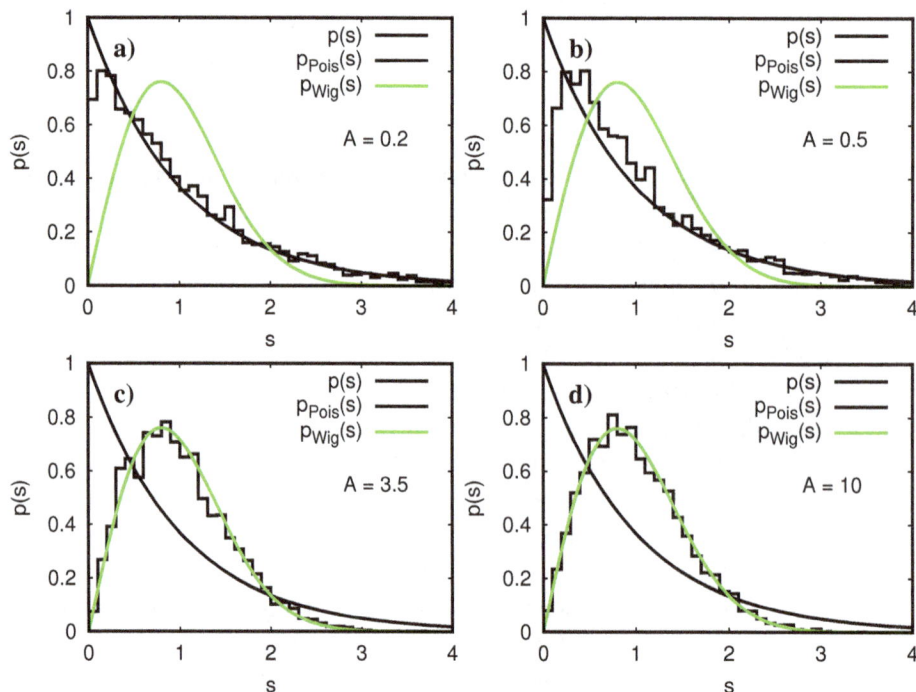

Figure 4. Histogram of unfolded level spacing statistics (blue line) for the exact energy eigenvalues E_m of H (using the lower half of the spectrum with $1 \leq m \leq d/2$). The different panels correspond to the Åberg parameter values $A = 0.2$ (**a**), $A = 0.5$ (**b**), $A = 3.5$ (**c**), $A = 10$ (**d**). The unfolding is done using the integrated density of states (9), where $q(E)$ is replaced by a fit polynomial of degree 5. The Poisson distribution $p_{\mathrm{Pois}}(s) = \exp(-s)$ (black line) and the Wigner surmise $p_{\mathrm{Wig}}(s) = \frac{\pi}{2} s \exp(-\frac{\pi}{4} s^2)$ (green line) are also shown for comparison.

A further confirmation that $A = 1$ is critical can be seen in Figure 5, which compares the dependence of the entropy S of exact eigenstates (lower half of the spectrum) on E_{1p} or E_{ex} with the theoretical thermalized entropy $S_{\mathrm{th}}(E)$. For the Åberg values $A = 0.2$ (and $A = 0.5$), the entropy S of all (most) states is significantly below its theoretical value S_{th}. Actually, the distribution of data points is considerably concentrated at smaller entropy values which is not so clearly visible in Figure. In particular, the average of the ratio of $S/S_{\mathrm{th}}(E_{1p})$ is 0.178 for $A = 0.2$ and 0.522 for $A = 0.5$. For the Åberg values $A = 3.5$ and $A = 10$, most or nearly all entropy values (for E_{1p}) are very close to the theoretical line with the average ratio $S/S_{\mathrm{th}}(E_{1p})$ being 0.990 for $A = 3.5$ and 0.998 for $A = 10$.

For $A = 3.5$, the states with lowest energies are not yet perfectly thermalized and the data points for E_{ex} and E_{1p} are still rather close. For $A = 10$, all states are well thermalized (when using the energy E_{1p}) while the data points for E_{ex} are quite outside the theoretical curve simply due to the overall increase of the width of the energy spectrum. This observation is also in agreement with the discussion of Figure 2. For smaller values $A < 0.2$ (not shown in Figure 5), we find that the data points are still closer to the E-axis while, for larger values $A > 10$, the data points are clearly on the theoretical curve for E_{1p} (but more concentrated on energy values closer to the center with larger entropy values and larger temperatures), while, for E_{ex}, according to (10), the overall width of the exact eigenvalue spectrum increases strongly and the data points are clearly outside the theoretical curve (except for a few states close to the band center).

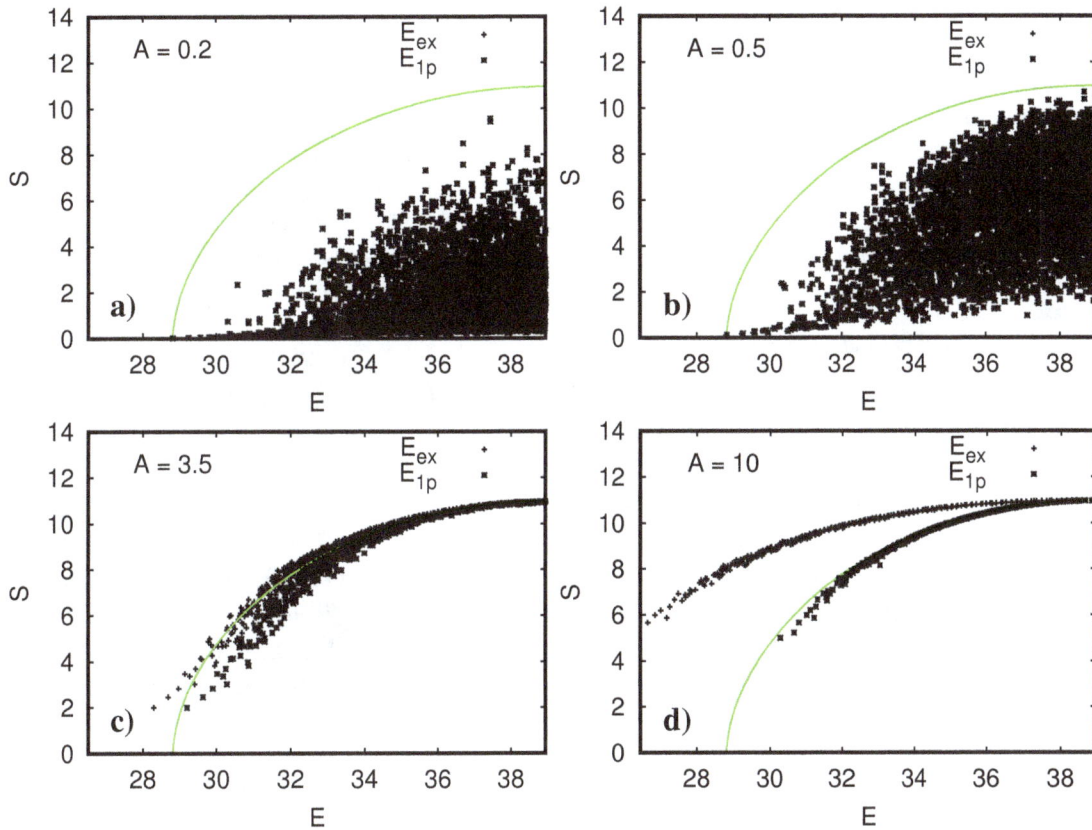

Figure 5. Dependence of the fermion entropy S on the effective one-particle total energy E_{1p} (blue cross symbols) and the exact many-body energy E_{ex} (red plus symbols). The green curve shows the theoretical entropy $S_{th}(E)$ obtained from the Fermi–Dirac occupation numbers as explained in the text. The different panels correspond to the Åberg parameter values $A = 0.2$ (**a**), $A = 0.5$ (**b**), $A = 3.5$ (**c**), $A = 10$ (**d**).

We note that the data points in Figure 5a,b significantly deviate from the theoretical thermalization curve since the Åberg criterion is not satisfied ($A < 1$). For $A = 3.5; 10$, the deviations are significantly reduced, but they are still more pronounced in a vicinity of the ground state in agreement with the relation (1).

In Figure 6, the occupation numbers n_k (averaged over several energy eigenvalues inside a given energy cell) are shown in the plane of energy E and orbital index k as color density plot for the Åberg parameter $A = 3.5$. The comparison with the theoretical occupation numbers $n(\varepsilon_k)$ (shown in the same way) provides further confirmation that, at $A = 3.5$, there is indeed already a quite strong thermalization of most eigenstates.

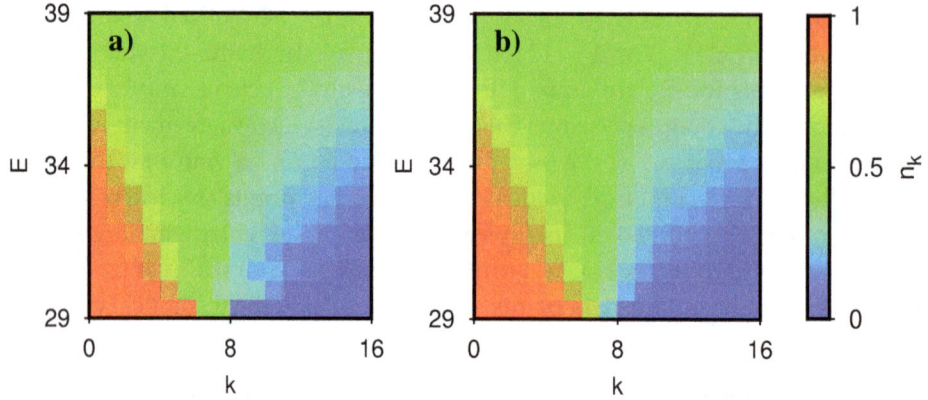

Figure 6. Color density plot of the orbital occupation number n_k in the plane of energy E and orbital index k. (**a**) n_k values of exact eigenstates of H with Åberg parameter $A = 3.5$; (**b**) thermalized Fermi–Dirac occupation number $n(\varepsilon_k)$, where β and μ are determined from (12) as a function of total energy E. The occupation number n_k is averaged over all eigenstates (**a**) or several representative values of E (**b**) inside a given energy cell. The energy interval $29 \leq E \leq 39$ corresponds roughly to the lower half of the spectrum (at $M = 16$, $L = 7$) for states with positive temperature and is similar to the energy interval used in Figures 4 and 5. The color bar provides the translation between n_k values and colors (red for maximum $n_k = 1$, green for $n_k = 0.5$ and blue for minimum $n_k = 0$).

3.6. Thermalization of Quantum Time Evolution

The question arises how or if a time dependent state $|\psi(t)> = \exp(-iHt)|\psi(0)>$, obeying the quantum time evolution with the Hamiltonian H and an initial state $|\psi(0)>$ being a non-interacting eigenstate $|n_M \cdots n_2 n_1>$ (with all $n_k \in \{0, 1\}$ and $\sum_k n_k = L$), evolves eventually into a thermalized state. We have computed such time dependent states using the exact eigenvalues and eigenvectors of H to evaluate the time evolution operator. As initial states, we have chosen four states (for $M = 16$, $L = 7$): (i) $|\phi_1> = |0000100000111111>$ where a particle at orbital 7 is excited from the non-interacting ground state (with all orbitals from 1 to 7 occupied) to the orbital 12, (ii) $|\phi_2> = |0010100000011111>$ where two particles at orbitals 6 and 7 are excited from the non-interacting ground state to the orbitals 12 and 14, (iii) $|\phi_3> = |0000011000110111>$ and (iv) $|\phi_4> = |1000100011001011>$. The states $|\phi_3>$ and $|\phi_4>$ are obtained from the exact eigenstate of H for $A = 0.35$ at level number $m = 123$ and 1354, respectively, by rounding the occupation numbers n_k to 1 (or 0) if $n_k > 0.5$ ($n_k < 0.5$) (states of top panels in Figure 2). The approximate energies (6) of these four states are $E \approx 30$ ($|\phi_1>$), $E \approx 32$ ($|\phi_2>$ and $|\phi_3>$) and $E \approx 35$ ($|\phi_4>$).

It is useful to express the time in multiples of the elementary quantum time step defined as:

$$\Delta t = \frac{t_H}{d} = \frac{1}{\sqrt{2\pi \sigma^2(A)}}, \tag{13}$$

where t_H is the Heisenberg time (at the given value of A), d the dimension of the Hilbert space and $\sigma(A)$ the width of the Gaussian density of states given in (10). The quantity Δt is the shortest physical time scale of the system (inverse of the largest energy scale) and obviously for $t \ll \Delta t$ the unitary evolution operator is close to the unit matrix multiplied by a uniform phase factor: $\exp(-iHt) \approx \exp(-iE_{mean}t)\,\mathbf{1}$ since the eigenvalues E_m of H satisfy $|E_m - E_{mean}| \lesssim \sigma(A)$. We expect that any signification deviation of $|\psi(t)>$ with respect to the initial condition $|\psi(0)>$ happens at $t \geq \Delta t$ (or later in case of very weak interaction). Furthermore, by analyzing the time evolution in terms of the ratio $t/\Delta t$ the results do not depend on the global energy scale of the spectral width. The longest time scale is the Heisenberg time $t_H \approx 10^4 \Delta t$ (since $d = 11440$ for $M = 16$, $L = 7$). Later, we also discuss intermediate time scales such as the inverse decay rate obtained from the Fermi golden rule.

To show graphically the time evolution, we compute the time dependent occupation numbers $n_k(t) = <\psi(t)|c_k^\dagger c_k|\psi(t)>$ and present them in a color density plot in the plane $(k, t/\Delta t)$. In addition, at the time value used last, we compute the effective total one-particle energy E_{1p} using the relation (11) (note that E_{1p} is not conserved with respect to the time evolution except for very weak interaction) and use this value to determine from (12) the inverse temperature β, chemical potential μ and the thermalized Fermi–Dirac filling factor $n(\varepsilon_k)$ at each k value for the orbital index. These values of ideally thermalized occupation numbers will be shown in an additional vertical bar (Note that this additional bar is not related to the usual color bar that provides the translation of colors to n_k values). The latter is shown in Figure 6 and applies also to all subsequent figures with color density plots for n_k values right behind the data for the last time values separated by a vertical white line. This presentation allows for an easy verification if the occupation numbers at the last time values are indeed thermalized or not.

In Figure 7, we show the time evolution for the initial state $|\phi_1>$ and the two Åberg parameter values $A = 1$ and $A = 3.5$ using a linear time scale with integer multiples of Δt and for $t \leq 2000\,\Delta t \approx t_H/6$. At $A = 1$, the occupation number n_{12} (of the excited particle) shows, at the beginning, a periodic structure, with an approximate period $400\,\Delta t$ for $t < 1000\,\Delta t$, and a modest decay for $t > 1000\,\Delta t$. At the same time, the first orbitals above the Fermi sea are slightly excited. At final $t = 2000\,\Delta t$, the state is clearly not thermalized. For $A = 3.5$, we see a very rapid partial decay of n_6 and n_{12} together with an increase of n_7. Furthermore, for n_k with $8 \leq k \leq 11$, there are later and more modest excitations with a periodic time structure. Here, the final state at $t = 2000\,\Delta t$ is also not thermalized, but it is closer to thermalization as for the case $A = 1$.

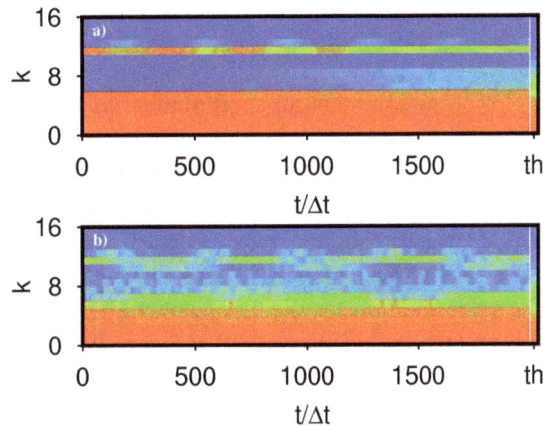

Figure 7. Color density plot of the orbital occupation number n_k in the plane of orbital index k and time t for the time dependent state $|\psi(t)>= \exp(-iHt)|\psi(0)>$ with initial condition $|\psi(0)>= |\phi_1>= |0000100000111111>$. The time values are integer multiples of the elementary quantum time step $\Delta t = t_H/d = 1/[\sqrt{2\pi}\,\sigma(A)]$ where t_H is the Heisenberg time (at the given value of A). The bar behind the vertical white line with the label "th" shows the theoretical thermalized Fermi–Dirac occupation numbers $n(\varepsilon_k)$ where β and μ are determined from (12) using the energy $E = E_{1p}$ of the state $|\psi(t)>$ at the last time value $t = 2000\,\Delta t$. The two panels correspond to the Åberg parameter $A = 1$ (**a**), $A = 3.5$ (**b**). For the translation of colors to n_k values, the color bar of Figure 6 applies.

The linear time scale used in Figure 7 is not very convenient since it cannot capture a rapid decay/increase of n_k well at small times and its maximal time value is also significantly limited below the Heisenberg time. Therefore, we use in Figures 8 and 9 a logarithmic time scale with $0.1\,\Delta t \leq t \leq 10^6\,\Delta t \approx 10^2\,t_H$. Note that, in these figures, the different n_k values for each cell are not time averaged but represent the precise values for certain, exponentially increasing, discrete time values (see caption of Figure 8 for the precise values). Therefore, in case of periodic oscillations of n_k, there will be, for larger time values, a quasi random selection of different time positions with respect to the period.

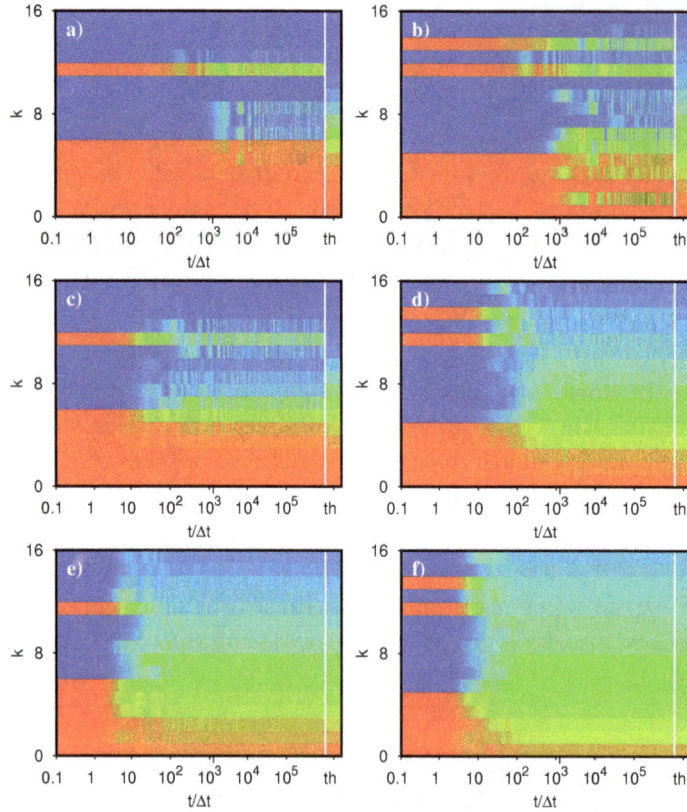

Figure 8. Color density plot of the orbital occupation number n_k in the plane of orbital index k and time t for the time dependent state $|\psi(t)\rangle = \exp(-iHt)|\psi(0)\rangle$. The time axis is shown in logarithmic scale with time values $t_n = 10^{(n/100)-1}\,\Delta t$ and integer $n \in \{0, 1, \ldots, 700\}$ corresponding to $0.1 \le t_n/\Delta t \le 10^6$. The elementary quantum time step Δt is the same as in Figure 7. The bar behind the vertical white line with the label "th" shows the theoretical thermalized Fermi–Dirac occupation numbers $n(\varepsilon_k)$ where β and μ are determined from (12) using the energy $E = E_{1\mathrm{p}}$ of the state $|\psi(t)\rangle$ at the last time value $t = 10^6\,\Delta t$. The additional longer tick below the t-axis right next to the tick for 10^3 gives the position of the maximal time value $t/\Delta t = 2000$ of Figure 7. The different panels correspond to the initial state $|\psi(0)\rangle = |\phi_1\rangle = |0000100000111111\rangle$ (**a,c,e**) or $|\psi(0)\rangle = |\phi_2\rangle = |0010100000011111\rangle$ (**b,d,f**) and Åberg parameter values $A = 1$ (**a,b**), $A = 3.5$ (**c,d**), $A = 10$ (**e,f**). For the translation of colors to n_k values, the color bar of Figure 6 applies.

In Figure 8, the time evolution for the initial states $|\phi_1\rangle$ and $|\phi_2\rangle$ is shown for the Åberg values $A = 1, 3.5, 10$. For $|\phi_1\rangle$ at $A = 1$ and $A = 3.5$, the observations of Figure 7 are confirmed with the further information that the absence of thermalization in these cases is also valid for time scales larger than $2000\,\Delta t$ up to $10^6\,\Delta t$ and for $A = 3.5$ the initial decay of n_6 and n_{12} happens at $t \approx 10\,\Delta t$. For $|\phi_1\rangle$ at $A = 10$, the decay starts at $t \approx 3\,\Delta t$ and an approximate thermalization happens at $t > 40\,\Delta t$. However, here there is still some time periodic structure and it would be necessary to do some time average to have perfect thermalization. For $|\phi_2\rangle$ at $A = 1$, the decay of excited orbitals 12 and 14 starts at $t \approx 100\,\Delta t$ and saturates at $t \approx 1000\,\Delta t$ at which time also orbitals 6 and 7 are excited. After this, there are very small excitations of orbitals 8, 9, 10 and maybe 13, 15. There is also some very modest decay of the Fermi sea orbitals 2, 4 and 5 at $t > 1000\,\Delta t$. The final state at $t = 10^6\,\Delta t$ is not thermalized even though some orbitals have n_k values close to thermalization. For $|\phi_2\rangle$ at $A = 3.5$, the decay of excited orbitals 12 and 14 starts at $t \approx 10\,\Delta t$ and, for $t > 300\,\Delta t$, there is thermalization (but requiring some time average as for $|\phi_1\rangle$ at $A = 10$). Interestingly, at intermediate times $10\,\Delta t < t < 100\,\Delta t$, the high orbitals 13 and 16 are temporarily slightly excited and decay afterwards rather quickly to their thermalized values. For $|\phi_2\rangle$ at $A = 10$, the decay of excited orbitals 12 and 14 starts even at $t \approx 3\,\Delta t$ and thermalization seems to set in at $t > 30\,\Delta t$.

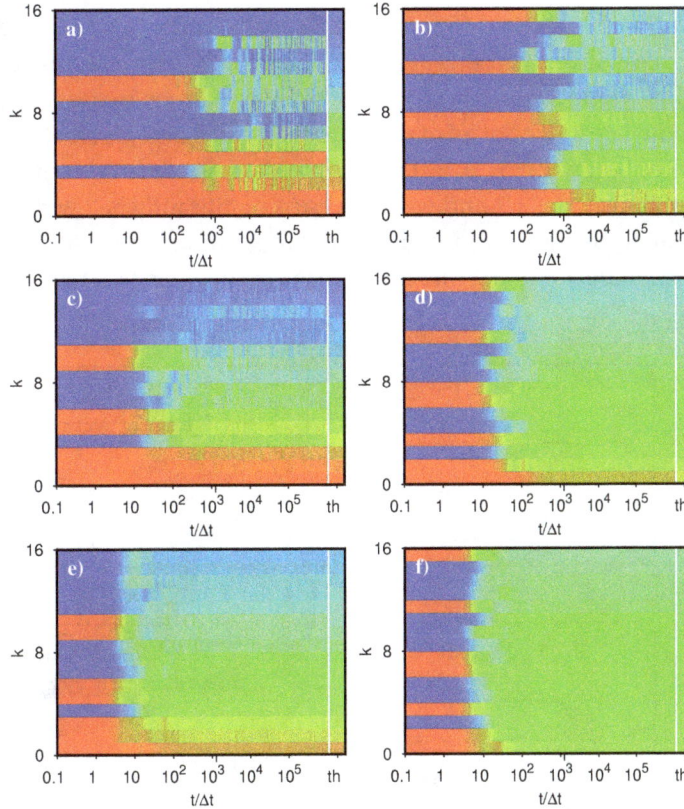

Figure 9. As in Figure 8 but for the initial states $|\psi(0)> = |\phi_3> = |0000011000110111>$ (**a,c,e**) and $|\psi(0)> = |\phi_4> = |1000100011001011>$ (**b,d,f**) (with the same A values as in Figure 8 for each row). These initial states can be obtained from the eigenstates of H for $A = 0.35$ at level numbers $m = 123$ or 1354, respectively, by rounding the occupation numbers to 1 (or 0) if $n_k > 0.5$ ($n_k < 0.5$) (see also top panels of Figure 2).

Figure 9 is similar to Figure 8 except for the initial states $|\phi_3>$ and $|\phi_4>$ which have occupation numbers $n_k \in \{0, 1\}$ obtained by rounding the n_k values of the two eigenstates visible in the two top panels of Figure 2. Here, the initial decay of excited orbitals starts roughly at $t \approx 300\,\Delta t$ ($t \approx (10 - 20)\,\Delta t$ or $t \approx (2 - 3)\,\Delta t$) for $A = 1$ ($A = 3.5$ or $A = 10$, respectively). There is no thermalization for both states at $A = 1$ (but some n_k values are close to thermalized values), approximate thermalization for $A = 3.5$ and $|\phi_3>$ and good thermalization for $A = 3.5$ and $|\phi_4>$ as well as $A = 10$ (both states).

Using the time dependent values $n_k(t)$, one can immediately determine the corresponding entropy $S(t)$ using (11). At $t = 0$, we have obviously $S(0) = 0$, since, for all four initially considered states, we have perfect occupation number values of either $n_k = 0$ or $n_k = 1$. Naturally, one would expect that the entropy increases with a certain rate and saturates then at some maximal value which may correspond (or be lower) to the thermalized entropy $S_{th}(E_{1p})$ (with E_{1p} determined for the state $|\psi(t)>$ at large times) depending if there is presence (or absence) of thermalization according to the different cases visible in Figures 8 and 9. However, in the absence of thermalization, we see that there may also exist periodic oscillations with a finite amplitude at very long time scales.

In Figure 10, we show the time dependent entropy $S(t)$ for the two initial states $|\phi_1>$, $|\phi_4>$ and the three values $A = 1$, $A = 3.5$ and $A = 10$ of the Åberg parameter. For $A = 10$, there is indeed a rather rapid saturation of the entropy of both states at a maximal value that is indeed close to the thermalized entropy $S_{th}(E_{1p})$. We note that E_{1p} is not conserved at strong interactions and that its initial value $E_{1p} \approx 30$ ($E_{1p} \approx 35$) at $t = 0$ evolves to $E_{1p} \approx 33.5$ ($E_{1p} \approx 37$) at large times for $|\phi_1>$ ($|\phi_4>$) corresponding roughly to $S \approx S_{th}(E_{1p}) \approx 9.2$ (10.8) visible as thin blue horizontal lines in Figure 10. For $A = 3.5$ (or $A = 1$), the thermalized entropy values, visible as thin green (red) lines, are lower as compared to the case $A = 10$ due to different final E_{1p} values. For $A = 3.5$ and $|\phi_4>$, there is also saturation of S to its thermalized value. For $A = 3.5$ and $|\phi_1>$, there seems to be an approximate saturation at a quite low value $S \approx 6$ but with periodic fluctuations in the range 6 ± 0.3. For $A = 1$ and $|\phi_4>$, there is a quite late and approximate saturation with some fluctuations that are visible for $t > 10^4 \Delta t$ and with $S \approx 10 \pm 0.2$. For $A = 1$ and $|\phi_1>$, there is a late periodic regime for $t > 10^3 \Delta$ with a quite large amplitude $S \approx 3 \pm 1$ and with $S_{max} \approx 4$ significantly below the thermalized entropy $S_{th}(E_{1p}) \approx 5.5$. The panels using a normal (instead of logarithmic) time scale with $t \leq 200 \Delta t$ miss completely the long time limits for $A = 1$ and might incorrectly suggest that there is an early saturation at quite low values of S.

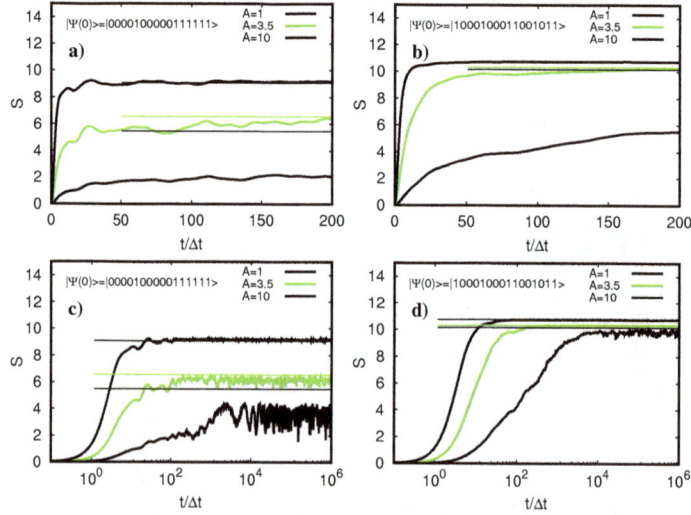

Figure 10.   Time dependence of the entropy S, computed by (11), of the state $|\psi(t)> = \exp(-iHt)|\psi(0)>$ for the Åberg parameter values $A = 1$ (red lines), $A = 3.5$ (green lines), $A = 10$ (blue lines) and initial states $|\psi(0)> = |\phi_1> = |0000100000111111>$ (**a,c**); $|\psi(0)> = |\phi_4> = |1000100011001011>$ (**b,d**); thick colored lines show numerical data of $S(t)$ and thin horizontal colored lines show the thermalized entropy $S_{th}(E_{1p})$ with E_{1p} being determined from $|\psi(t)>$ at $t = 10^6 \Delta t$; panels (**a,b**) use a linear time axis: $0 \leq t \leq 200 \Delta t$; panels (**c,d**) use a logarithmic time axis: $0.1 \Delta t \leq t \leq 10^6 \Delta t$; Δt is the elementary quantum time step (see also Figure 6).

The periodic (or quasi-periodic) time dependence of $n_k(t)$ or $S(t)$, for the cases with lower values of A and/or an initial state with lower energy, indicates that, for such states, only a small number $(2, 3, \ldots)$ of exact eigenstates of H contribute mostly in the expansion of $|\psi(t)>$ in terms of these eigenstates.

Figure 10 also shows that the initial increase of $S(t)$ is rather comparable between the two states for identical values of A even though the long time limit might be very different. Furthermore, a closer inspection of the data indicates that typically $S(t)$ is close to a quadratic behavior for $t \lesssim \Delta t$ but which immediately becomes linear for $t \gtrsim \Delta t$ similarly as the transition probabilities between states in the context of time dependent perturbation theory. To study the approximate slope in the linear regime, we define (For practical reasons, we decide to incorporate the quantum time step Δt in the definition of Γ_c, i.e., Γ_c is defined as the ratio of the initial slope $S'(t)$ over the global spectral bandwidth $\sim \sigma(A) \sim 1/\Delta t$) the quantity $\Gamma_c = dS(t)/d(t/\Delta t) = \Delta t S'(t)$ for $t = \zeta \Delta t$, where $\zeta \gtrsim 1$ is a numerical constant of order one. To determine Γ_c practically, we perform first the fit $S(t) = \bar{S}_\infty (1 - \exp[-\bar{\gamma}_1(t/\Delta t)])$ for $0 \leq t/\Delta t \leq 100$ and use the exponential decay rate $\bar{\gamma}_1$ to perform a refined fit $S(t) = S_\infty (1 - \exp[-\gamma_1(t/\Delta t) - \gamma_2(t/\Delta t)^2])$ for the interval $0 \leq t/\Delta t \leq 5/\bar{\gamma}_1$. From this, we determine $\Gamma_c = S_\infty \gamma_1$, which is rather close to $\bar{S}_\infty \bar{\gamma}_1$ for $A \leq 2$ but not for larger values of A where the decay time is reduced and not sufficiently large in comparison to the initial quadratic regime. Therefore, the quadratic term in the exponential is indeed necessary to obtain a reasonable fit quality. This procedure corresponds to an effective average of the value of ζ between 1 and roughly $1/\gamma_1$, which is indeed useful to smear out some oscillations in the initial increase of $S(t)$ for smaller values $A \leq 1$.

We note that, for many-body quantum systems, the exponential growth of entropy with time had been also discussed and numerically illustrated in [40] (see also related publications in References 25 and 26 there). Recently, such an exponential growth of entropy has been discussed in [62,70].

Figure 11 shows the dependence of these values of Γ_c on the parameter A for our four initial states. At first sight, one observes a behavior $\Gamma_c \propto A^2$ for $A \lesssim 2$ and a saturation for larger values of A. However, a more careful analysis shows that there are modest but clearly visible deviations with respect to the quadratic behavior in A (power law fits $\Gamma_c \propto A^p$ for $A \leq 2$ provide exponents close to $p \approx 1.75 - 1.85$) and it turns out that these deviations correspond to a logarithmic correction: $\Gamma_c = f(A) = (C_1 - C_2 \ln[g(A)]) g(A)$ with $g(A) = A^2$ (for fits with $A \leq 1$) or with $g(A) = A^2/(1 + C_3 A^2)$ (for fits with all A values).

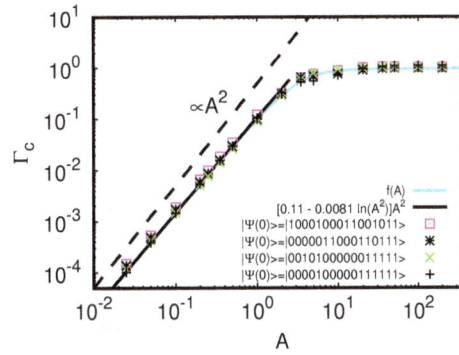

Figure 11.   Dependence of the initial slope $\Gamma_c = \Delta t\, S'(t)$ (at small time values $t \sim \Delta t$) of the time dependent entropy on the Åberg parameter A in double logarithmic scale. Δt is the elementary quantum time step (see also Figure 6). The practical determination of Γ_c is done using the fit $S(t) = S_\infty (1 - \exp[-\gamma_1(t/\Delta t) - \gamma_2(t/\Delta t)^2])$ which provides $\Gamma_c = S_\infty \gamma_1$. The different data points correspond to the four different initial states used in Figures 8 and 9. The dashed line corresponds to the power law behavior $\propto A^2$ and the light blue line corresponds to the fit $\Gamma_c = f(A) = (C_1 - C_2 \ln[g(A)]) g(A)$ with $g(A) = A^2/(1 + C_3 A^2)$ and fit values $C_1 = 0.107 \pm 0.009$, $C_2 = 0.0081 \pm 0.0023$, $C_3 = 0.092 \pm 0.017$ for the initial state $|\psi(0)>= |\phi_1>= |0000100000111111>$ corresponding to the red plus symbols. Fit values for the other initial states can be found in Table 2. The full black line corresponds to $f_0(A) = f(A)_{C_3=0} = [C_1 - C_2 \ln(A^2)] A^2$. The simpler fit $\Gamma_c = f_0(A)$ in the range $0.025 \leq A \leq 1$ provides the values $C_1 = 0.107 \pm 0.002$ and $C_2 = 0.0078 \pm 0.0005$ which are identical (within error bars) to the values found by the more general fit $\Gamma_c = f(A)$ for the full range of A values.

Table 2.   Values of the fit parameters $C_1, \ldots, C_5$ for the initial states $|\phi_1>, \ldots, |\phi_4>$ used for the analytical fits of Γ_c (Γ_F) in Figure 11 (see also Figure 12 and Figure 13).

Initial State	C_1	C_2	C_3	C_4	C_5
$\|\phi_1>= \|0000100000111111>$	0.107 ± 0.009	0.0081 ± 0.0023	0.092 ± 0.017	0.0048 ± 0.0001	0.0054 ± 0.0003
$\|\phi_2>= \|0010100000011111>$	0.100 ± 0.003	0.0103 ± 0.0011	0.069 ± 0.007	0.0068 ± 0.0001	0.0099 ± 0.0003
$\|\phi_3>= \|0000011000110111>$	0.110 ± 0.005	0.0130 ± 0.0016	0.080 ± 0.012	0.0076 ± 0.0001	0.0098 ± 0.0003
$\|\phi_4>= \|1000100011001011>$	0.103 ± 0.003	0.0210 ± 0.0007	0.018 ± 0.005	0.0094 ± 0.0001	0.0140 ± 0.0002

To understand this behavior, we write for sufficiently small times $n_k(t) \approx 1 - \delta n_k(t)$ (if $n_k(0) = 1$) or $n_k(t) \approx \delta n_k(t)$ (if $n_k(0) = 0$), where $\delta n_k(t)$ is the small modification of $n_k(t)$. Time dependent perturbation theory suggests that $\delta n_k(t) \sim (t/\Delta t)^2$ for $t \lesssim \Delta t$ and $\delta n_k(t) \approx a_k A^2 t/\Delta t$ for $t \gtrsim \Delta t$ such that still $\delta n_k(t) \ll 1$ with coefficients a_k dependent on k (and also on M, L) and satisfying a linear relation to ensure the conservation of particle number. Using (11) and neglecting corrections of order

δn_k^2, we obtain: $S \approx -\sum_k (\delta n_k \ln \delta n_k - \delta n_k)$ and $\Gamma_c = \Delta t\, S'(t) \approx -\Delta t \sum_k \delta n_k'(t) \ln \delta n_k(t)$ with $t = \xi\, \Delta t$. Since $\delta n_k'(t) \approx a_k A^2 / \Delta t$, we find indeed the behavior :

$$\Gamma_c = [C_1 - C_2 \ln(A^2)]\, A^2 \quad , \quad C_1 = -\sum_k a_k \langle \ln(a_k \xi) \rangle \quad , \quad C_2 = \sum_k a_k, \qquad (14)$$

where $\langle \cdots \rangle$ indicates an average over some modest values of $\xi \gtrsim 1$. The precise values of a_k may depend rather strongly on the orbital index k and the initial state (see also Figures 8 and 9), but the coefficients C_1, C_2 depend only slightly on the initial state (see Table 2). Furthermore, by replacing $A^2 \to g(A) = A^2 / (1 + C_3 A^2)$ to allow for a saturation at large A and with a further fit parameter C_3, it is possible to describe the numerical data by the more general fit $\Gamma_c = f(a)$ for the full range of A values.

3.7. Survival Probability and Fermi's Golden Rule

The knowledge of the time dependent states $|\psi(t)\rangle$ allows us also to compute the decay function $p_{\text{dec}}(t) = |\langle \psi(0)|\psi(t)\rangle|^2$ which represents the survival probability of the initial non-interacting eigenstate due to the influence of interactions. Again, for the very short time window $t \lesssim \Delta t$, we expect a quadratic decay: $1 - p_{\text{dec}}(t) \approx \langle (H - E_{\text{mean}})^2 \rangle t^2 \approx \text{const.}\, (t/\Delta t)^2$ with $\langle \cdots \rangle$ being the quantum expectation value with respect to $|\psi(0)\rangle$ and a numerical constant $\lesssim 1$ since $1/\Delta t$ represents roughly the spectral width of H. For $t \gtrsim \Delta t$, but such that $1 - p_{\text{dec}}(t) \ll 1$, we have, according to Fermi's golden rule: $1 - p_{\text{dec}}(t) = \Gamma_F\, (t/\Delta t)$, where Γ_F is the decay rate (Again, for practical reasons and similarly to Γ_c, we incorporate in the definition of Γ_F the time scale Δt, i.e., $\Gamma_F = \Delta t \times$ usual decay rate found in the literature and meaning that Γ_F is defined as the ratio of the usual decay rate over the global spectral bandwidth) of the state.

To determine Γ_F numerically, we apply the fit: $p_{\text{dec}}(t) = C \exp(-\Gamma_F t/\Delta t)$ in two steps. First, we use the interval $1 \le t/\Delta t \le 50$ and, if $5/\Gamma_F < 50$, corresponding to a rapid decay (which happens for larger values of A), we repeat the fit for the reduced interval $1 \le t/\Delta t \le 5/\Gamma_F$. The choice of the Amplitude $C \neq 1$ and the condition $t \ge \Delta t$ for the fit range allow for taking into account the effects due to the small initial window of quadratic decay. In Figure 12, we show two examples for the initial state $|\phi_1\rangle$ and the Åberg values $A = 3.5$ and $A = 10$. In both cases, the shown maximal time value $t_{\text{max}} = 50\, \Delta t$ (if $A = 3.5$) or $t_{\text{max}} \approx 13.5$ (if $A = 10$) defines the maximal time value for the fit range. For $A = 10$, the fit nicely captures the decay for $1 \le t/\Delta t \le 6$, while, for $A = 3.5$, there are also some oscillations in the decay function for which the fit procedure is equivalent to some suitable average in the range $1 \le t/\Delta t \le 30$. For very small values of A, the fit procedure also works correctly since it captures only the initial decay that is important if $p_{\text{dec}}(t)$ does not decay completely at large times and which typically happens in the perturbative regime $A \lesssim 1$.

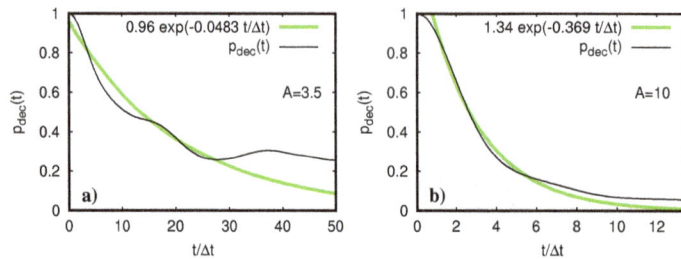

Figure 12. Decay function $p_{\text{dec}}(t) = |\langle \psi(0)|\psi(t)\rangle|^2$ obtained numerically from $|\psi(t)\rangle$ with the initial state $|\psi(0)\rangle = |0000100000111111\rangle$ (thin red line) and the fit $p_{\text{dec}}(t) = C \exp(-\Gamma_F t/\Delta t)$ (thick green line) for the two Åberg values $A = 3.5$ (**a**) and $A = 10$ (**b**). The fit values are $C = 0.959 \pm 0.011$, $\Gamma_F = 0.0483 \pm 0.0015$; (**a**)$C = 1.339 \pm 0.015$, $\Gamma_F = 0.369 \pm 0.005$; (**b**) corresponding to the decay times $\Gamma_F^{-1} = 20.7$ (**a**), 2.71; (**b**). Δt is the elementary quantum time step (see also Figure 6).

Figure 13 shows the dependence of Γ_F on A for the usual four initial states together with the fit $\Gamma_F = f(A) = C_4 A^2 / (1 + C_5 A^2)$ for the data with initial state $|\phi_1>$. The values of the parameters C_4, C_5 for this and the other initial states are given in Table 2. Here, the initial quadratic dependence $\Gamma_F \propto A^2$ is highly accurate (with no logarithmic correction). Similarly to Γ_c, there is only a slight dependence of the values of Γ_F and the fit values on the choice of initial state.

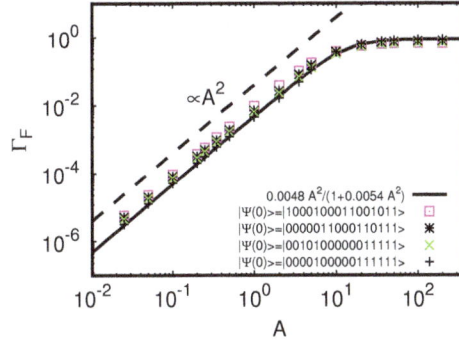

Figure 13. Dependence of the decay rate Γ_F corresponding to Fermi's gold rule on the Åberg parameter A in double logarithmic scale. The practical determination of Γ_F is done using the exponential fit function of Figure 12 for the numerically computed decay function $p_{\mathrm{dec}}(t)$. The different data points correspond to the four different initial states used in Figures 8 and 9. The dashed line corresponds to the power law behavior $\propto A^2$ and the full black line corresponds to the fit $\Gamma_F = f(A) = C_4 A^2 / (1 + C_5 A^2)$ and fit values $C_4 = 0.0048 \pm 0.0001$, $C_5 = 0.0054 \pm 0.0003$ for the initial state $|\psi(0)> = |\phi_1> = |0000100000111111>$ corresponding to the red plus symbols. Fit values for the other initial states can be found in Table 2. Δt is the elementary quantum time step (see also Figure 6).

Theoretically, we expect according to Fermi's Golden rule that: $\Gamma_F \approx (\Delta t) \, 2\pi V_{\mathrm{Fock}}^2 \, \rho_c(E)$, where $V_{\mathrm{Fock}}^2 = \mathrm{Tr}_{\mathrm{Fock}}(V^2)/(Kd) = \sigma_0^2 \, \alpha A^2 / K$ according to the discussion below (10) and $\rho_c(E)$ is the effective two-body density of states for states directly coupled by the interaction such that $\rho_c(E_{\mathrm{mean}}) = 1/\Delta_c$ (see discussion below (7)). We note that V_{Fock} is the typical interaction matrix element in Fock space which is slightly larger than V_{mean} (see the theoretical discussion above for the computation of the coefficient α used in (10) and Appendix A of [46]). The factor Δt is due to our particular definition (Again, for practical reasons and similarly to Γ_c, we incorporate in the definition of Γ_F the time scale Δt, i.e., $\Gamma_F = \Delta t \times$ usual decay rate found in the literature and meaning that Γ_F is defined as the ratio of the usual decay rate over the global spectral bandwidth) of decay rates. The expression of Γ_F is actually also valid for larger values of A provided we use the density of states $\rho_c(E)$ in the presence of interactions that provides an additional factor $1/\sqrt{1 + \alpha A^2}$ according to (10). Therefore, at the band center, we have: $2\pi \Delta t \rho_c = K/[\sigma_0(L) \sigma_0(L=2)(1+\alpha A^2)]$, which gives, together with (8):

$$\Gamma_F = \frac{\sigma_0(L)}{\sigma_0(L=2)} \left(\frac{\alpha A^2}{1 + \alpha A^2} \right) = \sqrt{\frac{L(M-L)}{2(M-2)}} \left(\frac{\alpha A^2}{1 + \alpha A^2} \right). \tag{15}$$

For $M = 16$ and $L = 7$, the square root factor is 1.5 and we have to compare $1.5\alpha \approx 0.0132$ with the values of C_4 in Table 1, which are somewhat smaller, probably due to a reduction factor for the energy dependent density of states since the energies of the initial states have a certain distance to the band center. Furthermore, according to (15), we have to compare C_5 with $\alpha \approx 0.00877$ which is not perfect but gives the correct order of magnitude. For both parameters, the numerical matching is quite satisfactory taking into account the very simple argument using the same typical value of the interaction matrix elements for all cases of initial states.

Finally, we mention that, for the three Åberg parameter values $A = 1$, $A = 3.5$, $A = 10$ used in Figures 8 and 9, we have typical decay times in units of Δt being $1/\Gamma_F \approx 300, 30, 3$, respectively (with some modest fluctuations depending on initial states). These values match quite well the observed time values at which the initially occupied orbitals start to decay (see the above discussion of Figures 8 and 9).

3.8. Time Evolution of Density Matrix and Spatial Density

We now turn to the effects of the many-body time evolution in position space (see for example Figure 1). For this, we compute the spatial density

$$\rho(x,y,t) = <\psi(t)|\Psi^{\dagger}(x,y)\,\Psi(x,y)|\psi(t)> \quad , \quad \Psi^{(\dagger)}(x,y) = \sum_{k} \varphi_k^{(*)}(x,y)\,c_k^{(\dagger)}, \tag{16}$$

where $\varphi_k(x,y)$ is the one-particle eigenstate of orbital k, with some examples shown in Figure 1. Here, $\Psi^{(\dagger)}(x,y)$ denotes the usual fermion field operators (in the case of continuous x, y variables) or standard fermion operators for discrete position basis states (when using a discrete grid for x and y positions as we did for the numerical solution of the non-interacting Sinai oscillator model in Section 2). The sum over orbital index k in (16) requires in principle a sum over a *full complete* basis set of orbitals with infinite number (case of continuous x, y values) or a very large number (case of discrete x-y grid) significantly larger than the very modest number of orbitals M we used for the numerical solution of the many-body Sinai oscillator.

However, we can simply state that, in our model, by construction, all orbitals with $k > M$ are never occupied such that, in the expectation value for $\rho(x,y,t)$, only the values $k \leq M$ are necessary. Taking this into account together with the fact that the one-particle eigenstates are real valued, we obtain the more explicit expression:

$$\rho(x,y;t) = \sum_{k,l=1}^{M} \varphi_k(x,y)\,\varphi_l(x,y)\,n_{kl}(t) \quad , \quad n_{kl}(t) = <\psi(t)|c_k^{\dagger}\,c_l\,|\psi(t)>, \tag{17}$$

where $n_{kl}(t)$ is the density matrix in orbital representation generalizing the occupation numbers $n_k(t)$ which are its diagonal elements. Due to the complex phases of $|\psi(t)>$ (when expanded in the usual basis of non-interacting many-body states), the density matrix is complex valued but hermitian: $n_{kl}^*(t) = n_{lk}(t)$. Therefore, its anti-symmetric imaginary part does not contribute in $\rho(x,y;t)$. We have numerically evaluated (17) and we present in Figure 14 color plots of the density matrix and the spatial density $\rho(x,y;t)$ for $A = 3.5$, the initial state $|\psi(0)> = |\phi_2>$ and four time values $t/\Delta t = 0.1, 30, 100, 1000$. Since the density $\rho(x,y;t)$ does not provide a lot of spatial structure, we also show in Figure 14 the density difference with respect to the initial condition $\Delta\rho(x,y;t) = \rho(x,y;t) - \rho(x,y;0)$ which reveals more of its structure (figures and videos for the time evolution of this and other cases are available for download at the web page [71]).

At the time $t/\Delta t = 0.1$, density matrix and spatial density are essentially identical to the initial condition at $t = 0$. For $\Delta\rho$, we see a non-trivial structure since there is a small difference with the initial condition and the color plot simply amplifies small maximal amplitudes to maximal color values (red/yellow for strongest positive/negative values even if the latter are small in an absolute scale). The density matrix is diagonal and its diagonal values are either 1 (for initially occupied orbitals) or 0 (for initially empty orbitals) and the spatial density simply gives the sum of densities due to the occupied eigenstates.

At $t/\Delta t = 30$, we see a non-trivial structure in the density matrix with a lot of non-vanishing values in certain off-diagonal elements. Furthermore, the orbitals 13 and 16 are also slightly excited (see also discussion of Figure 8) and there is a significant change of the spatial density.

Later, at $t/\Delta t = 100$, the number/values of off-diagonal elements in the density matrix is somewhat reduced, but they are still visible. Especially between orbitals 12 and 13 as well as 14 and 15, there is a rather strong coupling. Orbital 13 is now more strongly excited than the initially excited orbital 12. In addition, orbitals 14 and 15 are quite strong. The spatial density has become smoother and the structure of $\Delta\rho$ is roughly close to the case at $t/\Delta t = 30$ but with some significant differences.

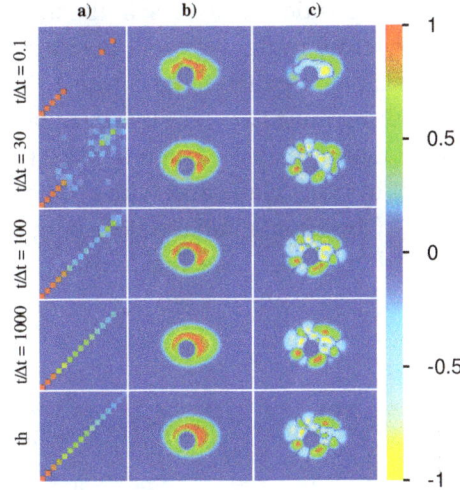

Figure 14. Time dependent density matrix $|n_{kl}(t)|$ (**a**), spatial density $\rho(x,y,t)$ (**b**), spatial density difference with respect to the initial condition $\Delta\rho(x,y,t) = \rho(x,y,t) - \rho(x,y,0)$ (**c**) all computed from $|\psi(t)>$ for the initial state $|\psi(0)>= |\phi_2>= |0010100000011111>$ and the Åberg parameter $A = 3.5$. Panels in column (**a**) correspond to (k,l) plane with $k,l \in \{1,\ldots,16\}$ being orbital index numbers. Panels in columns (**b**,**c**) correspond to the same rectangular domain in (x,y) plane as in Figure 1. The five rows of panels correspond to the time values $t/\Delta t = 0.1, 30, 100, 1000$ and the thermalized case (label "th") where the density matrix is diagonal with entries being the thermalized occupations numbers $n_{kk} = n(\varepsilon_k)$ at energy $E = 32.9$ (typical total one-particle energy of $|\psi(t)>$ for $t/\Delta t \geq 1000$). The numerical values of the color bar represent values of $|n_{kl}|$ (**a**), $(\rho/\rho_{max})^{1/2}$ (**b**), $\mathrm{sgn}(\Delta\rho)(|\Delta\rho|/\Delta\rho_{max})^{1/2}$ (**c**) where ρ_{max} or $\Delta\rho_{max}$ are maximal values of ρ or $|\Delta\rho|$, respectively.

Finally, at $t/\Delta t = 1000$, the density matrix seems be diagonal with values close to the thermalized values. There is a further increase of the density smoothness and $\Delta\rho$ has a similar but different structure as for $t/\Delta t = 100$ or $t/\Delta t = 30$.

Apparently, at intermediate times $20 \leq t/\Delta t \leq 100$, there are some quantum correlations between certain orbitals, visible as off-diagonal elements in the density matrix which disappear for later times. This kind of decoherence is similar to the exponential decay observed in [46] for the off-diagonal element of the 2×2 density matrix for a qubit coupled to a chaotic quantum dot or the SYK black hole. However, to study this kind of decoherence more carefully in the context here, it would be necessary to use as initial state a non-trivial linear combination of two non-interacting eigenstates and not to rely on the creation of modest off-diagonal elements for intermediate time scales as we see here.

The spatial density is globally rather smooth and typically quite well given by the "classical" relation $\rho(x,y;t) \approx \sum_k \varphi_k^2(x,y)\, n_k(t)$ in terms of the time dependent occupation numbers. Only for intermediate time scales with more visible quantum coherence (more off-diagonal elements $n_{kl}(t) \neq 0$), this relation is less accurate. However, at $A = 3.5$, the density still exhibits small but regular fluctuations in its detail structure as can be seen in the structure of $\Delta\rho$ for later time scales. A closer inspection of the data (for time values not shown in Figure 14) also shows that, even at long time scales, there are significant fluctuations of ρ when t is slightly changed by a few multiples of Δt.

In Figure 14, we also show for comparison the theoretical thermalized quantities where, in (17), the density matrix is replaced by a diagonal matrix with entries being the thermalized occupations numbers $n_{kk} = n(\varepsilon_k)$ at energy $E = 32.9$, which is the typical total one-particle average energy of $|\psi(t)>$ for the long time limit $t/\Delta t \geq 1000$ showing that, at $t/\Delta t = 1000$, the state is very close to thermalization but still with small significant differences (see also discussion of Figure 7 for this case).

We may also generalize the spatial density (16) to a spatial density correlator which we define as:

$$\rho_{\mathrm{corr}}(x,y;x_0,y_0;t) = <\psi(t)|\Psi^\dagger(x,y)\,\Psi(x_0,y_0)|\psi(t)> = \sum_{k,l=1}^{M} \varphi_k(x,y)\,\varphi_l(x_0,y_0)\,n_{kl}(t) \qquad (18)$$

depending on initial (x_0, y_0) and final position (x, y). As an illustration, we choose the fixed value $(x_0, y_0) = (1.22, 0.15)$ which is very close to the maximal position (center of the red area) of the one-particle ground state $\varphi_1(x, y)$ visible in panel (a) of Figure 1. The spatial density correlator is potentially complex with a non-vanishing imaginary part in case of non-vanishing off-diagonal matrix elements of $n_{kl}(t) \neq 0$ for $k \neq l$. In Figure 15, we present density plots of absolute value, real and imaginary part of $\rho_{\mathrm{corr}}(x, y; x_0, y_0; t)$ in (x, y) plane and with the given value $(x_0, y_0) = (1.22, 0.15)$ for the same parameters of Figure 14 (concerning initial state, Åberg parameter, time values and also thermalized case). However, for the thermalized case, the density matrix is diagonal by construction and the imaginary part of $\rho_{\mathrm{corr,th}}(x, y; x_0, y_0)$ vanishes (giving a blue panel due to zero values).

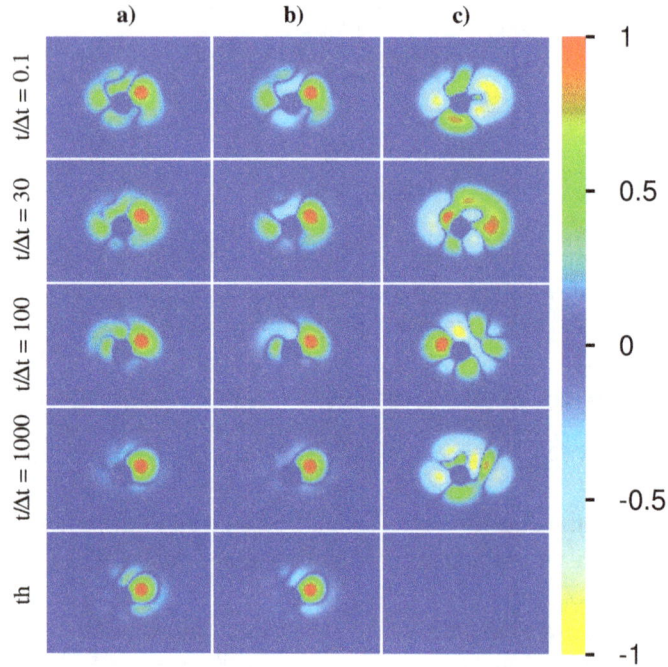

Figure 15. Time dependent spatial density correlator $\rho_{\mathrm{corr}}(x, y; x_0, y_0; t)$ shown in the same rectangular domain in (x, y) plane as in Figure 1. The columns correspond to absolute value (**a**), real part (**b**), imaginary part (**c**). The initial point is given by $(x_0, y_0) = (1.22, 0.15)$ which is very close to the maximal position (center of the red area) of the one-particle ground state $\varphi_1(x, y)$ visible in panel (**a**) of Figure 1. Initial state, Åberg parameter and meaning of row labels are as in Figure 14. The numerical values of the color bar represent values of $\mathrm{sgn}(u)|u/u_{\max}|^{1/2}$ where u is absolute value (**a**), real part (**b**), imaginary part (**c**), of ρ_{corr} and $u_{\max}$ is the maximal value of $|u|$. The data for thermalized case and imaginary part is completely zero (blue panel in bottom right corner) since the spatial density correlator is for the thermalized case purely real.

There are significant time dependent fluctuations of $\rho_{\mathrm{corr}}(x, y; x_0, y_0; t)$ for all time scales with real part and absolute value being dominated by rather strong maximal values for positions close to the initial position. However, the imaginary part (which vanishes at $t = 0$ and is typically smaller than the values in maximum domain of a real part) shows a more interesting structure since the color plot amplifies small amplitudes (in absolute scale). Apart from this, the absolute and real part values for positions outside the maximum domain (far away from the initial position) seem to decay for long timescales, which is also confirmed by the thermalized case. Even though the case for $t/\Delta t = 1000$ seems to be rather close to the thermalized case (for absolute value and real part), there are still differences that are more significant here as in Figure 14.

Globally, the obtained results show that the dynamical thermalization takes place well, leading to the usual Fermi–Dirac thermal distribution when the Åberg criterion is satisfied and interactions are sufficiently strong to drive the system into the thermal state.

4. Estimates for Cold Atom Experiments

We discuss here typical parameters for cold atoms in a trap following [72]. Thus, for sodium atoms, we have $\omega \approx \omega_{x,y,z} \approx 2\pi 10 Hz$ with $a_0 = \sqrt{\hbar/(m\omega)} = 6.5\mu m$ and oscillator level spacing $E_u = \hbar\omega \approx 0.5nK$ (nanoKelvin). The typical scattering length is $a_s \approx 3nm$ being small compared to a_0. The atomic density is $\rho_0 = 1/(a_0)^3 \approx 4 \times 10^9 cm^{-3}$. Since $a_s \ll a_0$, the two-body interaction is of δ-function type with $v(\mathbf{r_1} - \mathbf{r_2}) = (4\pi\hbar^2 a_s/m)\delta(\mathbf{r_1} - \mathbf{r_2})$ [66,72]. In our numerical simulations, $\delta(\mathbf{r})$ is replaced by a function $H(r_c - |\mathbf{r}|)/(C 2^d r_c^d)$ with a small r_c, volume C of the unit sphere in d dimensions and with the Heaviside function $H(x) = 1$ (or 0) for $x \geq 0$ and $H = 0$ for $x < 0$. Hence, our parameter U introduced in Section 3 corresponds to $U = (C 2^d r_c^d)(4\pi\hbar^2 a_s/m)$.

Below, we present the estimates for the dynamical thermalization border for excitations of fermionic atoms in a vicinity of their Fermi energy in a 3D Sinai oscillator following the lines of Equation (1). In such a case, the two-body interaction energy scale between atoms is $U_s = 4\pi\hbar^2\rho_0 a_s/m = 4\pi(a_s/a_0)\hbar\omega$ [66,72] so that $U_s/\hbar\omega \sim 6 \times 10^{-3}$. Compared to sodium, the mass of Li atoms is approximately three times smaller so that, for the same ω, we have $a_0 \approx 10\,\mu m$ and $U_s/\hbar\omega \approx 4 \times 10^{-3}$. We think that the scattering length can be significantly increased via the Feshbach resonance, allowing for reaching effective interaction values $U_s/\hbar\omega \sim 1$ being similar to the value $A \sim 3$ used in our numerical studies with the onset of dynamical thermalization.

Usually, a 3D trap with fermionic atoms can capture about $N_a \sim 10^5$ atoms with $\omega \approx \omega_x \sim \omega_y \sim \omega_z \sim 2\pi 10 Hz$. Following the result (1), it is interesting to determine the DTC border dependence on $N_a \gg 1$ for a Sinai oscillator with $r_d \sim 1\mu m \sim a_0/5$. We assume that, similar to a 2D case, the scattering on an elastic ball in the trap center leads to quantum chaos and chaotic eigenstates with $\ell \leq N_a$ components (e.g., in the basis of oscillator eigenfunctions). The Fermi energy of the trap is then $E_F = \hbar(N_a\omega_x\omega_y\omega_z)^{1/3} \approx \hbar\omega N_a^{1/3}$ [52,53]. Assuming that all these components have random amplitudes of a typical size $1/\sqrt{\ell}$, we then obtain an estimate for a typical matrix element of two-body interaction between one-particle eigenstates

$$U_2 \approx \alpha_s \hbar\omega/\ell^{3/2}\,, \quad \alpha_s = 4\pi(a_s/a_0), \quad a_0 = \sqrt{\hbar/m\omega}. \tag{19}$$

The derivation of this estimate is very similar to the case of two interacting particles in a disordered potential with localized eigenstates [73]. At the same time, in the vicinity of the Fermi energy E_F, we have the one-particle level spacing $\Delta_1 = dE_F/dN_a \approx \hbar\omega/(3N_a^{2/3})$. Hence, the effective conductance appears in in (1) is $g = \Delta_1/U_2 \approx \ell^{3/2}/(3\alpha_s N_a^{2/3})$. Thus, from (1), we obtain the dynamical thermalization border for excitation energy δE in a 3D Sinai oscillator trap with N_a fermionic atoms:

$$\delta E > \delta E_{ch} \approx \Delta_1 g^{2/3} \approx 2\ell\Delta_1/(\alpha_s^{2/3}N_a^{4/9}) \sim N_a^{5/9}\Delta_1/\alpha_s^{2/3} \sim \hbar\omega/(\alpha_s^{2/3}N_a^{1/9}) \sim E_F/(\alpha_s^{2/3}N_a^{4/9}). \tag{20}$$

It is assumed that $\delta E \ll E_F$. Here, the last three relations are written in an assumption that $\ell \sim N_a$. Thus, at large N_a values and not too small α_s, the critical energy border δE_{ch} for dynamical thermalization is rather small compared to E_F. However, still $\delta E_{ch} \gg \Delta_1$. Here, we used the maximal value for the number of components $\ell \sim N_a$. It is possible that, in reality, ℓ can be significantly smaller than N_a. However, the determination of the dependence $\ell(N_a)$ requires separate studies taking into account the properties of chaotic eigenstates and their spreading over the energy surface. This spreading can have rather nontrivial properties (see, e.g., [23]). This is confirmed by the results presented in Appendix A for the 2D case of Sinai oscillator showing the numerically obtained dependence of two-body matrix elements on energy for transitions in a vicinity of Fermi energy E_F (see Figure A1 there).

5. Conclusions

In this work, we demonstrated the existence of interaction induced dynamical thermalization of fermionic atoms in a Sinai oscillator trap if the interaction strength between atoms exceeds a critical border determined by the Åberg criterion [38,39,41]. This thermalization takes place in a completely isolated system in the absence of any contact with an external thermostat. In the context of the Loschmidt–Boltzmann dispute [1,2], we should say that formally this thermalization is reversible in time since the Schrodinger equation of the system has symmetry $t \to -t$. The classically chaotic dynamics of atoms in the Sinai oscillator trap breaks in practice this reversibility due to exponential growth of errors induced by chaos. In the regime of quantum chaos, there is no exponential growth of errors due to the fact that the Ehrenfest time scale of chaos is logarithmically short [17,19,26,28]. An example for the stability of time reversibility is given in [27,28]. In fact, the experimental reversal of atom waves in the regime of quantum chaos has been even observed with cold atoms in [74]. In view of this and the fact that the spectrum of atoms in the Sinai oscillator trap is discrete, we can say that dynamical thermalization will have obligatory revivals in time returning from the thermalized state (e.g., bottom panels in Figure 8) to the initial state (top panels in Figure 8). This is the direct consequence of the Poincare recurrence theorem [75]. However, the time for such a recurrence grows exponentially with the number of components contributing to the initial state (which is also exponentially large in the regime of dynamical thermalization) and thus, during such a long time scale, external perturbations (coming from outside of our isolated system, e.g., not perfect isolation) will break in practice this time reversibility.

We hope that our results will initiate experimental studies of dynamical thermalization with cold fermionic atoms in systems such as the Sinai oscillator trap.

Author Contributions: All authors equally contributed to all stages of this work.

Funding: This work was supported in part by the Programme Investissements d'Avenir ANR-11-IDEX-0002-02, reference ANR-10-LABX-0037-NEXT (project THETRACOM). This work was granted access to the HPC GPU resources of CALMIP (Toulouse) under the allocation 2018-P0110.

Acknowledgments: We are thankful to Shmuel Fishman for deep discussions of quantum chaos problems and related scientific topics during many years.

Conflicts of Interest: The authors declare no conflict of interest.

Appendix A. Two-Body Matrix Elements near the Fermi Energy

In this Appendix, we present numerical results for the dependence of the quantity $V_{\text{mean}}(\varepsilon) = \sqrt{\langle V_{ij,kl}^2 \rangle}$ as a function of ε where the average is done *only for orbitals with energies* ε_n *close to* ε (for $n \in \{i, j, k, l\}$), i.e.: $|\varepsilon_n - \varepsilon| \leq \Delta\varepsilon$ with $\Delta\varepsilon = 2$. We note that this is different from the quantity V_{mean} used in Section 3 where the average was done over all orbitals (up to a maximal number being M). The reason for the special average with orbital energies close to ε (which will be identified with the Fermi energy E_F) is that these transitions are dominant in the presence of the Pauli blockade near the Fermi level.

We remind readers that, according to the discussion of Sections 2 and 3, the matrix elements $V_{ij,kl}$ were computed for an interaction potential of amplitude U for $|\mathbf{r_1} - \mathbf{r_2}| < r_c$ (with the radius $r_c = 0.2 r_d = 0.2$) and being zero for $|\mathbf{r_1} - \mathbf{r_2}| \geq r_c$. Furthermore, they have been anti-symmetrized and a diagonal shift $V_{ij,ij} \to V_{ij,ij} - (1/M_2) \sum_{k<l} V_{kl,kl}$ was applied to ensure that the interaction matrix has a vanishing trace.

Due to this shift and the precise average procedure, there is a slight (purely theoretical) dependence on the maximal orbital number M for this average (there is a cut-off effect for ε close to the maximal orbital energy ε_M). Due to this, we considered two values of $M = 30$ and $M = 60$.

The numerically obtained dependence is shown in Figure A1 and is well described by the fit $V_{\text{mean}}/U = a/\varepsilon^b$ with $a = 1.56 \times 10^{-4}$ and $b = 0.78$. The small value of a is due to antisymmetry of two-particle fermionic states and, due to a small value of $r_c = 0.2 r_d$, which leads to a decrease of the effective interaction strength being proportional to r_c^2.

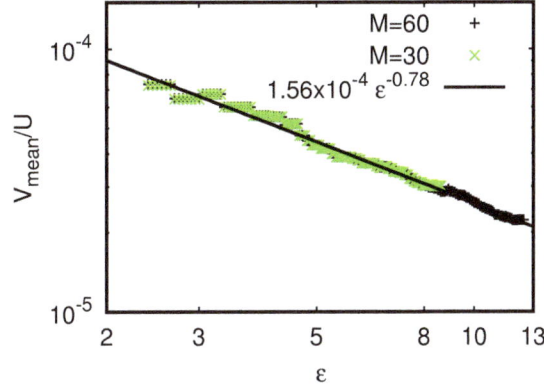

Figure A1. Dependence of the average two-body matrix element V_{mean} rescaled by the amplitude of interaction strength U on one-particle energy ε for two-body interaction transitions in a vicinity of Fermi energy $E_F = \varepsilon$; green symbols are for number of one-particle orbitals $M = 30$, red symbols are for $M = 60$; the blue line shows the fit $V_{\mathrm{mean}}/U = a/\varepsilon^b$ with $a = 0.000156 \pm 2 \times 10^{-6}$; $b = 0.781 \pm 0.005$.

We note that the Fermi energy is determined by the number of fermionic atoms N_a inside the 2D Sinai oscillators with $\varepsilon = E_F \approx \omega N_a^{1/2}$ assuming $\omega = \omega_x \approx \omega_y$. Therefore, we have $\varepsilon \propto M^{1/2} \sim N_a^{1/2}$, $\Delta_1 \sim \hbar\omega/N_a^{1/2}$ and $V_{\mathrm{mean}} \sim \alpha_s \hbar\omega/\ell^{3/2} \sim \alpha_s \hbar\omega/N_a^{b/2}$ (see (19)). Hence, the obtained exponent $b \approx 0.78$ corresponds to the number of one-particle components $\ell \sim N_a^{b/3} \sim N_a^{0.25} \sim n_x^{0.5}$. At the moment, we do not have a clear explanation for this numerical dependence. This dependence corresponds to $g = \Delta_1/V_{\mathrm{mean}} \sim \ell^{3/2}/(\alpha_s N_a^{1/2}) \sim 1/(\alpha_s N_a^{1/8})$. For such a dependence, we obtain that the DTC border in 2D takes place for an excitation energy $\delta E > \delta E_{ch} \sim g^{2/3}\Delta_1 \sim \hbar\omega/(\alpha_s^{2/3} N_a^{7/12})$. Thus, the thermalization can take place at rather low energy excitations above the Fermi energy with $\Delta_1 < \delta E \ll E_F$.

References

1. Loschmidt, J. *Über den Zustand des Wärmegleichgewichts eines Systems von Körpern mit Rücksicht auf die Schwerkraft*; II-73; Sitzungsberichte der Akademie der Wissenschaften: Wien, Austria, 1876; pp. 128–142.
2. Boltzmann, L. *Über die Beziehung eines Allgemeine Mechanischen Satzes zum Zweiten Haupsatze der Wärmetheorie*; II-75; Sitzungsberichte der Akademie der Wissenschaften: Wien, Austria, 1877; pp. 67–73.
3. Mayer, J.E.; Goeppert-Mayer, M. *Statistical Mechanics*; Wiley: New York, NY, USA, 1977.
4. Gousev, A.; Jalabert, R.A.; Pastawski, H.M.; Wisniacki, D.A. Loschmidt echo. *Scholarpedia* **2012**, *7*, 11687.
5. Arnold, V.; Avez, A. *Ergodic Problems in Classical Mechanics*; Benjamin: New York, NY, USA, 1968.
6. Cornfeld, I.P.; Fomin, S.V.; Sinai, Y.G. *Ergodic Theory*; Springer: New York, NY, USA, 1982.
7. Chirikov, B.V. A universal instability of many-dimensional oscillator systems. *Phys. Rep.* **1979**, *52*, 263. [CrossRef]
8. Lichtenberg, A.; Lieberman, M. *Regular and Chaotic Dynamics*; Springer: New York, NY, USA, 1992.
9. Sinai, Y.G. Dynamical systems with elastic reflections. Ergodic properties of dispersing billiards. *Uspekhi Mat. Nauk* **1970**, *25*, 141.
10. Gutzwiller, M.C. *Chaos in Classical and Quantum Mechanics*; Springer: New York, NY, USA, 1990.
11. Haake, F. *Quantum Signatures of Chaos*; Springer: Berlin, Germany, 2010.
12. Stockmann, H.-J. Microwave billiards and quantum chaos. *Scholarpedia* **2010**, *5*, 10243. [CrossRef]
13. Bohigas, O.; Giannoni, M.J.; Schmit, C. Characterization of chaotic quantum spectra and universality of level fluctuation. *Phys. Rev. Lett.* **1984**, *52*, 1. [CrossRef]
14. Wigner, E. Random matrices in physics. *SIAM Rev.* **1967**, *9*, 1. [CrossRef]
15. Mehta, M.L. *Random Matrices*; Elsvier Academic Press: Amsterdam, The Netherlands, 2004.
16. Ullmo, D. Bohigas-Giannoni-Schmit conjecture. *Scholarpedia* **2016**, *11*, 31721. [CrossRef]
17. Chirikov, B.V.; Izrailev, F.M.; Shepelyansky, D.L. Dynamical stochasticity in classical and quantum mechanics. *Math. Phys. Rev.* **1981**, *2*, 209–267.

18. Fishman, S.; Grempel, D.R.; Prange, R.E. Chaos, quantum recurrences, and Anderson localization. *Phys. Rev. Lett.* **1982**, *49*, 509. [CrossRef]

19. Chirikov, B.V.; Izrailev, F.M.; Shepelyansky, D.L. Quantum chaos: Localization vs. ergodicity. *Phys. D* **1988**, *33*, 77. [CrossRef]

20. Fishman, S. Anderson localization and quantum chaos maps. *Scholarpedia* **2010**, *5*, 9816. [CrossRef]

21. Anderson, P.W. Absence of diffusion in certain random lattices. *Phys. Rev.* **1958**, *109*, 1492. [CrossRef]

22. Frahm, K.M.; Shepelyansky, D.L. Quantum localization in rough billiards. *Phys. Rev. Lett.* **1997**, *78*, 1440. [CrossRef]

23. Frahm, K.M.; Shepelyansky, D.L. Emergence of quantum ergodicity in rough billiards. *Phys. Rev. Lett.* **1997**, *79*, 1833. [CrossRef]

24. Bohr, N. Über die Serienspektra der Element. *Zeitschrift für Physik* **1920**, *2*, 423. [CrossRef]

25. Ehrenfest, P. Bemerkung über die angenäherte Gültigkeit der klassischen Mechanik innerhalb der Quantenmechanik. *Zeitschrift für Physik* **1927**, *45*, 455. [CrossRef]

26. Shepelyanskii, D.L. Dynamical stochasticity in nonlinear quantum systems. *Theor. Math. Phys.* **1981**, *49*, 925. [CrossRef]

27. Shepelyansky, D.L. Some statistical properties of simple classically stochastic quantum systems. *Phys. D* **1983**, *8*, 208. [CrossRef]

28. Chirikov, B.; Shepelyansky, D. Chirikov standard map. *Scholarpedia* **2008**, *3*, 3550. [CrossRef]

29. Bohr, A.; Mottelson, B.R. *Nuclear Structure*; Benjamin: New York, NY, USA, 1969; Volume 1, p. 284.

30. Guhr, T.; Muller-Groeling, A.; Weidenmuller, H.A. Random-matrix theories in quantum physics: Common concepts. *Phys. Rep.* **1998**, *299*, 189. [CrossRef]

31. French, J.B.; Wong, S.S.M. Validity of random matrix theories for many-particle systems. *Phys. Lett. B* **1970**, *33*, 449. [CrossRef]

32. Bohigas, O.; Flores, J. Two-body random Hamiltonian and level density. *Phys. Lett. B* **1971**, *34*, 261. [CrossRef]

33. French, J.B.; Wong, S.S.M. Some random-matrix level and spacing distributions for fixed-particle-rank interactions. *Phys. Lett. B* **1971**, *35*, 5. [CrossRef]

34. Bohigas, O.; Flores, J. Spacing and individual eigenvalue distributions of two-body random Hamiltonians. *Phys. Lett. B* **1971**, *35*, 383. [CrossRef]

35. Thouless, D.J. Maximum Metallic Resistance in Thin Wires. *Phys. Rev. Lett.* **1977**, *39*, 1167. [CrossRef]

36. Imry, Y. *Introduction to Mesoscopic Physics*; Oxford University Press: Oxford, UK, 2002.

37. Akkermans, E.; Montambaux, G. *Mesoscopic Physics of Electrons and Photons*; Cambridge University Press: Cambridge, UK, 2007.

38. Åberg, S. Onset of chaos in rapidly rotating nuclei. *Phys. Rev. Lett.* **1990**, *64*, 3119. [CrossRef] [PubMed]

39. Åberg, S. Quantum chaos and rotational damping. *Prog. Part. Nucl. Phys.* **1992**, *28*, 11. [CrossRef]

40. Shepelyansky, D.L. Quantum chaos and quantum computers. *Phys. Scr.* **T2001**, *90*, 112. [CrossRef]

41. Jacquod, P.; Shepelyansky, D.L. Emergence of quantum chaos in finite interacting Fermi systems. *Phys. Rev. Lett.* **1997**, *79*, 1837. [CrossRef]

42. Shepelyansky, D.L.; Sushkov, O.P. Few interacting particles in a random potential. *Europhys. Lett.* **1997**, *37*, 121. [CrossRef]

43. Gornyi, I.V.; Mirlin, A.D.; Polyakov, D.G. Many-body delocalization transition and relaxation in a quantum dot. *Phys. Rev. B* **2016**, *93*, 125419. [CrossRef]

44. Gornyi, I.V.; Mirlin, A.D.; Polyakov, D.G.; Burin, A.L. Spectral diffusion and scaling of many-body delocalization transitions. *Ann. Phys.* **2017**, *529*, 1600360. [CrossRef]

45. Kolovsky, A.R.; Shepelyansky, D.L. Dynamical thermalization in isolated quantum dots and black holes. *EPL* **2019**, *117*, 10003. [CrossRef]

46. Frahm, K.M.; Shepelyansky, D.L. Dynamical decoherence of a qubit coupled to a quantum dot or the SYK black hole. *Eur. Phys. J. B* **2018**, *91*, 257. [CrossRef]

47. Landau, L.D.; Lifshitz, E.M. *Statistical Mechanics*; Wiley: New York, NY, USA, 1976.

48. Ermann, L.; Vergini, E.; Shepelyansky, D.L. Dynamics and thermalization a Bose–Einstein condensate in a Sinai oscillator trap. *Phys. Rev. A* **2016**, *94*, 013618. [CrossRef]

49. Davis, K.B.; Mewes, M.-O.; Andrews, M.R.; van Druten, N.J.; Durfee, D.S.; Kurn, D.M.; Ketterle, W. Bose–Einstein Condensation in a Gas of Sodium Atoms. *Phys. Rev. Lett.* **2015**, *75*, 3969. [CrossRef] [PubMed]

50. Anglin, J.A.; Ketterle, W. Bose Einstein condensation of atomic gases. *Nature* **2002**, *416*, 211. [CrossRef] [PubMed]

51. Ketterle, W. Nobel lecture: When atoms behave as waves: Bose–Einstein condensation and the atom laser. *Rev. Mod. Phys.* **2002**, *74*, 1131. [CrossRef]

52. Valtolina, G.; Scazza, F.; Amico, A.; Burchianti, A.; Recati, A.; Enss, T.; Inguscio, M.; Zaccanti, M.; Roati, G. Exploring the ferromagnetic behaviour of a repulsive Fermi gas through spin dynamics. *Nat. Phys.* **2017**, *13*, 704. [CrossRef]

53. Burchianti, A.; Scazza, F.; Amico, A.; Valtolina, G.; Seman, J.A.; Fort, C.; Zaccanti, M.; Inguscio, M.; Roati, G. Connecting dissipation and phase slips in a Josephson junction between fermionic superfluids. *Phys. Rev. Lett.* **2018**, *120*, 025302. [CrossRef] [PubMed]

54. Sachdev, S.; Ye, J. Gapless spin-fluid ground state in a random quantum Heisenberg magnet. *Phys. Rev. Lett.* **1993**, *70*, 3339. [CrossRef] [PubMed]

55. Kitaev, A. A Simple Model of Quantum Holography. *Video Talks at KITP Santa Barbara*, 7 April and 27 May 2015.

56. Sachdev, S. Bekenstein-Hawking entropy and strange metals. *Phys. Rev. X* **2015**, *5*, 041025. [CrossRef]

57. Polchinski, J.; Rosenhaus, V. The spectrum in the Sachdev-Ye-Kitaev model. *J. High Energy Phys.* **2016**, *4*, 1. [CrossRef]

58. Maldacena, J.; Stanford, D. Remarks on the Sachdev-Ye-Kitaev model. *Phys. Rev. D* **2016**, *94*, 106002. [CrossRef]

59. Garcia-Garcia, A.M.; Verbaarschot, J.J.M. Spectral and thermodynamic properties of the Sachdev-Ye-Kitaev model. *Phys. Rev. D* **2016**, *94*, 126010. [CrossRef]

60. Nandkishore, R.; Huse, D.A. Many-body localization and thermalization in quantum statistical mechanics. *Annu. Rev. Condens. Matter Phys.* **2015**, *6*, 15. [CrossRef]

61. Alessiom, L.D.; Kafri, Y.; Polkovnikov, A.; Rigol, M. From quantum chaos and eigenstate thermalization to statistical mechanics and thermodynamics. *Adv. Phys.* **2016**, *65*, 239. [CrossRef]

62. Borgonovi, F.; Izrailev, F.M.; Santos, L.F.; Zelevinsky, V.G. Quantum chaos and thermalization in isolated systems of interacting particles. *Phys. Rep.* **2016**, *626*, 1. [CrossRef]

63. Alet, F.; Laflorencie, N. Many-body localization: An introduction and selected topics. *Comptes Rendus Phys.* **2018**, *19*, 498. [CrossRef]

64. Gribakin, G.F.; Flambaum, V.V. Calculation of the scattering length in atomic collisions using the semiclassical approximation. *Phys. Rev. A* **1999**, *48*, 1998. [CrossRef]

65. Flambaum, V.V.; Gribakin, G.F.; Harabati, C. Analytical calculation of cold-atom scattering. *Phys. Rev. A* **1993**, *59*, 546. [CrossRef]

66. Busch, T.; Englert, B.-G.; Rzazewski, K.; Wilkens, M. Two cold atoms in a harmonic trap. *Found. Phys.* **1998**, *28*, 549. [CrossRef]

67. Kohler, T.; Goral, K.; Julienne, P. Production of cold molecules via magnetically tunable Feshbach resonances. *Rev. Mod. Phys.* **2006**, *78*, 1311. [CrossRef]

68. Landau, L.D.; Lifshitz, E.M. *Quantum Mechanics: Non-Relativistic Theory*; Pergamon Press: New York, NY, USA, 1977.

69. Flambaum, V.V.; Izrailev, F.M. Distribution of occupation numbers in finite Fermi systems and role of interaction in chaos and thermalization. *Phys. Rev. E* **1997**, *55*, R13. [CrossRef]

70. Borgonovi, F.; Izrailev, F.M.; Santos, L.F. Exponentially fast dynamics of chaotic many-body systems. *Phys. Rev. E* **2019**, *99*, 010101. [CrossRef] [PubMed]

71. Available online: http://www.quantware.ups-tlse.fr/QWLIB/fermisinaioscillator/ (accessed on 15 July 2019).

72. Ketterle, W.; Durfee, D.S.; Stamper-Kurn, D.M. Making, probing and understanding Bose–Einstein condensates. In *Proceedings of the International School of Physics "Enrico Fermi"*; Inguscio, M., Stringari, S., Wieman, C.E., Eds.; Course CXL; IOS Press: Amsterdam, The Netherlands, 1999; p. 67.

73. Shepelyansky, D.L. Coherent propagation of two interacting particles in a random potential. *Phys. Rev. Lett.* **1994**, *73*, 2607. [CrossRef] [PubMed]

74. Ullah, A.; Hoogerland, M.D. Experimental observation of Loschmidt time reversal of a quantum chaotic system. *Phys. Rev. E* **2012**, *83*, 046218. [CrossRef] [PubMed]

75. Poincare, H. Sur les equations de la dynamique et le probleme des trois corps. *Acta Math.* **1890**, *13*, 1.

8

SAXS Analysis of Magnetic Field Influence on Magnetic Nanoparticle Clusters

Fábio Luís de Oliveira Paula (ID)

Complex Fluid Group, Institute of Physics, University of Brasilia—UnB, Campus Darcy Ribeiro, Brasilia (DF) 70919-970, Brazil; fabioluis@fis.unb.br or fabioluis@unb.br

Abstract: In this work, we investigated the local colloidal structure of ferrofluid, in the presence of the external magnetic field. The nanoparticles studied here are of the core-shell type, with the core formed by manganese ferrite and maghemite shell, and were synthesized by the coprecipitation method in alkaline medium. Measures of Small Angle X-ray Scattering (SAXS) performed in the Brazilian Synchrotron Light Laboratory (LNLS) were used for the study of the local colloidal structure of ferrofluid, so it was possible to study two levels of structure, cluster and isolated particles, in the regimes with and without applied magnetic field. In the methodology used here there is a combination of the information obtained in the system with and without magnetic field application. In this way, it is possible to undertake a better investigation of the colloidal dispersion. The theoretical formalism used: (i) the unification equation proposed by Beaucage G.; (ii) the analysis of the radial distribution function $p(r)$ and (iii) theoretical calculation of the radius of gyration as a function of the moment of inertia of the spherical of n-nanoparticles.

Keywords: Beaucage model; $MnFe_2O_{4+\delta}@\gamma\text{-}Fe_2O_3$ nanoparticles; D11-SAXS1 beamline

PACS: 61.05.cf; 78.67.Bf; 47.65.Cb; 81.16.Be; 05.45.Df

1. Introduction

Ferrofluids are a colloidal suspension of dispersions of magnetic nanostructures in a specific carrier liquid [1–3]. Together with the original conjunction of liquid and magnetic properties, ferrofluids can be confined, displaced, deformed and controlled by application of an external magnetic field. The applications of ferrofluids in the area of condensed matter are the most diverse due to their properties. These properties have aroused the interest of researchers in the field of chemistry and physics of condensed matter and in the development of nanotechnologies [4], in biomedical [5] and industrial applications. Recent studies have intensified the local colloidal structure in ferrofluids, and have mainly used the technique of small angle scattering; Fu et al. (2016) [6] investigated the self-assembly of supercrystals field-induced colloidal structure of core-shell NPs of core-shell iron oxide dispersed in toluene by small angle neutron scattering (SANS); Rozynek et al. (2011) [7] the effect of magnetic field on the structure formation in an oil-based magnetic fluid with various concentrations of magnetite particles by SAXS; Campi et al. (2019) [8] analyzed nanoparticles clusters of nickelate perovskite by SμXRD (Scanning X-ray micro-Diffraction); Campi et al. (2019) [9] investigated hybrid nanoparticles diffusion and nanoscale aggregation, and observed how fractal dimensional changes leading to a mass surface fractal transition (SAXS); and Wandersman et al. (2015) [10] field induced anisotropic cooperativity in a magnetic colloidal glass. We can emphasize the study of the local colloidal structure with the objective of improving in the applications in the field of medicine, with specific use in hyperthermia in the treatment of cancer. In this context, we would like to highlight some works, pertinent to the influence of the colloidal organization of ferrofluids on the effect of hyperthermia: Myrovali et al. (2016) [11] has verified the role of chain formation to further optimize the heating efficiency in the hyperthermia of Fe_3O_4 magnetic particles when magnetically aligned; Abenojar et al. (2016) [12] effect of nanochain and nanocluster formation on magnetic hyperthermia properties; Serantes et al. (2014) [13] observed that a system with chain-like arrangement biomimicking magnetotactic bacteria has the superior heating performance, increasing more than 5 times in comparison with the randomly distributed system when aligned with the magnetic field; Martinez-Boubeta et al. (2013) [14] heat generation of iron-oxide nanoparticles for magnetic hyperthermia; Mehdaoui et al. (2013) [15] increase of magnetic hyperthermia efficiency due to dipolar interactions in magnetic nanoparticles; and Bañobre-López et al. (2013) [16] use of magnetic nanoparticles-basead hyperthermia in cancer therapy.

In order to investigate these properties in colloids, a Small Angle X-ray Scattering (SAXS) technique is very efficient. To analyze the scattering signal using a global dispersion function proposed by Beaucage G. [17,18], which unifies the local laws of Guinier and Porod in only one function that describes the form factor of a dispersion object in terms of radius of gyration and interfaces in the means. These models were successfully used to characterize a disordered cluster structure [19], a fractal mass of particles [20] as well as to distinguish between individual polydispersed particles and aggregates [21]. This work aims to investigate the local structure of magnetic colloids [22] under external magnetic field application, the colloids in question are composed of magnetic nanoparticles based on manganese ferrite type core-shell $MnFe_2O_{4+\delta}@\gamma$-$Fe_2O_3$ [23] and dispersed in aqueous medium. SAXS measurements were performed with external magnetic field applied [24] to determine the dimensions between nanoparticles and clusters [25], for future studies in hyperthermia [26] for biomedical applications.

2. Experimental

2.1. Sample Preparation

All the reagents used in this work are of analytical purity, purchased from SIGMA-ALDRICH. The magnetic nanoparticles (NPs) were synthesized according to the hydrothermal co-precipitation method according to the one proposed by Tourinho et al. [27,28]. The initial aqueous solutions of (0.5 M) $MnCl_2 \cdot 4H_2O$ and (0.5 M) $FeCl_3 \cdot 6H_2O$ were mixed in alkaline medium with $CH_3 \cdot NH_2 \cdot$ (methylamine) at 100 °C. Then, a treatment with $Fe(NO_3)_3$ at 100 °C was carried out to guarantee the stability of the particles and to avoid their degradation in acidic media [29]. This surface treatment induces iron enrichment of the nanoparticles, thus creating a superficial layer of maghemite γ-Fe_2O_3 that surrounds the particle core made of a nickel ferrite. The synthesized ferrofluid has volumetric fraction $\phi = 2.0\%$ and pH = 2.

2.2. Measurements

The chemical composition of NPs is checked by determination of Mn and Fe concentrations with atomic absorption spectroscopy (AAS) technique. Both measurements with a Thermo Scientific Spectrometer model S series AA, with specific lines are chosen for each metal (Mn (279.5 nm) and Fe (248.3 nm/372.0 nm)) to avoid interference effects.

In order to obtain the size distribution and polydispersity and aspects of the morphology of our nanoparticles, measurements were taken using an Electron Transmitting Electron Microscope (MET) JEOL 100CX2. The sample was diluted, then placed on the ultrasound for 1 h and then placed in the sample port until the liquid evaporated.

The magnetic characterization of ferrofluid is performed using the vibrating sample magnetometer (VSM; PPMS, Quantum Design model 6000). The magnetization curves are obtained as a function of the applied field (up to 9000 kA/m) at room temperature.

The colloidal characterization was performed SAXS measurements at the Brazilian Synchrotron Light Laboratory (LNLS), in the beamline D11A-SAXS1 UVX ring. With wavelength of $\lambda = 1.7556$ Å, it is q the scattering vector between $5.08 \times 10^{-3} < q < 0.124$ Å^{-1} ($q = 4\pi\sin(\theta/2)/\lambda$, θ is angle scattering). The detector used was MarCCD165 2D and the sample-detector distance was 2.250 m. The sample holder of quartz capillaries had a diameter of 1.5 mm and for the measurements with applied magnetic field two Neodymium magnets [7] were used, perpendicular to the X-ray beam. The data were corrected by the contribution of the empty sample holder and the scattering of water.

3. Theoretical Background

For magnetic characterization of the particles, let us consider an assemble of independent single-domain grains with a magnetic moment $\mu = \pi m_s d^3/6$; where m_s is the magnetization of the NPs. In the presence of an external field, the liquid matrix with the suspended particles behaves as an ideal superparamagnet and the ferrofluid magnetization is given by the Langevin law: $M = m_s \phi, L(\varsigma) = \coth \varsigma - \varsigma^{-1}, \varsigma = \mu_0 \mu H/k_B T$, with ϕ the ferrite volume fraction, k_B the Boltzmann constant T the temperature. This equation shows that at $H = 0$ the magnetization is zero. As the field H is turned on, the magnetic moments tend to align in the field direction so that at high fields, M saturates at $m_s \phi$. For polydisperse ferrofluids and considering particle size distributions, weighting Langevin with log-normal distribution.

The analysis of SAXS experimental data was first using the unified global equation proposed by Beaucage et al. [17,18]. This equation incorporates the Guinier and Porod boundaries and allows information on parameters such as radius of gyration, log-normal polydispersity index and the exponent associated with the interfacial regime, fractal dimensions of clusters. Thus, in Equation (1), the first level of structure corresponds to the larger scatter object, i.e., the primary clusters composed of some NPs, its size is comprised by means of the radius of gyration, Rg_1. The second level of structure is relative to the isolated spherical NPs, Rg_2.

$$I(q) = G_1 e^{-q^2 Rg_1^2/3} + B_1 e^{-q^2 Rg_2^2/3} h_1^{(-P_1)} + G_2 e^{-q^2 Rg_1^2/3} + B_2 h_2^{(-P_2)} \tag{1}$$

where $h_1 = q/[\mathrm{erf}(q[(Rg_1)]/\sqrt{6})]^3$, $h_2 = q/[\mathrm{erf}(q[(Rg_2)]/\sqrt{6})]^3$, where $\mathrm{erf}(x)$ is an error function, G_1 and G_2 are the pre-factors of Guinier, B_1 and B_2, the pre-factors of Porod, P_1 and P_2 are the indices of the power law, for the respective regions of analysis. The mean radius R_{part} and the polydispersion in size (σ) of individual spherical particles are determined by the following expressions.

$$R_{part} = \sqrt{\left(\frac{5}{3}\right)} Rg_2 \exp\left(7\sigma^2\right), \sigma = \sqrt{\frac{\ln\left(\frac{B_2 Rg_2^4}{1.62 G_2}\right)}{12}} \tag{2}$$

In the case of the correlated particles, we have the insertion of one more multiplicative term $S(q)$ to Equation (1). $S(q)$ is the structure factor, ξ is the mean of the distances between the correlated objects, and k is the degree of correlation [30,31].

$$S(q) = \frac{1}{1 + k\left(3\frac{\sin(q\xi)-(q\xi)\cos(q\xi)}{(q\xi)^3}\right)} \tag{3}$$

The function of distance distribution in pairs $p(r)$, corresponds to the Fourier transform of the signal of $I(q)$. The function reveals valuable information about the shape of the scatter object and the size, allowing a more intuitive interpretation of the intensity profile [32].

$$p(r) = r^2/(2\pi^2) \int I(q) \frac{\sin(qr)}{qr} q^2 dq \tag{4}$$

First principles in the theoretical radius of gyration of n-particles that make up the clusters takes into account the association of the moment of inertia of each spherical particle ($I = (2MR^2)/5$) with mean radius of R and mean mass M and associated with the theorem of parallel axes for the formulation of the spatial configuration of the clusters. The symbol I_{xx}, I_{yy}, I_{zz} are known as the moments of inertia of a 3D rigid body about the respective axes.

$$I_{xx} = \int \left(y^2 + z^2\right) dm, \tag{5a}$$

$$I_{yy} = \int \left(x^2 + z^2\right) dm, \tag{5b}$$

$$I_{zz} = \int \left(x^2 + y^2\right) dm, \tag{5c}$$

$$I_{xx} + I_{yy} + I_{zz} = 2MRg^2. \tag{5d}$$

For a system of more particles, we combine the Equation (5d) for each particle together with the theorem of parallel axes. For the case without magnetic field application. Equation (6b) describes the case $n = N_{clust} = 10$ particles, and $d_{H=0}$ is a mean distance between particle.

$$l \sum_{i=1}^{n=N_{clust}} \left(\frac{2}{5}MR_i^2 + \frac{Md_i^2}{4} + \frac{2}{5}MR_i^2 + \frac{Md_i^2}{4} + \frac{2}{5}MR_i^2 + \frac{Md_i^2}{4}\right) = 2(nM)Rg_{H=0}^2, \tag{6a}$$

$$Rg_{H=0}^2 \simeq \frac{3}{5}R_{H=0}^2 + \frac{9d_{H=0}^2}{10}. \tag{6b}$$

For the case with applied magnetic field, we do not have as much rotational freedom, since the objects are aligned with the magnetic field, in this way we have two of the three components with calculation equal to zero. Equation (7b) describes the a case $N_{clust}^{\perp,\parallel}$ of the $n = 10$ particles, and $d_{\parallel,\perp}$ is a mean distance between particles in the projections parallel and perpendicular to the field, the general mean distance between particles can be found as $d_{H \neq 0} = \sqrt{(d_{\parallel}^2 + d_{\perp}^2)}$.

$$l \sum_{i=1}^{n=N_{clust}^{\perp,\parallel}} \left(0 + 0 + \frac{2}{5}MR_{i,\perp,\parallel}^2 + \frac{Md_{i,\perp,\parallel}^2}{4}\right) = 2(nM)Rg_{\perp,\parallel}^2, \tag{7a}$$

$$Rg_{\perp,\parallel}^2 \simeq \frac{R_{\perp,\parallel}^2}{5} + \frac{7d_{\perp,\parallel}^2}{3}. \tag{7b}$$

4. Results and Discussions

The results show that in the chemical characterization performed, after the coprecipitation step, the stoichiometry of the core NPs does not correspond to that of an ideal ferrite. Thus, we write the chemical formula as $MnFe_2O_{4+\delta}$ ($Fd\bar{3}m$ space group symmetry) to obtain the stoichiometry of naked NPs, δ is the oxidation parameter. Other complex materials also have the similar behavior, Campi et al. (2019) [9] makes analysis of perovskite $La_2CuO_{4+\delta}$ (with $\delta = 0.1$ has orthorhombic $Fmmm$ space group symmetry) with space resolved SµXRD (Scanning X-ray micro-Diffraction). An excess of positive charges occurs due to the oxidation of Mn^{+2} to Mn^{+3} in the synthesis process, in order to balance exactly the extra positive charge that appears in the formula is inserted a δ parameter of 0.48 to adjust the electronegativity. In recent works by Martins et al. [23] and Moreira et al. [33], the mean valence of manganese in ferrites synthesized by coprecipitation was determined. Several analyzes were performed, including the X-ray Absorption Near Edge Structure (XANES) analysis with the Linear Combination Fitting (LFC) method of Mn_3O_4, Mn_2O_3 and MnO_2 oxides. It was found that there is more Mn^{+3} than other valence states Mn^{+2} and Mn^{+4}, resulting in an average oxidation of +3. Such analysis collaborates with the result of LFC was the absorption edge shift determined with the first derivative of the normalized X-ray absorption spectroscopy (XAS) signal. The Rietveld refinements of X-ray Powder Diffraction (XPD) and Neutron Powder Diffraction (NPD) by Martins and the Pair Distribution Function (PDF) made by Moreira, both performed for non-stoichiometric ferrites, showed an increase in the oxygen occupancy rate. The value of χ_m further undergoes a reduction of its value with the formation of the iron rich shell (γ-Fe$_2$O$_3$). Thus, we find a value of $\chi_m = 0.22$ much smaller than an ideal ferrite that is 0.33. Based on the non-stoichiometric core of the manganese ferrite and the magnetite shell and using the results of the χ_m, it was possible to determine the thickness of the shell ($th = 0.57$ nm). The calculation of the molar fraction is given by, $\chi_m = [Mn]/([Mn] + [Fe])$. Studies correlated to this mechanism of the core-shell model were reported in the works of Gomes et al. [34], Martins et al. [23], and Pilates et al. [26].

Figure 1a presents the TEM picture of the sample and shows that the NPs are roughly spherical. The NPs size distribution was estimated by measuring the size of about 300 particles, using a log normal law of size distribution which corresponds to the probability density that a particle has a diameter $d^{TEM} = 7.1$ nm, with a standard deviation $\sigma^{TEM} = 0.25$. Figure 1b presents the normalized room temperature magnetization curve for NPs; the full line is the best fit of Langevin, adjusting the experimental data, which allowed us to determine $m_s = 172$ kA/m, $d^{mag} = 6.8$ nm and $\sigma^{mag} = 0.26$.

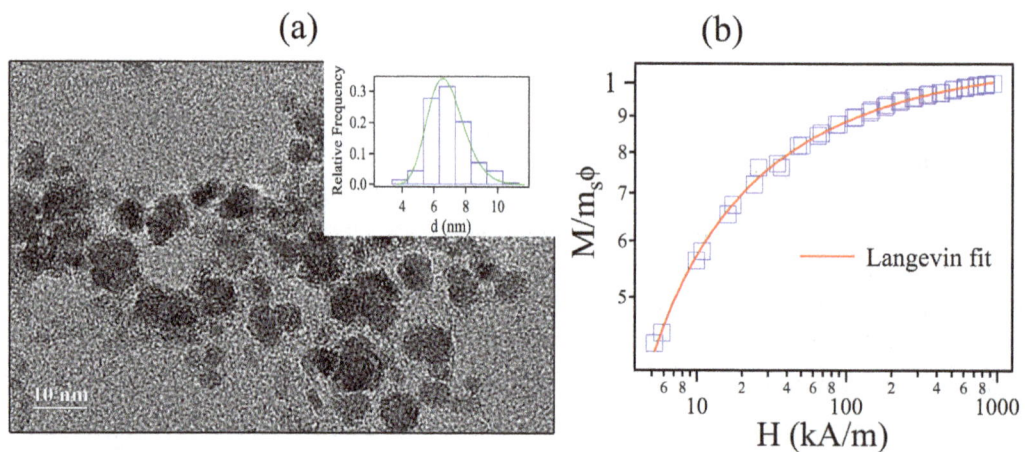

Figure 1. (a) TEM picture of the nanoparticles (NPs), the insert display with size distribution histogram adjusted with (—) log-normal distribution, and (b) Reduced magnetization of (□) sample as function of the applied magnetic filed. The solid line (—) is the fit to the data using the Langevin formalism.

Figure 2 shows the two-dimensional patterns ((a) 2D isotropic pattern; (b) 2D anisotropic pattern) at Small-Angle X-ray Scattering of the ferrofluid sample with manganese ferrite core-shell type (MnFe$_2$O$_{4+\delta}$@γ-Fe$_2$O$_3$) nanoparticles [23]. Figure 2b shows the external magnetic field of intensity is $H = 538$ mT applied perpendicular to the direction of propagation of the incident beam. The curve 1D, intensity scattering $I(q)$ is calculated after averaging in a constant radius, ring with the aid of FIT2D software [35]. In relation to the two-dimensional, anisotropic, we don't integrate the entire surface of the image to obtain the curves 1D, $I(q, H = 0)$, and yes, we delimit two distinct integration regions. These regions correspond to azimuthal averages over $\pm 5°$ in the image, in the perpendicular directions $I_\perp(q, H)$ and parallel $I_\parallel(q, H)$ to the applied magnetic field [36,37]. Angular sector was used in the integration of the 2D pattern in order not to have a radial intensity variation with the angle, since for larger angles we could have regions with intensity variations due to a radial anisotropy of the intensity in the integration area (Figure 2b).

In Figure 3 we show the small angle scattering curves, with the Guinier and Porod components for the two levels of structure as presented in Equation (1). In the Guinier region, all curves have a non-zero inclination, there is a behavior of globally attractive systems, probably associated to the presence of primary clusters of NPs. In case of a system without interaction, we would have a plateau in the region of Guinier, that is, $I(q) \sim q^{-0}$. In Figure 3c the curve of $I_\perp(q, H)$, the slope is less pronounced than $I_\parallel(q, H)$ (Figure 3b) and $I(q, H = 0)$ (Figure 3a), that is, it is a less attractive system. In $I_\perp(q, H)$ there is a correlation peak at $q^{peak} \sim 0.034$ Å^{-1}, characteristic of a globally repulsive system with a mean distance between the primary clusters of $\xi^{peak} \sim 181$ Å ($\xi^{peak} = 2\pi/q^{peak}$). In the curves with a large q we observe the law of power q^{-4} corresponding to the interface hard of nanoparticles

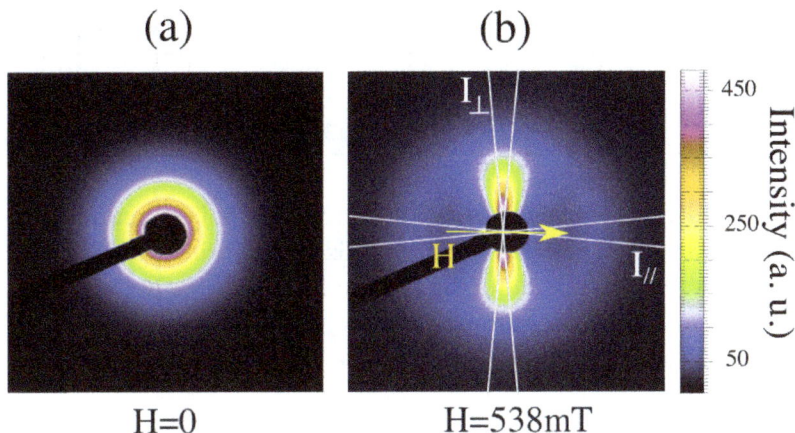

Figure 2. Two-dimensional X-ray scattering patterns for ferrofluid based on manganese ferrite nanoparticles. (**a**) isotropic pattern, with no applied magnetic field; (**b**) anisotropic pattern, Ferrofluid with external magnetic field application of $H = 538$ mT, the white lines limit are azimuthal averages over $\pm 5°$ in directions perpendicular and parallel to the magnetic field to obtain 1D curves shown in Figure 3b,c.

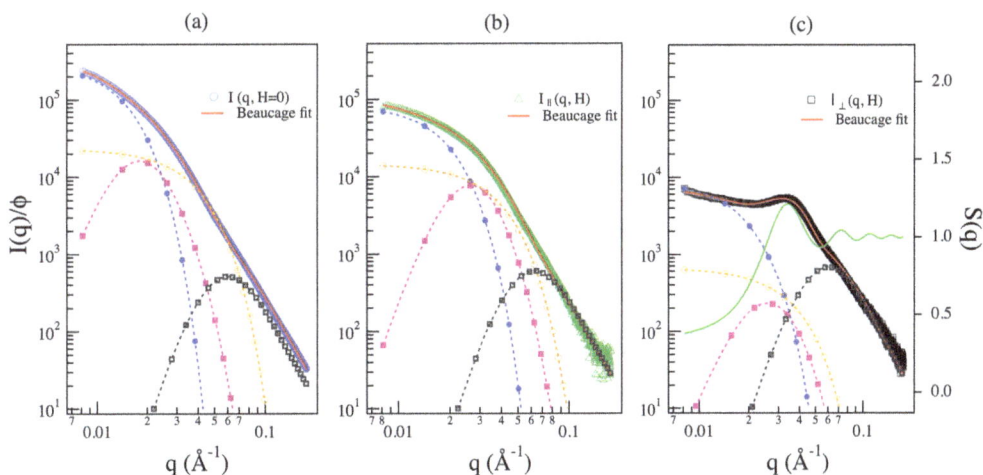

Figure 3. The X-ray scattering curves normalized by the volumetric fraction. Better fit results, following the (—) unified Beaucage expression described in the text. Contribution of the (-●-) Guinier and (-■-) components related to the first level of the structure corresponding to the largest object in the dispersion, (-⊖-) Guinier and (-⊟-) Porod components referring to the second structure level are relative to the isolated NPs, and (—) $S(q)$ is the structure factor. (**a**) $I(q, H = 0)$ scattering curve without external field application; (**b**) $I_{\parallel}(q, H)$ scattering curve obtained in the integration of the 2D pattern in the direction parallel to the magnetic field; (**c**) $I_{\perp}(q, H)$ scattering curve perpendicular to the magnetic field.

The number of NPs per cluster involved in the first structure level is $N_{clust} = (Rg_1/Rg_2)^{P_1}$, using the values Rg_1 and Rg_2 Table 1, which respectively describes the clusters and individual particles by the Beaucage model (Equation (1)). The values found were $N_{clust} \sim 10$, $N_{clust}^{\parallel} \sim 5.3$ and $N_{clust}^{\perp} \sim 1.8$, respectively, without field, field applied parallel and orthogonally. Only with the correlation of the values found of $N_{clust}^{\parallel}$ and $N_{clust}^{\perp}$ is it possible to obtain information about the cluster in the field action.

Table 1. Fitting parameters obtained the SAXS data analysed using the Beaucage model. The parameters Rg_1 and P_1 refer to the clusters; Rg_2 and P_2, the individual particles; k and ξ parameters of the structure factor and N_{clust} are related to the number of nanoparticles per clusters.

	Rg_1 [Å]	P_1	Rg_2 [Å]	P_2	k	ξ [Å]	N_{clust}
$I(q, H=0)$	105	2.7	46	3.95	-	-	~10
$I_\parallel(q, H)$	81.5	2.79	45	4.00	-	-	~5.3
$I_\perp(q, H)$	57.5	2.3	45	3.85	2.01	181	~1.8

In Figure 4, we have the schematic representation of the spatial configuration of the NPs, so we have that, the same quantitative of particles have reorganized, repositioning in the direction of the field, the fact that collaborates for such effect and the relation found $N_{clust}^{\parallel} \times N_{clust}^{\perp} \approx N_{clust}$. In Figure 4a,b the radial distribution analysis of the scattering curves obtained by means of Equation (4) allowed a maximum size for the scattering object; in the case, without field, we have $D_{max}^{H=0;p(r)} = 210$ Å.

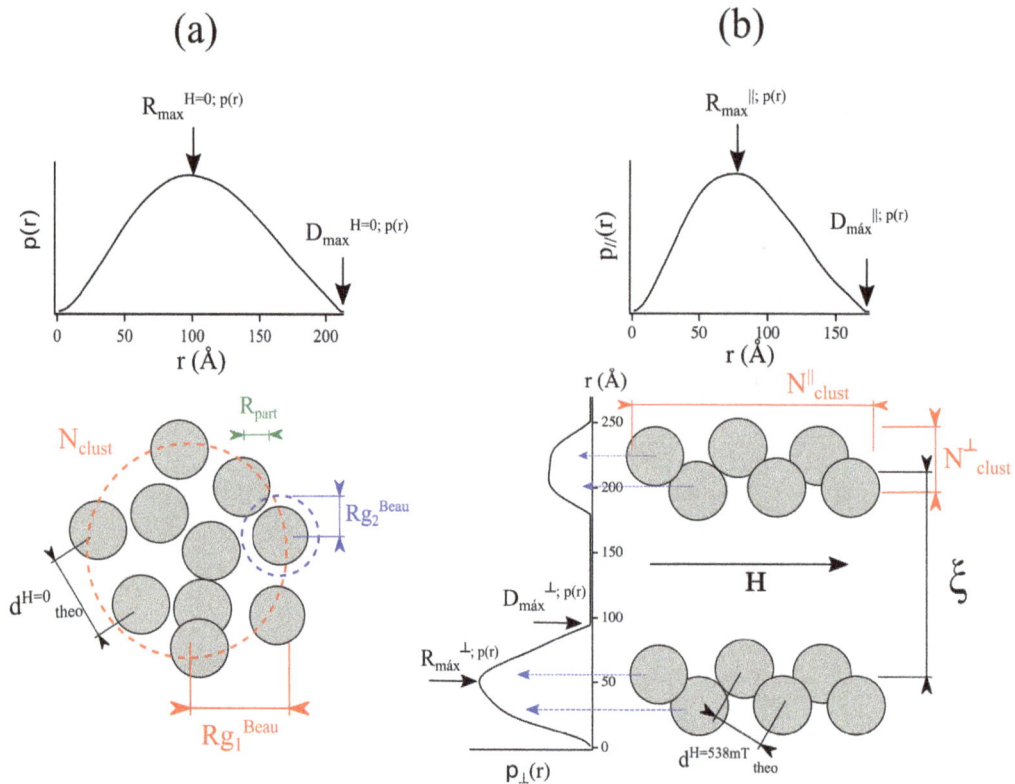

Figure 4. Parameter distribution function (—) corresponding to the Fourier transform to the Small Angle X-ray Scattering (SAXS) curves and the schematic drawing of the spatial configuration of the NPs, taking into account the information obtained by the Beaucage model and the curves of $p(r)$ and the theoretical radius of gyration ($d_{theo}^{H=538mT}$, Equation (7b); $d_{theo}^{H=0mT}$, Equation (6b) is and the theoretical distance calculated with and without field; $D_{max}^{H=0,\parallel,\perp;p(r)}$ determination the size of the scattered object, that $R_{max}^{H=0,\parallel,\perp;p(r)}$ is the average of the scatter object; $N_{clust}^{\parallel,\perp}$ and N_{clust} is the number of NPs in the cluster with and without field; $Rg_{1,2}^{Beau}$ is the radius of the gyration): (**a**) $H = 0$ T; (**b**) with field, the formation of chains of particles with a mean inter-cluster distance of ξ ~181 Å.

In the case with the applied magnetic field, we have a rearrangement of the spatial configuration of the object due to the action of the field, the particles are oriented in the direction of the field and, in the perpendicular direction, the objects have a distancing behavior for the formation of chains in the direction of the field. We then see that $D_{max}^{\|;p(r)} = 160$ Å and, in the direction perpendicular to the field, two regions were determined; the first $D_{max}^{\perp;p(r)} = 90$ Å and then another $D_{max}^{\perp;p(r)} = 70$ Å, with a range of $\xi^{p(r)} \sim 180$ Å .

For a better understanding, see the schematic illustration in Figure 4. The result $D_{max}^{H=0;p(r)} > D_{max}^{\|;p(r)}$, is given by the alignment of the dipole moments of the NPs, thus increasing the magnetic interaction, consequently decreasing the distance between particles.

In Figure 4a and Table 2, in a system without field, $p^{H=0}(r)$ is quasi-symmetric indicating an object scattered approximately in a sphere, this is due to the profile of the curve being almost symmetric, with radius of the clusters of $R_{clust}^{H=0;p(r)} \sim 90$ Å. In Figure 4b, the distribution $p^{\perp}(r)$ in the direction of the field, of $R_{clust}^{\|;p(r)} \sim 74$ Å, at $p^{\|}(r)$, in the direction of the field perpendicular to the field, the radius of the clusters is $R_{clust}^{\perp;p(r)} \sim 51$ Å. The analysis of the radial distribution perpendicular and parallel allows us to obtain the dimensions of the thickness and the length of the clusters, as represented schematically in Figure 4.

Table 2. $D_{max}^{H=0,\|,\perp;p(r)}$ is the maximum of the pair distribution, with $R_{clust}^{H=0,\|,\perp;p(r)}$ being the maximum point of $p(r)^{H=0,\|,\perp}$, respectively. The $Rg_{theo}^{H=0,\|,\perp}$ is the theoretical radius of gyration of the clusters that is calculated by the moments of inertia and $d_{theo}^{H=0,\|,\perp}$ and the distance between particles in the clusters.

	$D_{max}^{H=0,\|,\perp;p(r)}$ [Å]	$R_{clust}^{H=0,\|,\perp;p(r)}$ [Å]	$Rg_{theo}^{H=0,\|,\perp}$ [Å]	$d_{theo}^{H=0,\|,\perp}$ [Å]
$H = 0$	210	90	103.2	105
$\|$	160	74	84.0	53
$\perp$	90	51	58.2	37

We see a decrease in D_{max} for the signs $p^{\|}(r)$ and $p^{\perp}(r)$, this is due to the reorganization of the scattering objects before the field action. We also have that due to the limitation of the degrees of freedom for the scattering objects on the action of the magnetic field, it is possible to verify by the values of the radius of gyration by the Beaucage method (Table 2) is $Rg_1^{H=0;Beau} = 105 > Rg_1^{\|;Beau} = 81.5 > Rg_1^{\perp;Beau} = 57.5$ Å. To better understand this relationship, the calculation of the theoretical radius of gyration was performed using Equations (6b) and (7b). In the case of the system without field: $Rg_{theo}^{H=0} = 103.2$ Å and $d_{theo}^{H=0} = 105$ Å; and with field, we have due to the limitation, degrees of rotational freedom. Thus, two of these components of Equation (7a) are zero, $Rg_{theo}^{\|} = 84$ Å and $d_{theo}^{\|} = 53$ Å and $Rg_{theo}^{\perp} = 58.2$ Å and $d_{theo}^{\perp} = 37$ Å. The values determined for the radius of gyration by the method of Beaucage are in agreement with those determined by the theoretical calculation. The average distance between the particles with magnetic field is obtained by means of the projections of the distances obtained by the analysis of the signal in the parallel and perpendicular direction:

$$d_{theo}^{H=538mT} = \sqrt{(d_{theo}^{\|2} + d_{theo}^{\perp 2})} = \sqrt{(53^2 + 37^2)} = 64.63 \text{ Å}.$$

It is observed that $d_{theo}^{H=0} > d_{theo}^{H=538 \text{ mT}}$, this is due to the spatial reorganization of the scattering objects, since there is a greater alignment of the dipoles moments of the NPs towards the field, resulting in greater intraparticle interaction. The average radius of the individual particles $R_{part} = 35$ Å and the polydispersity $\sigma = 0.24$ were determined using the expressions for spherical scattering objects (slope in $I(q) \sim q^{-4}$) in the Beaucage Equation (2). The values for diameter and polydispersion in the techniques of morphological characterization (TEM) and magnetic (magnetization curve) are in good agreement with that determined using the SAXS technique. We see that the diameter $d_{part} = (2R_{part})$, $d^{mag}, d^{TEM} \approx 70$ Å, and polydispersion $\sigma_{part}, \sigma^{mag}, \sigma^{TEM} \approx 0.25$.

5. Conclusions

It was very promising to study the local structure of these dispersions using the SAXS technique, since it was possible to have access to the spatial organization of the nanoparticles in clusters in two regimes with and without applied magnetic field. In this sense, one can better measure the size of the clusters and the individual nanoparticles. In addition to information about the mean distance between particles and between the chains of particles when the field is applied. The calculation performed for theoretical radius of gyration for the clusters, even if simplistic, proved to be very consistent with the values found in the analyzes using the unified global equation and the radial distribution function. The results showed the behavior of a globally repulsive colloidal system in the direction perpendicular to the field and globally attractive in the direction of the magnetic field. For the repulsive system, we have that the distance between clusters is ξ^{peak}, $\xi^{P(r)}$, $\xi^{Beau} \approx 180$ Å. The analysis of the globally attractive curves helped us to understand the dimensions of the clusters, as to the form and quantity of particles involved in their formation. In this way, we find that the number of particles in the clusters follows the relation, $N_{clust}^{\parallel} \times N_{clust}^{\perp} \approx N_{clust}$ reaffirming the spatial reconfiguration of the particles on the action of the magnetic field. It was also evidenced the good agreement of the values for diameter and polydispersion in the techniques of morphological characterization (TEM) and magnetic with the technique of SAXS. We intend to study, in the light of this work, the relation of the volumetric fraction and the intensity of the magnetic field applied in the formation of the clusters.

Funding: This work was supported by Brazilian agencies, the National Council for Scientific and Technological Development through the National Institute of Science and Technology of Complex Fluids (INCT-FCx-CNPq—grant numbers 465259/2014-6 and 400849/2016-0), the Coordination for the Improvement of Higher Education Personnel (CAPES) and the Federal District Research Foundation (FAPDF—grant numbers 0193-001376/2016, 0193-001194/2016 and 0193.001569/2017).

Acknowledgments: This research used resources of the Brazilian Synchrotron Light Laboratory (LNLS), an open national facility operated by the Brazilian Centre for Research in Energy and Materials (CNPEM) for the Brazilian Ministry for Science, Technology, Innovations and Communications (MCTIC). The D11-SAXS1 beamline staff is acknowledged for the assistance during the experiments. Special thanks to R. Aquino and J. Depeyrot for technical assistance. Thanks to Aldenora M. O. Paula for textual review. Thanks to the post doctorate program CAPES/COFECUB, grant contract BEX 7288/14-0.

Conflicts of Interest: The author declares no conflict of interest.

Abbreviations

The following abbreviations are used in this manuscript:

SAXS   Small-Angle X-ray Scattering
LNLS   Brazilian Synchrotron Light Laboratory

References

1.   Papell, S.S. Low Viscosity Magnetic Fluid Obtained by the Colloidal Suspension of Magnetic Particles. U.S. Patent 3215572, 2 November 1965.

2.   Rosensweig, R.E.; Kaiser, R. *Study of Ferromagnetic Liquid, Phase I*; NTIS Rep. No. NASW-1219; NASA Office of Advanced Reseach and Technology: Washington, DC, USA, 1967.

3.   Rosensweig, R.E. *Ferrohydrodynamics*; Cambridge University Press: Cambridge, UK; London, UK, 1985.

4.   Beeran, A.E.; Fernandez, F.B.; Nazeer, S.S.; Jayasree, R.S.; John, A.; Anil, S.; Vellappally, S.; Al Kheraif, A.A.A.; Varma, P.R.H. Multifunctional nano manganese ferrite ferrofluid for efficient theranostic application. *Colloids Surf. B Biointerfaces* **2015**, *136*, 1089–1097. [CrossRef] [PubMed]

5.   Pankhurst, Q.A.; Thanh, N.T.K.; Jones, S.K.; Dobson, J. Progress in applications of magnetic nanoparticles in biomedicine. *J. Phys. D Appl. Phys.* **2009**, *22*, 224001. [CrossRef]

6.   Fu, Z.; Xiao, Y.; Feoktystov, A.; Pipich, V.; Appavou, M.; Su, Y.; Feng, E.; Jina, W.; Bruckel, T. Field-induced self-assembly of iron oxide nanoparticles investigated using small-angle neutron scattering. *Nanoscale* **2016**, *8*, 18541–18550. [CrossRef] [PubMed]

7. Rozynek, Z.; Jozefczak, A.; Knudsen, K.D.; Skumiel, A.; Hornowski, T.; Fossum, J.O.; Timko, M.; Kopcanshy, P.; Koneracka, M. Structuring from nanoparticles in oil-based ferrofluids. *Eur. Phys. J. E* **2011**, *34*. [CrossRef] [PubMed]

8. Campi, G.; Poccia, N.; Joseph, B.; Bianconi, A.; Mishra, S.; Lee, J.; Roy, S.; Agung Nugroho, A.; Buchholz, M.; Braden, M.; et al. Direct Visualization of Spatial inhomogeneity of Spin Stripes Order in $La_{1.72}Sr_{0.28}NiO_4$. *arXiv* **2019**, arXiv:1905.02124.

9. Campi, G.; Bianconi, A. Evolution of Complexity in Out-of-Equilibrium Systems by Time-Resolved or Space-Resolved Synchrotron Radiation Techniques. *Condens. Matter* **2019**, *4*, 32. [CrossRef]

10. Wandersman, E.; Chushkin, Y.; Dubois, E.; Dupuis, V.; Robert A.; Perzynski, R. Field induced anisotropic cooperativity in amagnetic colloidal glass. *Soft Matter* **2015**, *11*, 7165. [CrossRef] [PubMed]

11. Myrovali, E.; Maniotis, N.; Makridis, A.; Terzopoulou, A.; Ntomprougkidis, V.; Simeonidis, K.; Sakellari, D.; Kalogirou, O.; Samaras, T.; Salikhov, R.; et al. Arrangement at the nanoscale Effect on magnetic particle hyperthermia. *Sci. Rep.* **2016**, *6*, 37934. [CrossRef]

12. Abenojar, E.C.; Wickramasinghe, S.; Bas-Concepcion, J.; Samia, A.C.S. Structural effects on the magnetic hyperthermia properties of iron oxide nanoparticles. *Prog. Nat. Sci. Mater. Int.* **2016**, *26*, 440–448. [CrossRef]

13. Serantes, D.; Simeonidis, K.; Angelakeris, M.; Chubykalo-Fesenko, O.; Marciello, M.; Del Puerto Morales, M.; Baldomir, D.; Martinez-Boubeta, C. Multiplying Magnetic Hyperthermia. Response by Nanoparticle Assembling. *J. Phys. Chem. C* **2014**, *118*, 5927–5934. [CrossRef]

14. Martinez-Boubeta, C.; Simeonidis, K.; Makridis, A.; Angelakeris, M.; Iglesias, O.; Guardia, P.; Cabot, A.; Yedra, L.; Estradé, S.; Peiró, F.; et al. Learning from nature to improve the heat generation of iron-oxide nanoparticles for magnetic hyperthermia applications. *Sci. Rep.* **2013**, *3*, 1652. [CrossRef] [PubMed]

15. Mehdaoui, B.; Tan, R.P.; Meffre, A.; Carrey, J.; Lachaize, S.; Chaudret, B.; Respaud, M. Increase of magnetic hyperthermia efficiency due to dipolar interactions in low-anisotropy magnetic nanoparticles: Theoretical and experimental results. *Phys. Rev. B* **2013**, *87*, 174419. [CrossRef]

16. Bañobre-López, M.; Teijeiro, A.; Rivas, J. Magnetic nanoparticle-based hyperthermia for cancer treatment. *Rep. Pract. Oncol. Radiother.* **2013**, *18*, 397–400. [CrossRef] [PubMed]

17. Beaucage, G.; Schaefer, D.W. Structural studies of complex systems using small-angle scattering: A unified Guinier/power-law approach. *J. Non-Cryst. Solids* **1994**, *172–174*, 797–805. [CrossRef]

18. Beaucage, G. Approximations Leading to a Unified Exponential/Power-Law Approach to Small-Angle Scattering. *J. Appl. Cryst.* **1995**, *28*, 717. [CrossRef]

19. Beaucage, G. Small-Angle Scattering from Polymeric Mass Fractals of Arbitrary Mass-Fractal Dimension. *J. Appl. Cryst.* **1996**, *29*, 134. [CrossRef]

20. Beaucage, G. Determination of branch fraction and minimum dimension of mass-fractal aggregates. *Phys. Rev. E* **2004**, *70*, 031401. [CrossRef]

21. Beaucage, G.; Kammler, H.K.; Pratsinis, S.E. Particle size distributions from small-angle scattering using global scattering functions. *J. Appl. Cryst.* **2004**, *37*, 523. [CrossRef]

22. Paula, F.L.O.; Depeyrot, J.; Fossum, J.O.; Tourinho, F.A.; Aquino, R.; Knudsen, K.D.; da Silva, G.J. Small-angle X-ray and small-angle neutron scattering investigations of colloidal dispersions of magnetic nanoparticles and clay nanoplatelets. *J. Appl. Cryst.* **2007**, *40*, 269–273. [CrossRef]

23. Martins, F.H.; da Silva, F.G.; Paula, F.L.O.; Gomes, J.A.; Aquino, R.; Mestnik-Filho, J.; Bonville, P.; Porcher, F.; Perzynski, R.; Depeyrot, J. Local Structure of Core-Shell $MnFe_2O_{4+\delta}$ Based Nanocrystals: Cation Distribution and Valence States of Manganese Ions. *J. Phys. Chem. C* **2017**, *121* 8982–8991. [CrossRef]

24. Paula, F.L.O.; da Silva, G.J.; Aquino, R.; Depeyrot, J.; Fossum, J.O.; Knudsen, K.; Tourinho, F.A. Gravitational and Magnetic Separation in Self-Assembled Clay-Ferrofluid Nanocomposites. *Braz. J. Phys.* **2009**, *39*, 163–170. [CrossRef]

25. Castro, L.L.; da Silva, M.; Bakuzis, A.; Miotto, R. Aggregate formation on polydisperse ferrofluids: A Monte Carlo analysis. *J. Magn. Magn. Mater.* **2005**, *293*, 553–558. [CrossRef]

26. Vanessa, P.; Cabreira, R.G.; Gomide, G.S.; Coppola, P.; Silva, F.G.; Paula, F.L.O.; Perzynski, R.; Goya, F.G.; Aquino, R.; Depeyrot, J. Core/Shell Nanoparticles of Non-Stoichiometric Zn-Mn and Zn-Co Ferrites as Thermosensitive Heat Sources for Magnetic Fluid Hyperthermia. *J. Phys. Chem. C* **2018**, *122*, 3028. [CrossRef]

27. Tourinho, F.A.; Franck, R.; Massart, R.; Perzynski, R. Synthesis and magneitc properties of managanese and cobalt ferrite ferrite ferrofluids. *Prog. Colloid Polym. Sci.* **1989**, *79*, 128–134.

28. Tourinho, F.A.; Franck, R.; Massart, R. Aqueous ferrofluids based on manganese and cobalt ferrites. *J. Mater. Sci.* **1990**, *25*, 3249–3254. [CrossRef]

29. Gomes, J.A.; Sousa, M.H.; Tourinho, F.A.; Aquino, R.; Depeyrot, J.; Dubois, E.; Perzynski, R. Synthesis of Core-Shell Ferrite Nanoparticles for Ferrofluids: Chemical and Magnetic Analysis. *J. Phys. Chem. C* **2008**, *112*, 6220–6227. [CrossRef]

30. Zaioncz, S.; Dahmouche, K.; Soares, B.G. SAXS Characterization of New Nanocomposites Based on Epoxy Resin/Siloxane/MMA/Acrylic Acid Hybrid Materials. *Macromol. Mater. Eng.* **2010**, *295*, 243–255. [CrossRef]

31. Beaucage, G.; Ulibarri, T.; Black, E.P.; Schaefer, D.W. *Hybrid Organic-Inorganic Composites*; Mark, J.E., Lee, C.Y.-C.; Bianconi, P.A., Eds.; ACS Symposium Series 585; American Chemical Society: Washington, DC, USA, 1985; p. 97.

32. Svergun, D.I.; Koch, M.H.; Timmins, P.A.; May, R.P. *Small Angle X-Ray and Neutron Scattering from Solutions of Biological Macromolecules*; Oxford University Press: Oxford, UK, 2013.

33. Moreira, A.F.L.; Paula, F.L.O.; Depeyrot, J. Evidence of Structural Distortions in Mixed Mn-Zn ferrite. *IOSR J. App. Phys.* **2019**, *11*, 36–44. [CrossRef]

34. Gomes, J.A.; Azevedo, G.M.; Depeyrot, J.; Mestnik-Filho, J.; Paula, F.L.O.; Tourinho, F.A.; Perzynski, R. Structural, Chemical, and Magnetic Investigations of Core-Shell Zinc Ferrite Nanoparticles. *J. Phys. Chem. C* **2012**, *116*, 24281–24291. [CrossRef]

35. Available online: http://www.esrf.eu/computing/scientic/FIT2D (accessed on 21 July 2014).

36. Mériguet, G.; Cousin, F.; Dubois, E.; Boué, F.; Cebers, A.; Farago, B.; Perzynski, R. What Tunes the Structural Anisotropy of Magnetic Fluids under a Magnetic Field? *J. Phys. Chem. B* **2006**, *110*, 4378–4386. [CrossRef]

37. Robbes, A.S.; Cousin, F.; Meneau, F.; Dalmas, F.; Boué, F.; Jestin, J. Nanocomposite Materials with Controlled Anisotropic Reinforcement Triggered by Magnetic Self-Assembly. *Macromolecules* **2011**, *44*, 8858–8865. [CrossRef]

9

Direct Visualization of Spatial Inhomogeneity of Spin Stripes Order in La$_{1.72}$Sr$_{0.28}$NiO$_4$

Gaetano Campi [1,*] [ID], Nicola Poccia [2,*], Boby Joseph [3] [ID], Antonio Bianconi [1,4] [ID], Shrawan Mishra [5,6], James Lee [5], Sujoy Roy [5], Agustinus Agung Nugroho [7] [ID], Marcel Buchholz [8], Markus Braden [8], Christoph Trabant [9], Alexey Zozulya [10,11] [ID], Leonard Müller [10], Jens Viefhaus [10] [ID], Christian Schüßler-Langeheine [9], Michael Sprung [10] and Alessandro Ricci [4,10,*]

1   Institute of Crystallography, CNR, via Salaria Km 29.300, 00015 Monterotondo, Roma, Italy
2   Institute for Metallic Materials, Leibniz IFW Dresden, 01069 Dresden, Germany
3   Elettra Sincrotrone Trieste. Strada Statale 14 - km 163.5, AREA Science Park, I-34149 Basovizza, Trieste, Italy
4   Rome International Center for Materials Science Superstripes RICMASS, Via dei Sabelli 119A, 00185 Roma, Italy
5   Advanced Light Source, Lawrence Berkeley National Laboratory, Berkeley, CA94720, USA
6   School of Materials Science and Technology, Indian Institute of Technology, Banaras Hindu University, Varanasi 221005, India
7   Faculty of Mathematics and Natural Sciences Intitut Teknologi Bandung, Jl. Ganesha 10 Bandung, Jawa Barat 40132, Indonesia
8   II. Physikalisches Institut, Universität zu Köln, Zülpicher Str. 77, 50937 Köln, Germay
9   Helmholtz-Zentrum Berlin für Materialien und Energie GmbH, Institute Methods and Instrumentation in Synchrotron Radiation Research, Albert-Einstein-Str. 15, 12489 Berlin, Germany
10 Deutsches Elektronen-Synchrotron DESY, Notkestraße 85, D-22607 Hamburg, Germany
11 European X-ray Free-Electron Laser Facility GmbH Holzkoppel 4, 22869 Schenefeld, Germany
*   Correspondence: gaetano.campi@ic.cnr.it (G.C.); n.poccia@ifw-dresden.de (N.P.); phd.alessandro.ricci@gmail.com (A.R.)

Abstract: In several strongly correlated electron systems, the short range ordering of defects, charge and local lattice distortions are found to show complex inhomogeneous spatial distributions. There is growing evidence that such inhomogeneity plays a fundamental role in unique functionality of quantum complex materials. La$_{1.72}$Sr$_{0.28}$NiO$_4$ is a prototypical strongly correlated perovskite showing spin stripes order. In this work we present the spatial distribution of the spin order inhomogeneity by applying micro X-ray diffraction to La$_{1.72}$Sr$_{0.28}$NiO$_4$, mapping the spin-density-wave order below the 120 K onset temperature. We find that the spin-density-wave order shows the formation of nanoscale puddles with large spatial fluctuations. The nano-puddle density changes on the microscopic scale forming a multiscale phase separation extending from nanoscale to micron scale with scale-free distribution. Indeed spin-density-wave striped puddles are disconnected by spatial regions with negligible spin-density-wave order. The present work highlights the complex spatial nanoscale phase separation of spin stripes in nickelate perovskites and opens new perspectives of local spin order control by strain.

Keywords: nanoscale phase separation; spin stripes; nickelates; quantum complex materials

1. Introduction

The complex organization of different orders seems to have a fundamental role in the mechanism governing the emergence of unique functionalities in quantum materials [1]. In cuprate perovskites, the stripes phases have been the object of interest for decades [1–3], while in this last decade new scanning X-ray diffraction methods have been developed due to the ability to focus X-ray synchrotron radiation to micron and sub-micron spots. These methods have made it possible to obtain visualization of spatial topological inhomogeneity of charge density wave order in doped cuprate perovskites [4,5]. Short range generalized Wigner charge density waves have been found to be spatially inhomogeneous with the formation of "striped charge puddles" anti-correlated with competing puddles of "striped dopants rich clusters" [4–6]. These experimental results have opened a new era in the long-standing research of complexity in doped strongly correlated perovskites, since they have falsified popular stripes theories which for decades have assumed a homogeneous spatial distribution of spin stripes and charge-stripes. In this work we focus on the spin stripes phase in doped nickelate perovskites. In order to determine the role that the spatial distribution of ordered phases in cuprates plays for the superconductivity, it is instructive to study a non-superconducting reference system like the layered nickelates [7]. Keeping this idea in mind, we push forward the investigation of the spatial distribution of spin-density-wave stripes ordering (SDW-stripes) in $La_{2-x}Sr_xNiO_4$ nickelates.

It is well known that spin stripes appear in layered nickelates [7] in the doping interval $0.15 \leq x \leq 0.5$ [7]. In the doping range $0.25 < x < 0.3$ magnetic stripes and charge stripes can be easily investigated separately. In $La_{2-x}Sr_xNiO_4$ the spin-order scattering exhibits peaks in the k-space for $(1-\varepsilon; 0; l)$ with odd and even l, whereas charge-order scattering always peaks at $(2\varepsilon; 0; l)$ with odd l, [7,8] where the notation refers to the commonly used orthorhombic unit cell. We have investigated a nominal single crystal $La_{1.72}Sr_{0.28}NiO_4$ to get direct visualization of the inhomogeneity of spin stripes incommensurate order in the bulk structure from nano to micro scale.

A large number of studies on the spin stripe order in $La_{2-x}Sr_xNiO_4$ were performed by traditional neutron scattering [9–13] probing the spin ordering with low energy neutrons, which have been confirmed by muon spectroscopy [14]. Several authors have focused on both magnetic and charge order [15–20] phenomena in nickelates. Resonant elastic X-ray scattering REXS [16–18] has been used to detect spin order directly via magnetic contrast but it has no spatial resolution. Electron diffraction and hard X-ray diffraction (XRD), or non-resonant-X-ray-magnetic-scattering (NXMS) of nickelates and related magnetic materials [20–25] have been used to probe the associated tiny lattice distortions related with polaron ordering or generalized Wigner CDW and magneto-elastic strain effects [26–28]. These last methods could have spatial resolution, therefore, they open new perspectives to unveil open puzzles on the complexity of the nature of stripes in $La_{2-x}Sr_xNiO_4$ nickelates. These experiments are needed to test theories i) proposing spin stripes in strongly correlated systems including orbital and polaronic degree of freedom [26–28] and ii) the theory predicting a frustrated phase separation controlled by strain, in a strongly correlated multiband system tuned to a Lifshitz transition, in the frame of the multiband Hubbard model [29].

The accumulated data on the $La_{2-x}Sr_xNiO_4$ system [7–21] enable us to present a temperature doping phase diagram of this nickelate system in panel (a) of Figure 1. The red area indicates the antiferromagnetic order (AFM) dominating the lower Sr-doping x given by the percentage of Sr for La substitutions in the La_2O_2 plane, which is assumed to give the number of injected doped electronic holes per Ni atom in the NiO_2 plane [7]. The blue and the green areas correspond to the observation of charge-density-wave order (CDW-stripes) and magnetic spin-density-wave order (SDW-stripes), respectively. The SDW-stripes modulation wave-vector direction for $0.15 < x < 0.5$ extends in real space diagonally to the Ni-O bond direction along the b-direction of the orthorhombic unit cell. For samples with tetragonal crystal symmetry, as the one studied here, the stripe order itself breaks the rotational symmetry of the ab-plane and therefore spin stripes were assumed to show two different orientations with a 90-degree rotation around the c-axis with equal probability. The spin stripes lead to superstructure peaks either in the neutron diffraction pattern [9–13] and X-ray diffraction [16–21]. Those at the lowest momentum transfer occur at wave-vectors $(1-\varepsilon, 0, 0)$ in the orthorhombic lattice for SDW-stripes order, where ε is a temperature dependent incommensurability value, which is well separated from different charge stripe wave-vectors in the k-space.

While neutron scattering has provided for decades k-space information on spin ordering averaged over large crystal area, the present X-ray investigation reported here provides spin ordering information on illuminated spots in micron sized samples using a focused X-ray synchrotron radiation beam which allows us for the first time to detect spatial inhomogeneous spin order in a nickelate single perovskite micro-crystal. The results show large spatial fluctuations of the spin order with fractal structure indicated by power-law distribution of the spin order parameter. The particular fractal distribution is assigned to quantum criticality near an electronic topological transition or Lifshitz transition as predicted by multiband Hubbard model for strongly correlated two band systems where the strain controls the energy shift between the two bands [29].

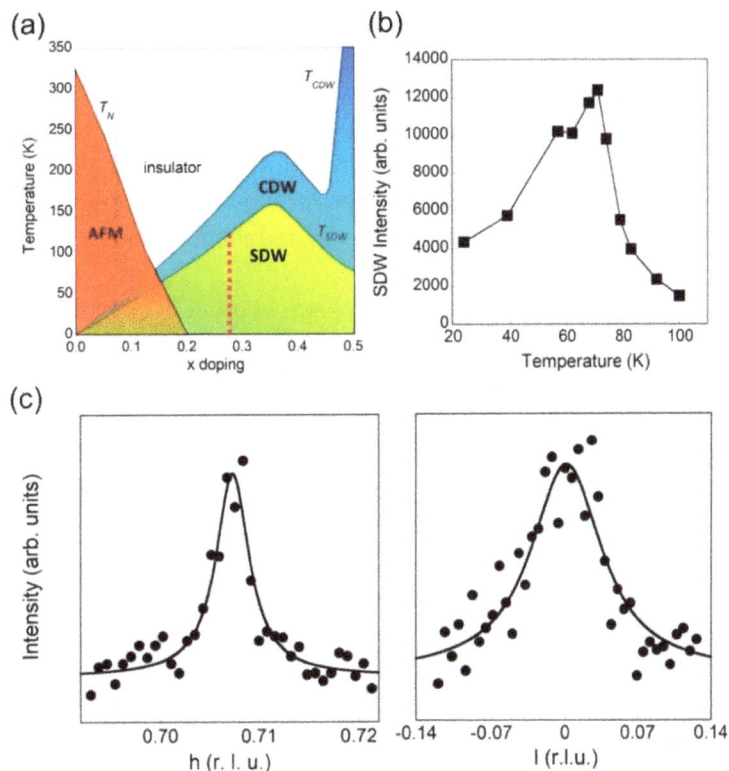

Figure 1. Phase diagram of nickelate systems and Spin-Density-Wave order. (**a**) Temperature-doping Phase diagram of the nickelate systems. In red the insulating Antiferromagnetic order (AFM), in blue the charge-density-wave order (CDW-stripes) and in green the spin-density-wave order (SDW-stripes). The red dotted line represents the temperature range where the sample of this work has been studied; (**b**) The intensity evolution of the SDW-stripes peak as a function of temperature; (**c**) SDW-stripes peak profile along the a* (left panel) and c* (right-panel) crystallographic directions, recorded at 30 K. The solid lines correspond to Lorentzian profiles fitted to the data, giving the in-plane and the out-of-plane correlation lengths around the average values of about 20 nm and 2 nm, respectively.

2. Materials and Methods

Single-crystalline La$_{1.72}$Sr$_{0.28}$NiO$_4$ was grown by floating zone technique. The seed and feed rods were prepared from polycrystalline powder obtained by solid state reaction of La$_2$O$_3$, SrCO$_3$ and NiO with an excess of NiO. The reaction was performed at 1200 °C for 20 h with intermediate grinding. The rods were densified at 1500 °C for 5 h in air.

Micro X-ray diffraction measurements of SDW-stripes order in the sample were carried out at beamline P10 of PETRA III (DESY, Hamburg, Germany) using an energy of 8 KeV. The scattering signal was detected at a sample to detector distance of 5 m using the large horizontal scattering set-up of beamline P10 including an evacuated flight path. A PILATUS 300K detector (DECTRIS, Baden-Daettwil, Switzerland) was used to record the X-rays scattered by the sample. By employing a focused beam with a diameter of about 1 μm and translating the sample, we mapped the spatial distribution of the $(1-\varepsilon,0,0)$ peak intensity over different areas of about 40×80 μm^2 in steps of 1 μm in both directions resulting in 3321 diffraction images. The scanning was performed translating the sample along the 80 horizontal lines. The exposure time for each X ray diffraction frame was of 5 s providing a X ray flux on each spot on the sample of about 10^9 count per second (cps). The time needed to collected a map was around 5 hours.

3. Results and Discussion

Scanning micro X-ray Diffraction (SμXRD) has been demonstrated to be a powerful tool in unraveling material inhomogeneity in superconductors at the micro and nanoscale, and has been successfully applied to the cuprate systems doped by oxygen interstitials: $HgBa_2CuO_{6+y}$, known as Hg1201 [4], La_2CuO_{4+y}, known as La124, [5,6] $Bi_2Sr_2CaCu_2O_{8+y}$, known as Bi2212 [30], $YBa_2Cu_3O_{6+y}$ known as Y123 [31–33], to iron-based superconductors [34] and to cobaltates materials opening a new era for our knowledge of quantum complex materials at nanoscale [35–79]. Here we have investigated a single crystal $La_{2-x}Sr_xNiO_4$ with Sr doping $x = 0.28$ with the spin modulation wave-vector $\varepsilon = 0.29$ consistent with the empirical relation for spin stripes wave-vector ε close to the percentage of the number of holes x per Ni sites. The red dotted line in Figure 1a represents the temperature range where the sample of this work has been studied.

In order to probe the spatial evolution of the SDW-stripes order in real space we used a micron-size X-ray beam probing the local SDW-stripes via the corresponding intensity of the magnetic so-called SDW-stripe superlattice peak at (0.71,0) in the orthorhombic (h,k) plane which is well separated from the so-called CDW-stripe superlattice reflection at (0.56,0).

The temperature variation of the intensity of the SDW-stripes reflection peak as a function of temperature is shown in panel (b) of Figure 1. The SDW-stripes peak is identifiable for temperatures below $T_{SDW} = 120$ K in agreement with previous works [7]. On further cooling, its integrated intensity increases, reaches a maximum around $T^* = 65$ K, followed by an intensity reduction when the temperature is further decreased as shown in Figure 1b which confirms the previous X-ray scattering results [16]. The determined temperature dependence of the SDW-stripes intensity is similar to the one reported for nickelates of different doping levels [16]. Thus, we can assume the observed behavior to be a general characteristic for SDW-stripes order. This behavior resembles what has been predicted for incommensurate cuprate stripes to occur at low temperatures, when a freezing to the lattice potential disturbs the long-range order [36–38].

The XRD diffraction peak profile of SDW-stripes shows a large anisotropy in the k-space shown in panel (c) of Figure 1. Line cuts through the SDW-stripes peak along the h-direction and l-direction, are presented in the left and right panel of Figure 1c, respectively. The solid lines correspond to results of Lorentzian profiles fitted to the data. The width of the two Lorentzian profiles gives the correlation lengths in the NiO_2 plane direction much larger than in the out-of-plane direction reflecting the quasi two-dimensionality of the magnetic interactions [7,35] in the NiO_2 atomic layer modules. In fact, this system can be described as a hetero-structure at atomic limit made of weakly interacting atomic NiO_2 layers separated by $La_{2-x}Sr_xO_2$ spacing layers. The NiO_2 layers have a sizeable compressive strain tuned by Sr doping, because of the lattice mismatch between the antiferromagnetic striped layers and the spacer layers which is a key variable controlling spin and charge ordering [51,52].

We have found an intrinsic inhomogeneous spatial organization of SDW-stripes forming stripy domains organized in micrometric stripes with different spatial arrangement as a function of temperature. In Figure 2 we report the micro X-ray diffraction maps probing the spatial inhomogeneity of the spin stripes order. Three maps were collected by scanning the same sample area at 30 K, 50 K and $T^* = 70$ K temperatures. In the spatial maps the spin-stripes peak intensity is plotted in a logarithm scale with red color areas. We observe in panels (a, b, c) of Figure 2 the formation of SDW rich regions in the sample corresponding to very high spin-stripes diffraction intensity with the shape of microscale stripes.

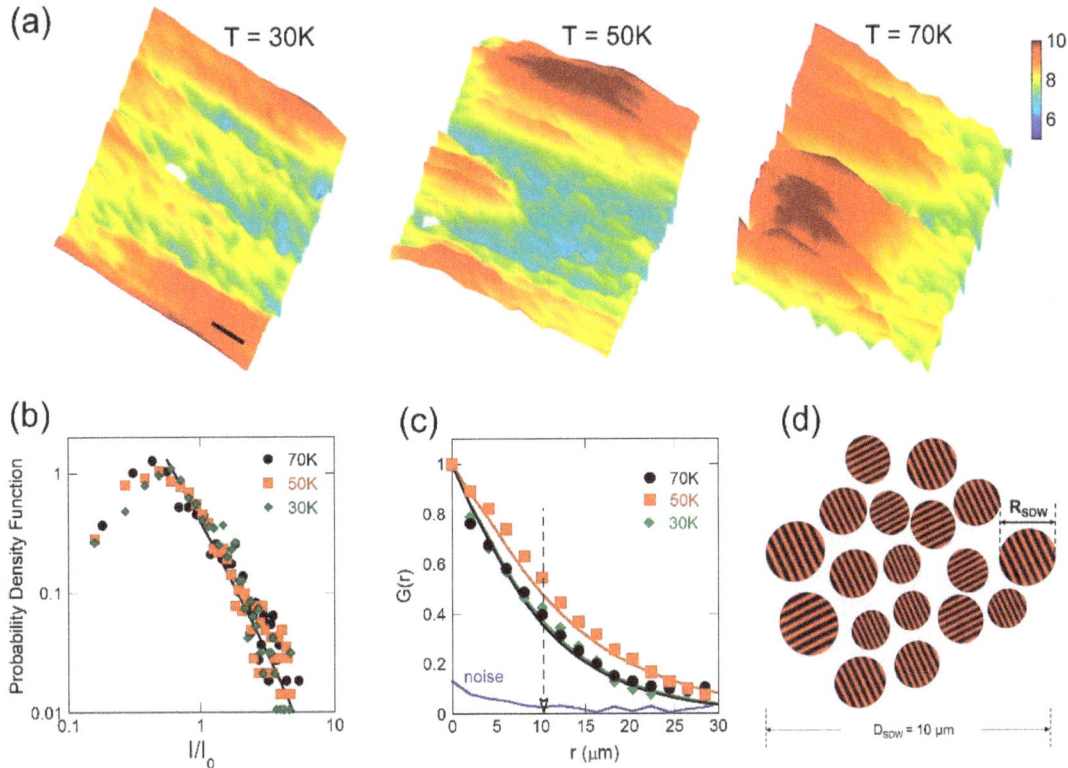

Figure 2. Spatial inhomogeneous stripes order. (**a**) Maps of spin stripes order showing phase segregation with spin stripes aggregates (red area) separated by regions (green-blue) with no spin stripes order signal. Each single pixel of the presented maps has been obtained recording the intensity of the XRD reflections probing SDW-stripes in a specific x,y position of the sample. l. In order to reconstruct the spatial maps, the sample has been scanned over an area of about 40×80 μm^2 in steps of 1 μm in both directions. The black scale bar shown in the upper frame collected at 30 K corresponds to 10 μm. Red areas show SDW-stripes domains of the probed SDW-stripes forming puddles of the order of about 10 micrometers. Blue areas are representative of SDW-stripes domains where the SDW-stripes signal is not detected; (**b**) Probability density functions of the intensity of the spin stripes signal calculated from spatial maps at 30 K, 50 K and T* = 70 K. The distributions show evident fat-tails characterized by a power-law behavior with a critical exponent of 2.1 (solid black line) which rescale on the same curve with different values of the cut-off (**c**). The radial correlation function G(r) calculated from the three spatial maps. The blue line represents the spatial correlation obtained for a random distribution of stripes XRD peak intensities obtained by just shuffling the data. The G(r) intensities decay exponentially, on the noise level, at D$_{SDW}$ indicating the size of a typical domain of 10 microns made by aggregation of individual nanoscale stripes puddles in the NiO$_2$ in the ab-plane; (**d**) Pictorial view of the stripes puddles of about R$_{SDW}$ ≈ 20 nm, given by the correlation length extracted from the width of the diffraction reflection lines (see Figure 1), forming aggregates of about D$_{SDW}$ ≈ 10 micron size below the critical temperature T*.

The microscale spin rich stripes are separated by blue or green areas where the spin diffraction signal is the noise level. The two regions are separated by the yellow interface domains between red micro-scale stripes with spin modulations, and blue-green regions with no spin stripe modulation. The red spin stripes as a function of temperature shows a maximum density at T* = 70 K and by decreasing the temperature some of the red color spin stripes disappear and are replaced by the increasing blue-green regions with missing SDW.

Panel (b) in Figure 2b shows the statistical analysis of the distribution of the SDW-stripes intensity, I, in terms of the probability density function $P(x)$ where $x = I/I_0$ and I_0 is the average intensity of the map. The intensity distributions strongly deviate from a Gaussian distribution and show extended fat tails which can be fitted by an exponentially truncated power-law behavior

$$P(x) = x^{-\alpha}\exp(-x/x_\tau)$$

with a critical exponent $\alpha = 2.1 \pm 0.2$ and cut-off $x_\tau = 7 \pm 0.5$ at $T^* = 70$ K shown by a solid line in panel (b). Similar behavior has been reported for the distribution of the oxygen interstitials and the CDW order accompanied by local lattice distortions in the active layers of cuprates and related materials [4–6,30–34]. This result underlines a spatial organization of the spin stripes order in "scale-free" or fractal-like geometries showing long-range power-law correlations common to systems showing fractal geometry which are quantified by the experimentally determined critical exponent α and the cut-off x_τ. This physical state appears in systems "tuned" near a quantum critical point which is a feature of striped quantum complex matter phase in perovskite materials [4–6,30–34].

For investigating the spatial distribution of the microscale SDW-strip we have calculated the radial distribution function G(r) of the XRD reflection intensity in the spatial maps at the three temperatures reported in panel (c) in Figure 2. All the G(r) curves show a similar exponential decay falling on the noise level at distances with $D_{SDW} = 10$ μm. We associate this length to the average size of the microscale SDW-strip domain. The average size R_{SDW} of the nano-spin-puddles in the ab-plane deduced by the diffraction profile width in Figure 1 is of the order or 20 nm, therefore, each illuminated 1 μm spot in our scanning mode provides the average value over about 2.5×10^3 nano-spin-puddles. A pictorial view of spin-rich domain of radius DSDW = 10 μm hosting about 2.5×10^5 nano-spin-puddles of radius R_{SDW} is shown in panel (d) of Figure 2.

4. Conclusions

In this paper we have provided experimental evidence for spatial phase separation of magnetic stripes order in nickelates predicted for two bands strongly correlated systems near a Lifshitz transition [29]. In cuprates and related matter the coexistence of spin ordered, charge ordered and lattice ordered puddles [39–41] have pointed towards the possibility of quantum complex fluids at the interfaces spanning filamentary hyperbolic spaces [42] as it has been found in functional biological matter [43,44]. Moreover, it is possible that the doped charges in nickelates form polaron [45] anisotropic aggregates [46] with associated tilts [47] making quantum networks [48] of polarons. Therefore, the phase separation reported in this work could be assigned also to the liquid-striped liquid phase separation in liquids of anisotropic polarons similar to the liquid-striped liquid phase separation in water [49,50]. The anisotropy of polaron clusters in nickelates is assigned here to misfit strain [51,52] and orbital degrees of freedom [53–55]. The detection of the complex magnetic structures in strongly correlated electron systems by X- ray diffraction [56,57] can be used to support the association of the spin signal to polaronic distortions. The new mesoscopic phase separation with scale free spatial correlation for spin stripes order found here in nickelates is in agreement with previous indications [53–59] and it provides the experimental smoking gun evidence that the spin ordering in spin stripes phase in nickelates is near a quantum critical point. A similar spatial fractal landscape has been found in cuprates [46–52] and in other oxides near a quantum phase transition as in VO_2 [60–63], in ruthenates [64,65], and in diborides [66,67]. The observed scale free phase separation in nickelates is in agreement with the predictions by of the multiband Hubbard model of frustrated phase separation in strongly correlated two bands systems [29,68]. In this regime the strain manipulation provides a key physical variable [51,52] to drive the system near a particular quantum critical point at Lifshitz transitions or topological electronic transitions. Further experimental work is needed to clarify if hyperbolic space correlations predicted by theory [69,70] and observed in correlated metallic cuprates [4,42] are present or not in the stripes spatial landscape in nickelates. Finally, we have reported new information on the quantum complex scenario near criticality in nickelates which opens new venues to applications and developments of new magnetic and electronic devices. In fact, in the proximity of a topological Lifshitz transition it is possible to control novel macroscopic functionalities by a weak external stimulus such as stress [25–36,71–75], current density [64,65] or photon illumination dose [5,76–79].

Author Contributions: A.R. conceived the project and designed all the experiments. C.S.-L., M.S. and G.C. contributed to the planning of the experiments. The samples were grown by A.A.N., and characterized by M.B. (Marcel Buchholz), C.T., C.S.-L. and M.B. (Markus Braden); preliminary measurements using soft X-ray have been performed at ALS (Berkeley) by A.R., S.M., J.L., L.M. and S.R. and at P04-PETRA III (DESY) by M. B. (Marcel Buchholz), C.S.-L. and J.V.; micro X-ray diffraction experiments have been carried out at P10-PETRA III (DESY) by A.R., G.C., N.P., B.J., A.Z. and M.S.; data analysis has been done by A.R., G.C., A.B., N.P., M.B. ((Markus Braden)), C.S.-L., A.R. and M.S. discussed the results and worked on the interpretation of the data. The manuscript has been written by A.R., G.C and A.B. collecting feedback from all the authors.

Funding: This research was supported by the Helmholtz Virtual Institute: Dynamic Pathways in Multidimensional Landscapes. A.R. and G.C. acknowledge the Stephenson Distinguished Visitor Program by DESY. N.P. acknowledges the Italian Ministry for Education and Research and Marie Curie IEF project for partial financial support. A.A.N. acknowledges funding from Ministry of Research, Technology and Higher Education through Hibah WCU-ITB. Work in Cologne was supported by the DFG through SFB608 and by the German Ministry for Science and Education through contract 05K10PK2. Moreover, support by the DFG within SFB 925 and by the Superstripes-onlus is gratefully acknowledged. Work at the ALS, LBNL was supported by the Director, Office of Science, Office of Basic Energy Sciences, of the US Department of Energy (Contract No. DE-AC02-05CH11231).

Conflicts of Interest: The authors declare no conflict of interest.

References

1. Dagotto, E. Complexity in strongly correlated electronic systems. *Science* **2005**, *309*, 257–262. [CrossRef] [PubMed]

2. Vojta, M. Lattice symmetry breaking in cuprate superconductors: Stripes, nematics and superconductivity. *Adv. Phys.* **2009**, *58*, 699–820. [CrossRef]

3. Bianconi, A.; Saini, N.L. *Stripes and Related Phenomena*; Springer Science Business Media: Berlin, Germany, 2001.

4. Campi, G.; Bianconi, A.; Poccia, N.; Barbo, G.; Arrighetti, G.; Innocenti, D.; Karpinski, J.; Zhigadlo, D.N.; Kazakov, S.M.; Burghammer, M.; et al. Inhomogeneity of charge-density-wave order and quenched disorder in a high-T_c superconductor. *Nature* **2015**, *525*, 359–362. [CrossRef] [PubMed]

5. Poccia, N.; Ricci, A.; Campi, G.; Fratini, M. Optimum inhomogeneity of local lattice distortions in La_2CuO_{4+y}. *Proc. Natl. Acad. Sci. U.S.A.* **2012**, *109*, 15685–15690. [CrossRef] [PubMed]

6. Fratini, M.; Poccia, N.; Ricci, A.; Campi, G.; Burghammer, M.; Aeppli, G.; Bianconi, A. Scale-free structural organization of oxygen interstitials in La_2CuO_{4+y}. *Nature* **2010**, *466*, 841–844. [CrossRef] [PubMed]

7. Ulbrich, H.; Braden, M. Neutron scattering studies on stripe phases in non-cuprate materials. *Phys. C Supercond.* **2012**, *481*, 31–45. [CrossRef]

8. Lee, S.H.; Cheong, S.W. Melting of quasi-two-dimensional charge stripes in $La_{5/3}Sr_{1/3}NiO_4$. *Phys. Rev. Lett.* **1997**, *79*, 2514. [CrossRef]

9. Yoshizawa, H.; Kakeshita, T.; Kajimoto, R.; Tanabe, T.; Katsufuji, T.; Tokura, Y. Stripe order at low temperatures in $La_{2-x}Sr_xNiO_4$ with $0.289 \lesssim x \lesssim 0.5$. *Phys. Rev. B* **2000**, *61*, R854. [CrossRef]

10. Kajimoto, R.; Kakeshita, T.; Yoshizawa, H.; Tanabe, T.; Katsufuji, T.; Tokura, Y. Hole concentration dependence of the ordering process of the stripe order in $La_{2-x}Sr_xNiO_4$. *Phys. Rev. B* **2001**, *64*, 144432. [CrossRef]

11. Lee, S.H.; Cheong, S.W.; Yamada, K.; Majkrzak, C.F. Charge canted spin order in $La_{2-x}Sr_xNiO_4$ (x = 0.275,1/3). *Phys. Rev. B* **2001**, *63*, 060405. [CrossRef]

12. Boothroyd, A.T.; Freeman, P.G.; Prabhakaran, D.; Enderle, M.; Kulda, J. Magnetic order and dynamics in stripe-ordered $La_{2-x}Sr_xNiO_4$. *Phys. B Condens. Matter* **2004**, *345*, 1–5. [CrossRef]

13. Freeman, P.G.; Christensen, N.B.; Prabhakaran, D.; Boothroyd, A.T. The temperature evolution of the out-of-plane correlation lengths of charge-stripe ordered $La_{1.725}Sr_{0.275}NiO_4$. *J. Phys. Conf. Ser.* **2010**, *200*, 012037. [CrossRef]

14. Chow, K.H.; Pattenden, P.; Blundell, S.J.; Hayes, W. Muon-spin-relaxation studies of magnetic order in heavily doped $La_{2-x}Sr_xNiO_{4+\delta}$. *Phys. Rev. B* **1996**, *53*, R14725. [CrossRef] [PubMed]

15. Schüßler-Langeheine, C.; Schlappa, J.; Tanaka, A.; Hu, Z.; Chang, C.F.; Schierle, E.; Benomar, M.; Ott, H.; Weschke, E.; Kaindl, G.; et al. Spectroscopy of stripe order in $La_{1.8}Sr_{0.2}NiO_4$ using resonant soft X-ray diffraction. *Phys. Rev. Lett.* **2005**, *95*, 156402. [CrossRef] [PubMed]

16. Schlappa, J.; Chang, C.F.; Schierie, E.; Tanaka, A. Static and fluctuating stripe order observed by resonant soft X-ray diffraction in $La_{1.8}Sr_{0.2}NiO_4$. *arXiv* **2009**, arXiv:0903.0994v1.

17. Coslovich, G.; Huber, B.; Lee, W.S.; Chuang, Y.D.; Zhu, Y.; Sasagawa, T.; Hussain, Z.; Bechtel, H.A.; Martin, M.C.; Shen, Z.X.; et al. Ultrafast charge localization in a stripe-phase nickelate. *Nature Cmmun.* **2013**, *4*, 2643. [CrossRef]

18. Wilkins, S.B.; Hatton, P.D.; Liss, K.D.; Ohler, M.; Katsufuji, T.; Cheong, S.W. High-resolution high energy X-ray diffraction studies of charge ordering in CMR manganites and nickelates. *Int. J. Mod. Phys. B* **2000**, *14*, 3753–3758. [CrossRef]

19. Chen, C.H.; Cheong, S.W.; Cooper, A.S. Charge modulations in $La_{2-x}Sr_xNiO_{4+y}$ ordering of polarons. *Phys. Rev. Lett.* **1993**, *71*, 2461. [CrossRef]

20. Ghazi, M.E.; Spencer, P.D.; Wilkins, S.B.; Hatton, P.D.; Mannix, D.; Prabhakaran, D. Incommensurate charge stripe ordering in in $La_{2-x}Sr_xNiO_4$ for x= (0.33, 0.30, 0.275). *Phys. Rev. B* **2004**, *70*, 144507. [CrossRef]

21. Du, C.H.; Ghazi, M.E.; Su, Y.X.; Pape, I. Critical Fluctuations and quenched Disordered Two-Dimensional Charge Stripes in $La_{5/3}Sr_{1/3}NiO_4$. *Phys. Rev. Lett.* **2000**, *84*, 3911. [CrossRef]

22. Johnson, R.D.; Mazzoli, C.; Bland, S.R.; Du, C. Magnetically induced electric polarization reversal in multiferroic $TbMn_2O_5$: Terbium spin reorientation studied by resonant X-ray diffraction. *Phys. Rev. B* **2011**, *83*, 054438. [CrossRef]

23. Vecchini, C.; Bombardi, A.; Chapon, L.C.; Lee, N.; Radaelli, P.G.; Cheong, S.W. Magnetic phase diagram and ordered ground state of $GdMn_2O_5$ multiferroic studied by X-ray magnetic scattering. *J. Phys. Conf. Ser.* **2014**, *519*, 012004. [CrossRef]

24. Pincini, D.; Boseggia, S.; Perry, R.; Gutmann, M. Persistence of antiferromagnetic order upon La substitution in the Mott insulator Ca_2RuO_4. *Phys. Rev. B* **2018**, *98*, 014429. [CrossRef]

25. Price, N.W.; Vibhakar, A.M.; Johnson, R.D.; Schad, J.; Saenrang, W.; Bombardi, A.; Chmiel, F.P.; Eom, C.B.; Radaelli, P.G. Strain engineering a multiferroic monodomain in thin-film $BiFeO_3$. *Phys. Rev. Appl.* **2019**, *11*, 024035. [CrossRef]

26. Caprara, S.; Sulpizi, M.; Bianconi, A.; Castro, C. Single-particle properties of a model for coexisting charge and spin quasicritical fluctuations coupled to electrons. *Phys. Rev. B* **1999**, *59*, 14980. [CrossRef]

27. Carlson, E.W.; Yao, D.X.; Campbell, D.K. Spin waves in striped phases. *Phys. Rev. B* **2004**, *70*, 064505. [CrossRef]

28. Raczkowski, M.; Fresard, R.; Oles, A.M. Microscopic origin of diagonal stripe phases in doped nickelates. *Phys. Rev. B* **2006**, *73*, 094429. [CrossRef]

29. Bianconi, A.; Poccia, N.; Sboychakov, A.O.; Rakhmanov, A.L.; Kugel, K.I. Intrinsic arrested nanoscale phase separation near a topological Lifshitz transition in strongly correlated two-band metals. *Supercond. Sci. Technol.* **2015**, *28*, 024005. [CrossRef]

30. Poccia, N.; Campi, G.; Fratini, M.; Ricci, A. Spatial inhomogeneity and planar symmetry breaking of the lattice incommensurate supermodulation in the high-temperature superconductor $Bi_2Sr_2CaCu_2O_{8+y}$. *Phys. Rev. B* **2011**, *84*, 100504. [CrossRef]

31. Ricci, A.; Poccia, N.; Campi, G.; Coneri, F.; Caporale, A.S.; Innocenti, D.; Burghammer, M.; Zimmermann, M.V.; Bianconi, A. Multiscale distribution of oxygen puddles in 1/8 doped $YBa_2Cu_3O_{6.67}$. *Sci. Rep.* **2013**, *3*, 2383. [CrossRef]

32. Ricci, A.; Poccia, N.; Campi, G.; Coneri, F.; Barba, L.; Arrighetti, G.; Polentarutti, M.; Burghammer, M.; Sprung, M.; Zimmermann, M.V.; et al. Networks of superconducting nano-puddles in 1/8 doped $YBa_2Cu_3O_{6.5+y}$ controlled by thermal manipulation. *New J. Phys.* **2014**, *16*, 053030. [CrossRef]

33. Campi, G.; Ricci, A.; Poccia, N.; Barba, L.; Arrighetti, G.; Burghammer, M.; Caporale, A.S.; Bianconi, A. Scanning micro-X-ray diffraction unveils the distribution of oxygen chain nanoscale puddles in $YBa_2Cu_3O_{6.33}$. *Phys. Rev. B* **2013**, *87*, 014517. [CrossRef]

34. Ricci, A.; Poccia, N.; Joseph, B.; Innocenti, D. Direct observation of nanoscale interface phase in the superconducting chalcogenide $K_xFe_{2-y}Se_2$ with intrinsic phase separation. *Phys. Rev. B* **2015**, *91*, 020503. [CrossRef]

35. Drees, Y.; Li, Z.W.; Ricci, A.; Rotter, M.; Schmidt, W.; Lamago, D.; Sobolev, O.; Rutt, U.; Gutowski, O.; Sprung, M.; et al. Hour-glass magnetic excitations induced by nanoscopic phase separation in cobalt oxides. *Nature Commun.* **2014**, *5*, 573. [CrossRef]

36. Mu, Y.; Ma, Y. Self-organizing stripe patterns in two-dimensional frustrated systems with competing interactions. *Phys. Rev. B* **2003**, *67*, 014110. [CrossRef]

37. Schmalian, J.; Wolynes, P.G. Stripe glasses: Self-generated randomness in a uniformly frustrated system. *Phys. Rev. Lett.* **2000**, *85*, 836. [CrossRef]

38. Bogner, S.; Scheidl, S. Pinning of stripes in cuprate superconductors. *Phys. Rev. B* **2001**, *64*, 054517. [CrossRef]

39. Poccia, N.; Chorro, M.; Ricci, A.; Xu, W.; Marcelli, A.; Campi, G.; Bianconi, A. Percolative superconductivity in $La_2CuO_{4.06}$ by lattice granularity patterns with scanning micro X-ray absorption near edge structure. *Appl. Phys. Lett.* **2014**, *104*, 221903. [CrossRef]

40. Bianconi, A.; Di Castro, D.; Bianconi, G.; Pifferi, A.; Saini, N.L.; Chou, F.C.; Johnston, D.C.; Colapietro, M. Coexistence of stripes and superconductivity: Tc amplification in a superlattice of superconducting stripes. *Phys. C: Supercond.* **2000**, *341*, 1719–1722. [CrossRef]

41. Chen, X.; Schmehr, J.L.; Islam, Z.; Porter, Z.; Zoghlin, E.; Finkelstein, K.; Ruff, J.P.C.; Wilson, S.D. Unidirectional spin density wave state in metallic $(Sr_{1-x}La_x)_2IrO_4$. *Nat. Commun.* **2018**, *9*, 103. [CrossRef]

42. Campi, G.; Bianconi, A. High-Temperature superconductivity in a hyperbolic geometry of complex matter from nanoscale to mesoscopic scale. *J. Supercond. Nov. Magn.* **2016**, *29*, 627–631. [CrossRef]

43. Campi, G.; Di Gioacchino, M.; Poccia, N.; Ricci, A.; Burghammer, A.; Ciasca, G.; Bianconi, A. Nanoscale correlated disorder in out-of-equilibrium myelin ultrastructure. *ACS Nano* **2017**, *12*, 729–739. [CrossRef]

44. Campi, G.; Cristofaro, F.; Pani, G.; Fratini, M.; Pascucci, B.; Corsetto, P.A.; Weinhausen, B.; Cedola, A.; Rizzo, A.M.; Visai, L.; et al. Heterogeneous and self-organizing mineralization of bone matrix promoted by hydroxyapatite nanoparticles. *Nanoscale* **2017**, *9*, 17274–17283. [CrossRef]

45. Zaanen, J.; Littlewood, P.B. Freezing electronic correlations by polaronic instabilities in doped La_2NiO_4. *Phys. Rev. B* **1994**, *50*, 7222. [CrossRef]

46. Kusmartsev, F.V.; Di Castro, D.; Bianconi, G.; Bianconi, A. Transformation of strings into an inhomogeneous phase of stripes and itinerant carriers. *Phys. Lett. A* **2000**, *275*, 118–123. [CrossRef]

47. Gavrichkov, V.A.; Shanko, Y.; Zamkova, N.G.; Bianconi, A. Is There Any Hidden Symmetry in Stripe Structure of Perovskite High-Temperature Superconductors? *J. Phys. Chem. Lett.* **2019**, *10*, 1840–1844. [CrossRef]

48. Bianconi, G. Quantum statistics in complex networks. *Phys. Rev. E* **2002**, *66*, 056123. [CrossRef]

49. Innocenti, D.; Ricci, A.; Poccia, N.; Campi, G.; Fratini, M.; Bianconi, A. A model for liquid-striped liquid phase separation in liquids of anisotropic polarons. *J. Supercond. Nov. Magn.* **2009**, *22*, 529–533. [CrossRef]

50. Campi, G.; Innocenti, D.; Bianconi, A. CDW and similarity of the Mott insulator-to-metal transition in cuprates with the gas-to-liquid-liquid transition in supercooled water. *J. Supercond. Nov. Magn.* **2015**, *28*, 1355–1363. [CrossRef]

51. Agrestini, S.; Saini, N.L.; Bianconi, G.; Bianconi, A. The strain of CuO_2 lattice: The second variable for the phase diagram of cuprate perovskites. *J. Phys. A Math. Gen.* **2003**, *36*, 9133. [CrossRef]

52. Bianconi, A.; Agrestini, S.; Bianconi, G.; Di Castro, D.; Saini, N.L. A quantum phase transition driven by the electron lattice interaction gives high Tc superconductivity. *J. Alloy. Compd.* **2001**, *317*, 537–541. [CrossRef]

53. Lichtenstein, A.I.; Fleck, M.; Oles, A.M.; Hedin, L. Dynamical mean-field theory of stripe ordering. In *Stripes and Related Phenomena*; Bianconi, A., Saini, N.L., Eds.; Springer: Boston, MA, USA, 2002; pp. 101–109.

54. Raczkowski, M.; Oleś, A.M. Competition between vertical and diagonal stripes in the Hartree-Fock approximation. *AIP Conf. Proc.* **2003**, *678*, 293–302.

55. Oles, A.M. Charge and orbital order in transition metal oxides. *Acta Phys. Polon. A* **2010**, *118*, 212. [CrossRef]

56. Beale, T.A.W.; Wilkins, S.B.; Johnson, R.D.; Prabhakaran, D.; Boothroyd, A.T.; Steadman, P.; Dhesi, S.S.; Hatton, P.D. Advances in the understanding of multiferroics through soft X-ray diffraction. *Eur. Phys. J. Spec. Top.* **2012**, *208*, 99–106. [CrossRef]

57. Radaelli, P.G.; Dhesi, S.S. The contribution of Diamond light source to the study of strongly correlated electron systems and complex magnetic structures. *Philos. Trans. R. Soc. A Math. Phys. Eng. Sci.* **2015**, *373*, 20130148. [CrossRef]

58. Mattoni, G.; Zubko, P.; Maccherozzi, F.; van der Torren, A.J.H.; Boltje, D.B.; Hadjimichael, M.; Manca, N.; Catalano, S.; Gibert, M.; Liu, Y.; et al. Striped nanoscale phase separation at the metal–insulator transition of heteroepitaxial nickelates. *Nat. Commun.* **2016**, *7*, 13141. [CrossRef]

59. Varignon, J.; Grisolia, M.N.; Íñiguez, J.; Barthelemy, A.; Bibes, M. Complete phase diagram of rare-earth nickelates from first-principles. *NPJ Quantum Mater.* **2017**, *2*, 21. [CrossRef]

60. Liu, M.; Sternbach, A.J.; Wagner, M.; Slusar, T.V.; Kong, T.; Budko, S.L.; Salinporn, K.; Qazibash, M.M.; McLeod, A.; Fei, Z.; et al. Phase transition in bulk single crystals and thin films of VO_2 by nanoscale infrared spectroscopy and imaging. *Phys. Rev. B* **2015**, *91*, 245155. [CrossRef]

61. Bianconi, A. Multiplet splitting of final-state configurations in X-ray-absorption spectrum of metal VO_2: Effect of core-hole-screening, electron correlation and metal-insulator transition. *Phys. Rev. B* **1982**, *26*, 2741. [CrossRef]

62. Marcelli, A.; Coreno, M.; Stredansky, M.; Xu, W. Nanoscale phase separation and lattice complexity in VO_2: The metal–insulator transition investigated by XANES via Auger electron yield at the vanadium L_{23}-edge and resonant photoemission. *Condens. Matt.* **2017**, *2*, 38. [CrossRef]

63. Gioacchino, D.; Marcelli, A.; Puri, A.; Zou, C. Metastability phenomena in VO_2 thin films. *Condens. Matt.* **2017**, *2*, 10. [CrossRef]

64. Zhang, J.; McLeod, A.S.; Han, Q.; Chen, X.; Bechtel, H.A.; Yao, Z.; Gilbert, S.N.; Ciavatti, T.; Tao, T.H.; Aronson, M.; et al. Nano-resolved current-induced insulator-metal transition in the Mott insulator Ca_2RuO_4. *Phys. Rev. X* **2019**, *9*, 011032. [CrossRef]

65. Luo, Y.; Pustogow, A.; Guzman, P.; Dioguardi, A.P.; Thomas, S.M.; Ronning, F.; Kikugawa, N.; Sokolov, D.A.; Jerzebeck, A.P.; Mackenzie, A.P.; et al. Normal state O^{17} NMR studies of Sr_2RuO_4 under uniaxial stress. *Phys. Rev. X* **2019**, *9*, 021044.

66. Bauer, E.; Paul, C.; Berger, S.; Majumdar, S.; Michor, H.; Giovannini, M.; Saccone, A.; Binaconi, A. Thermal conductivity of superconducting MgB_2. *J. Phys. Condens. Matt.* **2001**, *13*, L487. [CrossRef]

67. Agrestini, S.; Metallo, C.; Filippi, M.; Simonelli, L.; Gampi, G.; Sanipoli, C.; Liarokapis, E.; De Negri, S.; Giovannini, M.; Saccone, A.; et al. Substitution of Sc for Mg in MgB_2: Effects on transition temperature and Kohn anomaly. *Phys. Rev. B* **2004**, *70*, 134514. [CrossRef]

68. Kugel, K.I.; Rakhmanov, A.L.; Sboychakov, A.O.; Kusmartsev, F.V.; Poccia, N.; Bianconi, A. A two-band model for the phase separation induced by the chemical mismatch pressure in different cuprate superconductors. *Supercond. Sci. Technol.* **2008**, *22*, 014007. [CrossRef]

69. Bianconi, G.; Rahmede, C. Emergent hyperbolic network geometry. *Sci. Rep.* **2017**, *7*, 41974. [CrossRef]

70. Bianconi, G.; Ziff, R.M. Topological percolation on hyperbolic simplicial complexes. *Phys. Rev. E* **2018**, *98*, 052308. [CrossRef]

71. Kagan, M.Y.; Bianconi, A. Fermi-Bose mixtures and BCS-BEC crossover in high-Tc superconductors. *Condens. Matt.* **2019**, *4*, 51. [CrossRef]

72. Yanagisawa, T. Mechanism of high-temperature superconductivity in correlated-electron systems. *Condens. Matt.* **2019**, *4*, 57. [CrossRef]

73. Jurkutat, M.; Erb, A.; Haase, J. T_c and other cuprate properties in relation to planar charges as measured by NMR. *Condens. Matt.* **2019**, *4*, 67. [CrossRef]

74. Caprara, S. The ancient romans' route to charge density waves in cuprates. *Condens. Matt.* **2019**, *4*, 60. [CrossRef]

75. Yamamoto, S.; Fujiwara, T.; Hatsugai, Y. Electronic structure of charge and spin stripe order in $La_{2-x}Sr_xNiO_4$ (x=1/3,1/2). *Phys. Rev. B* **2007**, *76*, 165114. [CrossRef]

76. Campi, G.; Castro, D.D.; Bianconi, G.; Agrestini, A.; Saini, N.L.; Oyanagi, H.; Bianconi, A. Photo-Induced phase transition to a striped polaron crystal in cuprates. *Phase Transit.* **2002**, *75*, 927–933. [CrossRef]

77. Campi, G.; Dell'Omo, C.; Di Castro, D.; Agretini, S.; Filippi, M.; Bianconi, G.; Barba, L.; Cassetta, A.; Colapirtro, M.; Saini, N.L.; et al. Effect of temperature and X-ray illumination on the oxygen ordering in $La_2CuO_{4.1}$ Superconductor. *J. Supercond.* **2004**, *17*, 137–142. [CrossRef]

78. Poccia, N.; Bianconi, A.; Campi, G.; Fratini, M.; Ricci, A. Size evolution of the oxygen interstitial nanowires in La_2CuO_{4+y} by thermal treatments and X-ray continuous illumination. *Superconductor Sci. Technol.* **2012**, *25*, 124004. [CrossRef]

79. Campi, G.; Ricci, A.; Poccia, N.; Fratini, M.; Bianconi, A. X-rays Writing/Reading of charge density waves in the CuO_2 plane of a simple cuprate superconductor. *Condens. Matt.* **2017**, *2*, 26. [CrossRef]

10

Thermodynamic Stability, Thermoelectric, Elastic and Electronic Structure Properties of ScMN₂-Type (M = V, Nb, Ta) Phases Studied by ab initio Calculations

Robert Pilemalm [1],*[ID], Leonid Pourovskii [2,3][ID], Igor Mosyagin [4], Sergei Simak [1] and Per Eklund [1],*[ID]

[1] Department of Physics, Chemistry and Biology (IFM), Linköping University, SE-581 83 Linköping, Sweden; sergeis@ifm.liu.se

[2] Centre de Physique Théorique, Ecole Polytechnique, CNRS, Université Paris-Saclay, Route de Saclay, FR-91128 Palaiseau, France; leonid@cpht.polytechnique.fr

[3] Collège de France, 11 place Marcelin Berthelot, FR-75005 Paris, France

[4] Materials Modeling and Development Laboratory, NUST "MISIS", RU-119991 Moscow, Russia; igor.mosyagin@gmail.com

* Correspondence: robert.pilemalm@liu.se (R.P.); per.eklund@liu.se (P.E.)

Abstract: ScMN₂-type (M = V, Nb, Ta) phases are layered materials that have been experimentally reported for M = Ta and Nb, but they have up to now not been much studied. However, based on the properties of binary ScN and its alloys, it is reasonable to expect these phases to be of relevance in a range of applications, including thermoelectrics. Here, we have used first-principles calculations to study their thermodynamic stability, elastic, thermoelectric and electronic properties. We have used density functional theory to calculate lattice parameters, the mixing enthalpy of formation and electronic density of states as well as the thermoelectric properties and elastic constants (c_{ij}), bulk (B), shear (G) and Young's (E) modulus, which were compared with available experimental data. Our results indicate that the considered systems are thermodynamically and elastically stable and that all are semiconductors with small band gaps. All three materials display anisotropic thermoelectric properties and indicate the possibility to tune these properties by doping. In particular, ScVN₂, featuring the largest band gap exhibits a particularly large and strongly doping-sensitive Seebeck coefficient.

Keywords: ScTaN₂; inverse MAX phase; thermoelectric properties; density functional theory

1. Introduction

MAX phases, where M is a transition metal, A is an A-group element and X is carbon and/or nitrogen, comprise a family of more than 70 compounds. Since the mid-1990s, there has been extensive research on MAX phases due to their unique combination of metallic and ceramic properties, manifested by their unusual combination of properties such as excellent thermal and electrical conductivities, ductility, resistance to thermal shock and oxidation. MAX phases can be used for a variety of applications, such as high temperature structural applications, protective coatings, sensors, low friction surfaces and electrical contacts [1–6]. The stoichiometry of a ternary MAX phase is $M_{n+1}AX_n$, where n is 1,2 or 3 and the three different stoichiometries are referred to as 211 (n = 1), 312 (n = 2) and 413 (n = 3) [7,8].

A related structure to a 211 MAX phase is the ScTaN₂- and ScNbN₂-type structure. These phases have been observed experimentally [9–11], and a basic characterization of structure and some properties has been made. However, theoretical studies on these structures and how they relate to their physical properties are limited. Furthermore, $ScVN_2$ is expected to exist based on thermodynamic stability calculations [12], but it has not been observed experimentally.

The structure of $ScTaN_2$, $ScNbN_2$ and $ScVN_2$ can generally be described as the $ScMN_2$-type structure [12], which has space group $P6_3/mmc$ (#194) and comprises of alternating layers of $ScN_{6/3}$ octrahedra and $MN_{6/3}$ prisms. Sc occupies the 2a positions, M the 2d positions and N the 4f positions [11]. Table 1 shows experimentally characterized $ScTaN_2$ and $ScNbN_2$. Niewa et al. have also visualized the structure [11]. It can be noted that their positions are the inverse positions of a corresponding 211 MAX phase [1]. That is, the positions occupied by the M atoms in the 211 MAX structure correspond to the N positions in $ScMN_2$. Because of this relationship to a MAX phase, we term this structure "inverse MAX phase" (in analogy with, e.g., inverse perovskites).

Table 1. Atomic positions data and experimental lattice parameters (powder X-ray diffraction) of $ScTaN_2$ [11] and $ScNbN_2$ [9].

	ScTaN$_2$			
a (Å)		3.0534		
c (Å)		10.5685		
Atom	Site	x	y	z
Sc	2a	0	0	0
Ta	2d	1/3	2/3	3/4
N	4f	1/3	2/3	0.1231
	ScNbN$_2$			
a (Å)		3.0633		
c (Å)		10.5702		
Atom	Site	x	y	z
Sc	2a	0	0	0
Nb	2d	1/3	2/3	3/4
N	4f	1/3	2/3	0.1250

Furthermore, ScN is a semiconductor showing promise for thermoelectric applications due to its suitable thermal and electrical properties [12–17]. However, the thermal conductivity is relatively high $(8–12 \ Wm^{-1}K^{-1})$ [13,18], leading to a rather low thermoelectric figure of merit (ZT). Approaches to decrease the thermal conductivity include alloying [12,19–21], making artificial superlattices [22–25], or nanostructuring, as also exemplified for CrN [26]. In analogy with artificial superlattices, inherently nanolaminated materials like $ScTaN_2$, $ScNbN_2$ and $ScVN_2$ could be of interest for this purpose. Assessing this, however, requires determining their structural and electronic properties, e.g., are they metallic or semiconductors? This motivates the present theoretical study of the material properties of the inverse MAX phases, in order to screen their possibility for thermoelectrics.

2. Computational Details

We consider the enthalpy of formation in order to estimate phase stability of $ScMN_2$, where M is Ta, Nb or V, which is calculated as

$$\Delta H = H_{ScMN_2} - H_{ScN} - H_{MN}. \tag{1}$$

Each enthalpy, H, is considered at zero pressure and the energy of each phase is taken at its equilibrium volume. For relaxation, the energy and force tolerance was 0.0001 eV and 0.001 eV/Å, respectively. To determine the thermodynamic stability, we consider as competing phases the experimentally known cubic (NaCl, B1) binary nitrides, as no other (ternary) phases are known in these systems. For the case of VN, it should be noted that, according to previous theoretical calculations [27], the energy at 0 K for WC-type VN is lower than that of the experimentally observed B1 VN phase. However, the cubic phase is expected to be stabilized by atomic vibration at higher temperatures, rendering it more stable than the WC-type phase. It is therefore more relevant to consider it as a competing phase here [28,29].

The calculations were performed using density functional theory (DFT) as implemented in the Vienna Ab initio Simulation Package (VASP) [30–33]. Projector augmented wave basis sets [34] were used with a cutoff energy of 650 eV and the exchange–correlation potential was modeled with the generalized gradient approximation according to Perdew, Burke and Ernzerhof (PBE-GGA) [35]. For all systems, eight atom unit cells were used and $8 \times 8 \times 8$ k-point mesh for energy calculations, while, for elastic calculations, $25 \times 25 \times 11$ k-points mesh was used and an energy cutoff of 650 eV. The elastic tensor was determined in VASP from strain–stress relationships after the introduction of finite distortions in the lattice [36]. The magnitude of the strains was on the order of 0.015 Å.

Electronic density of states (DOS) was calculated with a plane wave cutoff of 650 eV, $17 \times 17 \times 5$ k-point mesh and a plane. The tetrahedron method with Blöchl correction [37] for integration over the Brillouin zone was used in all DOS calculations. The value of the level broadening was 0.2 eV. The calculations of DOS were performed with both the PBE-GGA functional with a $24 \times 24 \times 24$ k-points mesh and the Heyd–Scuseria–Ernzerhof hybrid functional (HSE06) [38] with a $17 \times 17 \times 5$ k-point mesh, respectively.

Transport properties were evaluated using an ab initio approach combining the linearized augmented plane wave (LAPW) method as implemented in the Wien2k package [39] with the semiclassical BoltzTraP [40] code for evaluating the conductivity and Seebeck tensors. We first computed the electronic structures of $ScMN_2$ by Wien2k using the lattice structures optimized previously by VASP. In these calculations, we employed the parameter $R_{MT}K_{max} = 7$ as well as a dense mesh 10^5 k-points in the Brillouin zone. Our test calculations with $R_{MT}K_{max} = 9$ showed little sensitivity of transport properties of this parameter. The transport integrals were subsequently evaluated from the converged Kohn–Sham band structure obtained by Wien2k using the BoltzTraP package [40].

3. Results and Discussion

Table 2 shows the calculated enthalpy of formation and calculated elastic constants of all three phases. For $ScTaN_2$, the enthalpy of formation is –1.067 eV; for $ScNbN_2$, it is −0.84 eV, and, for $ScVN_2$, it is −0.22 eV. It can be noted that the enthalpy of formation, which is calculated with respect to binary rock salt B1 structures, is negative for all three systems. Generally, this is expected to be representative also for the Gibbs free energy i.e., when the effect of temperature is accounted for, as it is known to be valid for a wide range of layered phases [41]. The reservation should be stated that the energy of the competing B1 VN phase is likely underestimated, since it is stabilized by vibrational effects at higher temperature, as mentioned in the Computational details section. In addition, the elastic constants fulfill the elastic stability criteria $C_{11} > |C_{12}|$, $2C^2_{13} < C_{33}(C_{11} + C_{12})$, $C_{44} > 0$ and $C_{11} − C_{12} > 0$, which are necessary and sufficient conditions for a hexagonal structure [42]. Thus, it can be concluded that all three phases are thermodynamically stable relative to the considered competing phases. Because the conditions of elastic stability are also met, all three phases can exist.

Table 2. Calculated enthalpy of formation, elastic constants, lattice parameters and fractional coordinate z of $ScTaN_2$, $ScNbN_2$ and $ScVN_2$.

Parameter	$ScTaN_2$	$ScNbN_2$	$ScVN_2$
ΔH_1 [eV]	−1.07	−0.84	−0.22
C_{11} [GPa]	551	522	480
C_{12} [GPa]	158	152	145
C_{13} [GPa]	143	130	121
C_{33} [GPa]	552	546	572
C_{44} [GPa]	196	189	167
a [Å]	3.0791	3.0824	2.9774
c [Å]	10.6254	10.6060	10.2591
z	0.1238	0.1237	0.1313

Table 2 furthermore reports the predicted hexagonal lattice parameters calculated with the energy relaxed with respect to volume of each phase. For $ScNbN_2$, the lattice parameters are, $a = 3.0824$ Å and $c = 10.6060$ Å; for $ScVN_2$, $a = 2.9774$ Å and $c = 10.2591$ Å; and, for $ScTaN_2$, $a = 3.0791$ Å and $c = 10.6254$ Å. For the first and the third case, these results can be compared with earlier experimental determination based on X-ray powder diffraction in Table 1. The difference in lattice parameter of $ScTaN_2$ and $ScNbN_2$ is expected since the use of the PBE-GGA functional tends to overestimate lattice parameters [43,44].

Table 3 shows the bulk modulus (B) and the shear modulus (G) estimated by both the Voigt (subscript V) [45] and the Reuss (subscript R) [45,46] approximation. The difference between the two approximations is that Voigt's assumes that the strain throughout the polycrystalline aggregate is uniform, while Reuss' assumes uniform stress instead [45]. In the case of a hexagonal lattice, the Voigt shear modulus (G_V) and the Voigt bulk modulus (B_V) are:

$$G_v = \frac{1}{15}(2c_{11} + c_{33} - c_{12} - 2c_{13}) + \frac{1}{5}\left(2c_{44} + \frac{c_{11} - c_{12}}{2}\right), \tag{2}$$

$$B_V = \frac{2}{9}(c_{11} + c_{12} + 2c_{13} + c_{33}/2). \tag{3}$$

Table 3. Elastic (E), shear (G) and bulk (B) moduli estimated with Voigt (subscript V) and Reuss (subscript R) approximations.

Parameter	$ScTaN_2$	$ScNbN_2$	$ScVN_2$
G_V [GPa]	197	190	179
B_V [GPa]	283	268	256
E_V [GPa]	479	460	436
G_R [GPa]	197	189	175
B_R [GPa]	283	268	255
E_R [GPa]	479	460	432
G_R/B_R	0.70	0.71	0.70
G_V/B_V	0.70	0.71	0.70

The Reuss shear modulus (G_R) and the Voigt bulk modulus (B_R) are:

$$G_R = \frac{15}{4(2s_{11} + s_{33}) - 4(s_{12} + 2s_{13}) + 3(2s_{44} + s_{66})}, \tag{4}$$

$$B_R = \frac{1}{(2s_{11} + s_{33}) + 2(s_{12} + 2s_{13})}. \tag{5}$$

For the calculations of the Young's modulus (E), we have, in both cases, used the following relation [45]:

$$E = \frac{9BG}{3B + G}. \tag{6}$$

To obtain a crude estimate of whether the materials can be expected to be ductile or brittle, the empirical Pugh's criterion can be used, which suggests that, if $G/B < 0.5$, the material tends to be ductile [47]. In the case of the three studied material systems, none of them meets this criterion regardless if Reuss' or Voigt's approximation is used, which can be seen from Table 3. This implies that the materials are not expected to be ductile.

The elastic moduli of $ScTaN_2$, $ScNbN_2$ and $ScVN_2$ can be compared to the elastic modulus of the MAX phase Ti_2AlC that has a value of 277 GPa [48], which is a typical value for a MAX phase. As can be seen from Table 3, the three studied inverse MAX phases have elastic moduli in the interval from 432 to 479 GPa, which implies that the inverse MAX phases (at least in these three cases) are stiff materials and much stiffer than the regular MAX phases.

Figures 1–3 show the electronic density of states (DOS) calculated for all three phases with the GGA-PBE functional as well as with the HSE06 hybrid functional [38]. The latter functional was used because the GGA functionals tend to underestimate bandgaps compared to experimental values [49], while hybrid functionals generally give more accurate bandgaps. For all three cases, when using the GGA-PBE functional, it was not obvious if the apparent bandgaps between occupied and empty states are real bandgaps or just pseudogaps. Using hybrid functionals, in contrast, it is clear that all three material systems exhibit small bandgaps. Other than the bandgaps, the structures of the density of states are very similar for PBE and HSE06. The bandgaps presented here are based on the calculations with hybrid functionals. In Appendix A, the energy band structures of the three phases are shown.

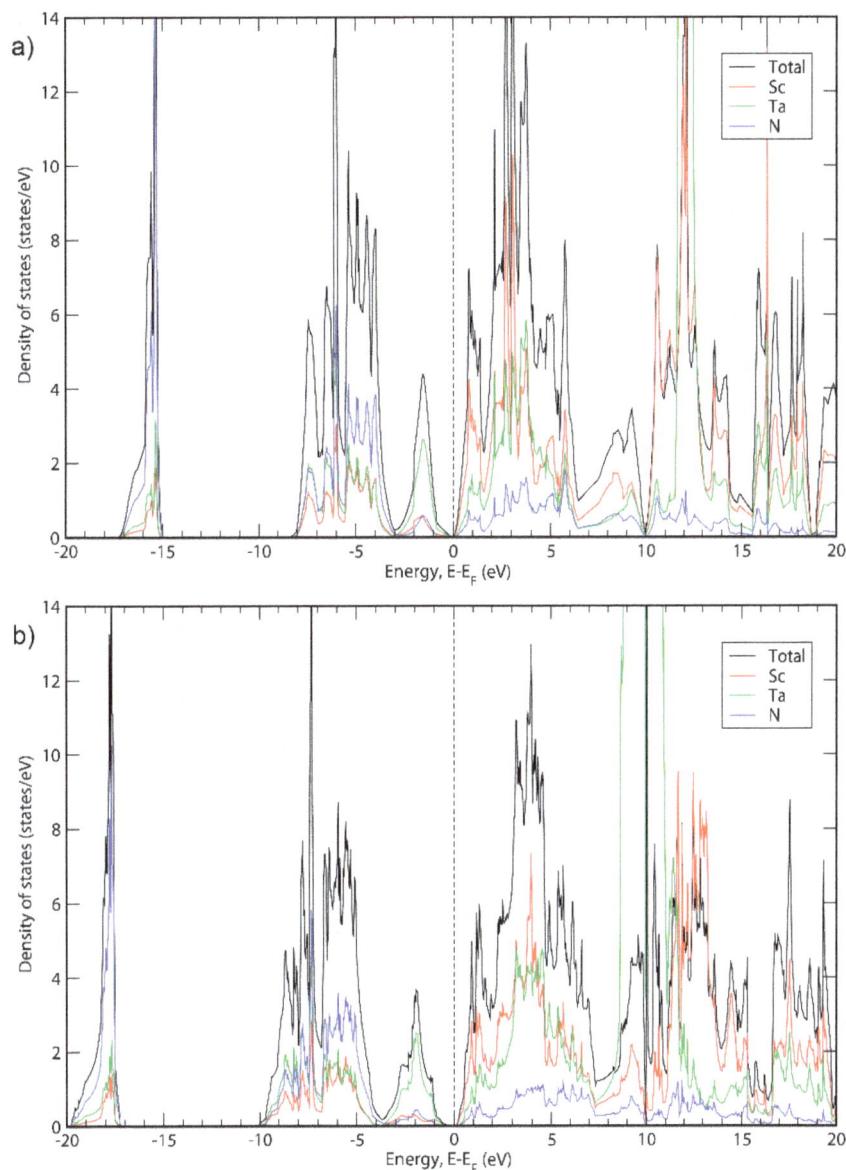

Figure 1. Total DOS and DOS projections for $ScTaN_2$ calculated with (**a**) GGA; (**b**) HSE06.

It can be seen in Figures 1–3 that the basic structure of the DOS is similar for all three material systems. The projected DOS indicated that the bands in the first peak of all three systems is mainly due to N. Furthermore, the main states around the top of the valence band are because of the presence of Ta, Nb or V for $ScTaN_2$, $ScNbN_2$ and $ScVN_2$, respectively, while the main states around the bottom of the conduction band are in all three cases due to a mix of states originating from Sc and the M metal. For higher-level energies in the conduction band (above 5 eV), the contribution to the DOS for all three systems is mainly from the Sc states. These results are consistent with $ScTaN_2$, where it has previously been shown that the first peak is mainly due to N (2p) states and the peak to the right of the chemical potential is due to Ta (5d) mixed with Sc (3d) states and the peak to the left of it is mainly due to Ta (5d) states [11].

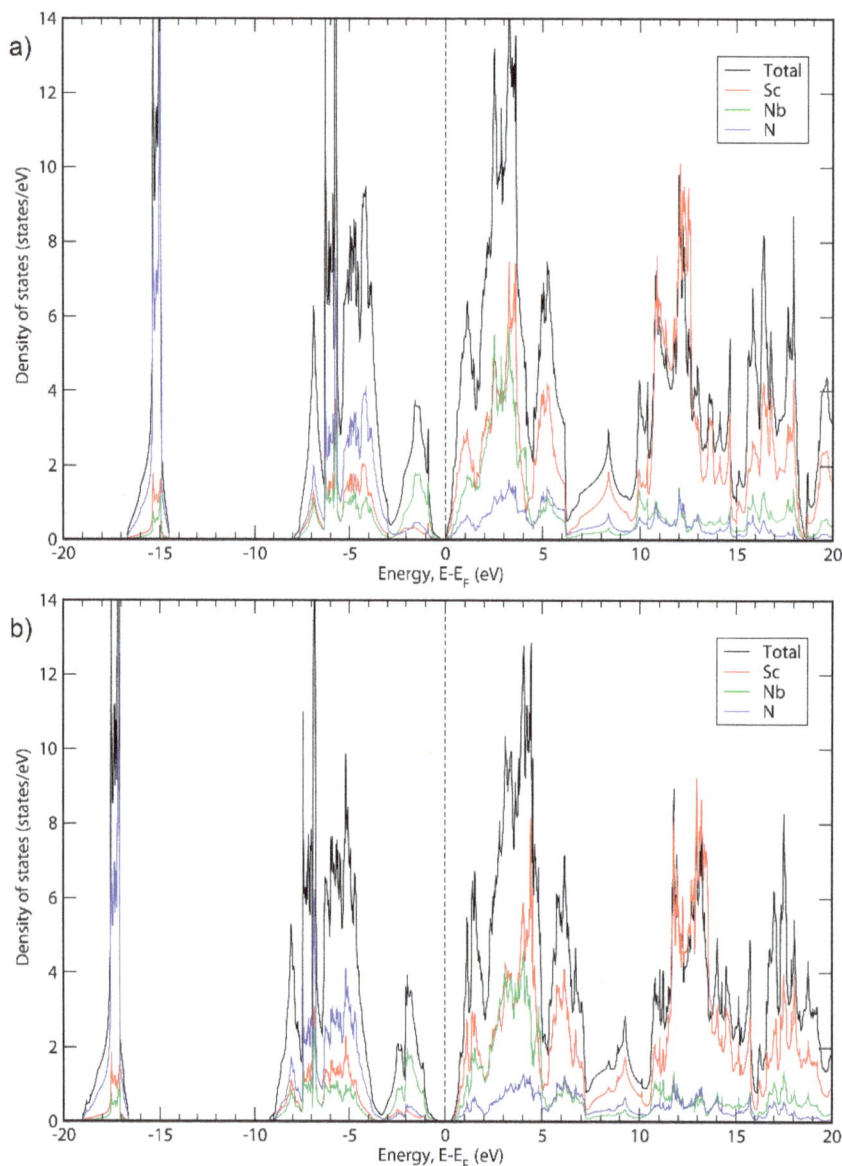

Figure 2. Total DOS and DOS projections for $ScNbN_2$ calculated with (**a**) GGA; (**b**) HSE06.

Figure 1 shows the total DOS of $ScTaN_2$ and the projected DOS of its elements calculated with the GGA-PBE functional (Figure 1a) and with the HSE06 functional (Figure 1b). The energy is adjusted so that the 0 eV corresponds to the highest occupied state. The bandgap calculated with HSE06 is 0.139 eV. Figure 2 shows the total DOS of $ScNbN_2$ and the projected DOS of its elements calculated with the GGA-PBE functional (Figure 2a) and with the HSE06 functional (Figure 2b), showing a bandgap of 0.350 eV (HSE06). Figure 3 shows the total DOS of $ScVN_2$ and the projected DOS of its elements calculated with the GGA-PBE functional (Figure 3a) and with the HSE06 functional (Figure 3b), showing a bandgap of 0.550 eV (HSE06). It can thus be concluded from here that $ScTaN_2$, $ScNbN_2$ and $ScVN_2$ are all narrow-bandgap semiconductors, where $ScTaN_2$ has the smallest bandgap and $ScVN_2$ has the largest. The formation of a semiconductor bandgap in these compounds can be explained by a splitting of the d shell of M^{3+} ion in the trigonal prismatic crystal field. The lowest level of z^2 symmetry is then completely filled in the case of d^2 occupancy while the excited crystal-field levels are empty. The smallest bandwidth of $3d$ band in V^{3+} then leads to the largest band gap in $ScVN_2$; the gap is progressively reduced with the d-band width increasing along the $V \rightarrow Nb \rightarrow Ta$ sequence.

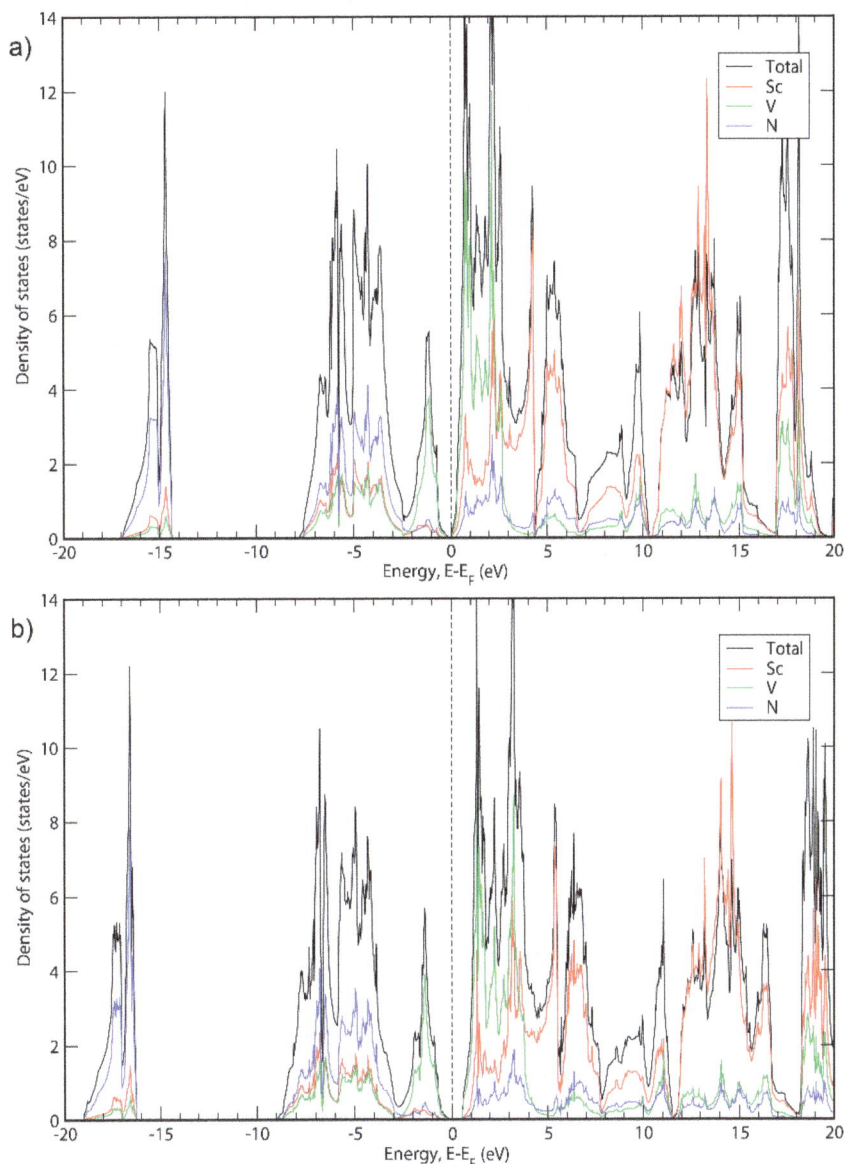

Figure 3. Total DOS and DOS projections for $ScVN_2$ calculated with (**a**) GGA; (**b**) HSE06.

Experimentally, temperature-dependent transport measurements on $ScTaN_2$ showed a metallic-like behavior by measurement of resistivity as a function of temperature, where it increased with temperature [11], but with high absolute values of resistivity. This is indicative of a highly doped degenerate semiconductor, typical also for ScN [13,14,17,19,50].

Previous calculations by Niewa et al. indicated a pseudogap in $ScTaN_2$, but with the local density approximation (LDA), which underestimates bandgaps just like GGA. This is consistent with the here determined narrow bandgaps in $ScMN_2$ (M = V, Nb, Ta), showing that all these materials are narrow-bandgap semiconductors. The bandgap is lower than the one in ScN (0.9 eV [51]). One can therefore expect quasimetallic behavior of the resistivity for $ScMN_2$ with impurities and/or dopants (or antisite defects as observed in Ref. [11]) analogously with the case of ScN [16,50]. Thus, the electronic properties and electrical conductivity of the $ScMN_2$ phases are at least qualitatively similar to those of ScN.

In order to obtain a qualitative estimate of the thermoelectric properties, we performed BoltzTraP calculations using the optimized lattice structures. The BoltzTraP code [40] implements the semiclassical Boltzmann theory of transport in the framework of the linearized augmented plane wave (LAPW) method Wien-2k [39]. Namely, by employing the approximation of a constant direction-independent relaxation time τ, the conductivity tensor can be written as

$$\sigma_{\alpha\beta}\left(i, \vec{k}\right) = e^2 \tau v_\alpha\left(i, \vec{k}\right) v_\beta\left(i, \vec{k}\right), \tag{7}$$

where $v_\alpha\left(i, \vec{k}\right)$ is the group velocity alone the direction $\alpha(= x, y, z)$ for the Kohn–Sham band i and the vector $\vec{k}$ of the Brillouin zone (BZ). The energy-projected conductivity tensor is then defined as

$$\sigma_{\alpha\beta}(\varepsilon) = \sum_{i,k} \sigma_{\alpha\beta}\left(i, \vec{k}\right) \frac{\delta\left(\varepsilon - \varepsilon_{i,k}\right)}{d\varepsilon}, \tag{8}$$

where $\varepsilon_{i,k}$ is the Kohn–Sham band energy for the band i and the BZ point $\vec{k}$, δ is the Dirac delta function. The transport tensors for given temperature T and chemical potential μ are moments of the energy-projected conductivity tensors multiplied by the energy derivative of Fermi function $f'_{\mu,T}(\varepsilon)$:

$$\sigma_{\alpha\beta}(T, \mu) = -\frac{1}{\Omega} \int \sigma_{\alpha\beta}(\varepsilon) f'_{\mu,T}(\varepsilon) d\varepsilon, \tag{9}$$

$$v_{\alpha\beta}(T, \mu) = -\frac{1}{\Omega} \int \sigma_{\alpha\beta}(\varepsilon)(\varepsilon - \mu) f'_{\mu,T}(\varepsilon) d\varepsilon, \tag{10}$$

where $\sigma_{\alpha\beta}(T, \mu)$ is the electric conductivity, and Ω is the unit cell volume. The Seebeck coefficient tensor is given by

$$S_{\alpha\beta} = \sum_{\gamma} \left(\sigma^{-1}\right)_{\gamma\alpha} v_{\gamma\beta}. \tag{11}$$

One immediately sees that, under the assumption of an energy and direction-independent relaxation time, the Seebeck coefficient is independent of τ and one may obtain its actual value by Equation (11). This approximation of energy and direction-independent relaxation time is employed by the BoltzTraP code [40]. The electrical conductivity is then proportional to τ. The estimation of relaxation time is not an objective of the present work; we thus calculated the ratio σ/τ in order to determine the impact of band-structure evolution along the ScMN$_2$ series on the electrical conductivity.

To capture the nonmetallic behavior of ScMN$_2$ compounds in these calculations we employed the modified Becke–Johnson (mBJ) exchange-correlation potential [52], which is known to accurately reproduce semiconducting band gaps with a modest computational cost. Our DFT-mBJ calculations predict ScVN$_2$ and ScNbN$_2$ to be semiconducting with narrow band gaps of 0.135 eV and 0.046 eV, respectively. ScTaN$_2$ is found to be metallic. The systematically smaller band gaps are thus predicted with the mBJ as compared to those obtained using the HSE06 hybrid functional. ScMN$_2$ are apparently located at a metallic-semiconductor threshold with the value of the bandgap being very sensitive to the choice of exchange-correlation potential. However, the qualitative evolution towards a more metallic behavior upon the isoelectronic substitution V → Nb → Ta is reproduced by both approaches. Therefore, our calculations can be expected to capture a valid qualitative picture of the evolution of thermoelectric properties upon this substation, despite some uncertainty regarding the value of the bandgap.

Figure 4a shows the Seebeck tensor components S_{xx} and S_{zz} at 290 K as a function of chemical potential μ for ScTaN$_2$, ScNbN$_2$ and ScVN$_2$. The dependence vs. μ has been computed in the rigid band approximation using the band structure obtained for the undoped case by self-consistent calculations as described in the previous paragraph. The curves indicate that all three material systems are anisotropic. For instance, at the chemical potential 0, S_{xx} for all three materials is positive, while S_{zz} is negative. This behavior—with an opposite sign of the Seebeck coefficient along the x-axis and z-axis—is analogous to that observed for the (metallic) Ti_3SiC_2 and Ti_3GeC_2 MAX phases, where the macroscopic Seebeck coefficient in randomly oriented bulk samples sum up to zero over a wide temperature range [3,53–55]. The opposite signs of S_{xx} and S_{zz} in the undoped materials can be understood by noticing that the sign of Seebeck coefficient for a given direction α is determined by the shape of $\sigma_{\alpha\alpha}(\varepsilon)$ in the vicinity of μ—see Equations (10) and (11). Taking into account the fact that $f'_{\mu,T}(\varepsilon) < 0$, one sees that a positive (negative) Seebeck coefficient originates from a positive (negative) first moment of $\sigma_{\alpha\alpha}(\varepsilon)$ for $|\varepsilon - \mu|$ being of the order of temperature. As is noted in our analysis of Figure 3, the z^2 orbital forms the top of valence band, while other d orbitals mainly contribute to the empty d states just at the bottom of conduction one (see Figure 3 and the corresponding discussion of DOS). The z^2 orbital should mostly contribute to the zz element of conductivity tensor (7), while the rest of d orbitals contributing more to the conductivity in the xy plane. Therefore, from Equations (7) and (8), one expects $\sigma_{xx}(\varepsilon)$ to be larger at the bottom of the conduction band than at the top of valence one; the opposite behavior is expected for $\sigma_{zz}(\varepsilon)$. Such behaviors are indeed observed in our calculated $\sigma_{xx}(\varepsilon)$ and $\sigma_{zz}(\varepsilon)$, resulting in positive $v_{xx}(T,0)$ and negative $v_{zz}(T,0)$, respectively, and thus in $S_{xx} > 0$ and $S_{zz} < 0$.

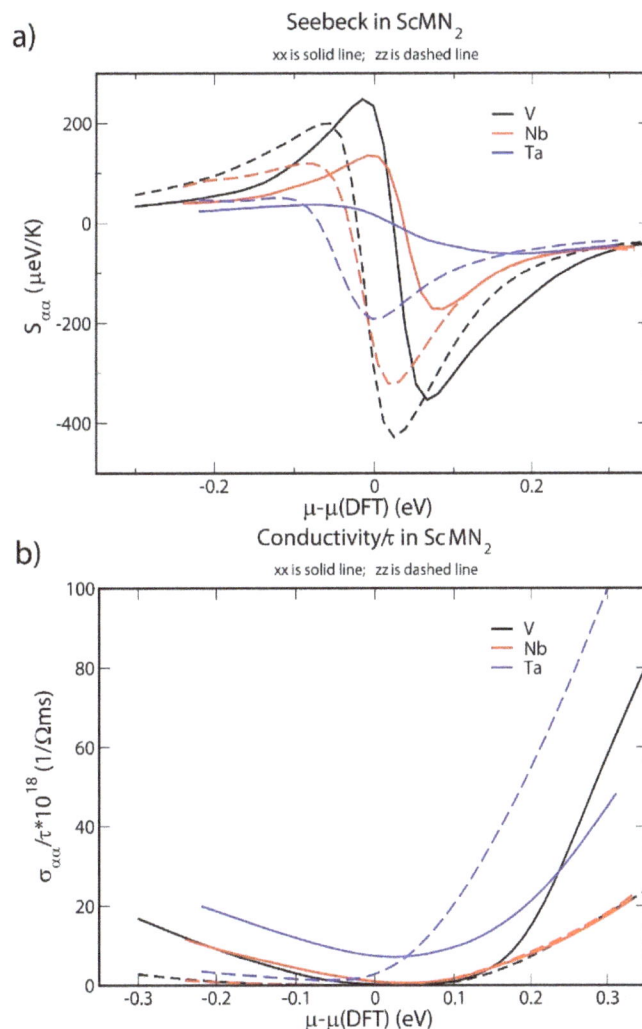

Figure 4. (a) Seebeck tensor components S_{xx} and S_{zz} as a function of chemical potential for $ScMN_2$, where M is Ta, Nb or V; (b) σ_{xx}/τ and σ_{zz}/τ as a function of chemical potential for $ScMN_2$, where M is Ta, Nb or V.

The maximum absolute values of the Seebeck components are high (in particular, more than 400 µV/K in ScVN$_2$), suggesting that all three materials can be candidates for thermoelectric applications. The doping level (as indicated by the chemical potential) is predicted to change the sign of Seebeck coefficients, with a particular large sensitivity to the doping is again found in ScVN$_2$. This indicates that both n- and p-type thermoelectric behavior could be obtained by an appropriate choice of dopants. Again, this is in analogy with ScN, which is n-type, but can be rendered p-type by sufficient doping with e.g., Mg [16,20,56].

Figure 4b shows the ratio of the conductivity-tensor components σ_{xx} and σ_{zz} to the relaxation time τ at 290 K as a function of chemical potential. The conductivity of undoped ScVN$_2$ is low, as expected for a semiconductor. For ScNbN$_2$, the predicted value of the gap is roughly of the same magnitude as the room-temperature energy, resulting in a noticeable thermally activated σ_{xx} but not σ_{zz}. Strong anisotropy can thus be noted also in this context. Furthermore, the electrical conductivity components, σ_{xx} and σ_{zz}, increase at chemical potential levels above 0 for all three material systems. That is, the electrical conductivity of the systems can also be increased by doping, and a tradeoff optimizing the power factor is possible.

4. Conclusions

We have used DFT calculations to investigate the mixing enthalpies, elastic properties, DOS, lattice parameters and thermoelectric properties of ScTaN$_2$, ScNbN$_2$ and ScVN$_2$. The evaluated mixing enthalpy of formation and of elastic properties indicate that all three systems are thermodynamically as well as elastically stable.

The DOS calculations show that all three systems have small band gaps suggesting that they are narrow-bandgap semiconductors. The band gaps are smaller than, for instance, the band gap of ScN and a quasimetallic behavior can be expected. Finally, the three systems are found to have anisotropic thermoelectric properties and the results indicate that their thermoelectric properties can be tuned by doping.

Author Contributions: Conceptualization, P.E.; Supervision, P.E. and S.S., Software and computational design, R.P., I.M., and L.P.; Formal Analysis, R.P. and L.P.; Writing-Original Draft Preparation, R.P.; Writing-Review & Editing, P.E., L.P. and R.P.; Visualization, R.P. and L.P.; Project Administration, R.P.; Funding Acquisition, P.E., S.S. and L.P.

Funding: This research was funded by the Swedish Research Council (VR) through project Grant No. 2016-03365, the Knut and Alice Wallenberg Foundation through the Academy Fellows Program, the Swedish Government Strategic Research Area in Materials Science on Functional Materials at Linköping University (Faculty Grant SFO-Mat-Liu No. 2009 00971) and the European Research Council through Grant No. ERC-319286-QMAC.

Acknowledgments: The calculations were performed using computer resources provided by the Swedish National Infrastructure for Computing (SNIC) at the National Supercomputer Centre (NSC). We would also like to thank Weine Olovsson at the NSC for giving practical guidance about the calculations.

Conflicts of Interest: The authors declare no conflict of interest. The funders had no role in the design of the study; in the collection, analyses, or interpretation of data; in the writing of the manuscript, and in the decision to publish the results.

Appendix A

Figures A1–A3 show the energy band structure of ScTaN$_2$, ScNbN$_2$ and ScVN$_2$ calculated with PBE-GGA. It can be noted that one or two bands in each graph intersect the energy level 0 at the Γ-point, corresponding to a pseudogap as described in the main text. In Figure A1, the M-point also intersects the energy level at zero. However, the GGA-PBE functional underestimates the bandgap. The difference, if the bands were to be calculated with a hybrid functional, is that the bands would be displaced from the zero level, opening a bandgap (with values as stated in the main text) at the Γ-point.

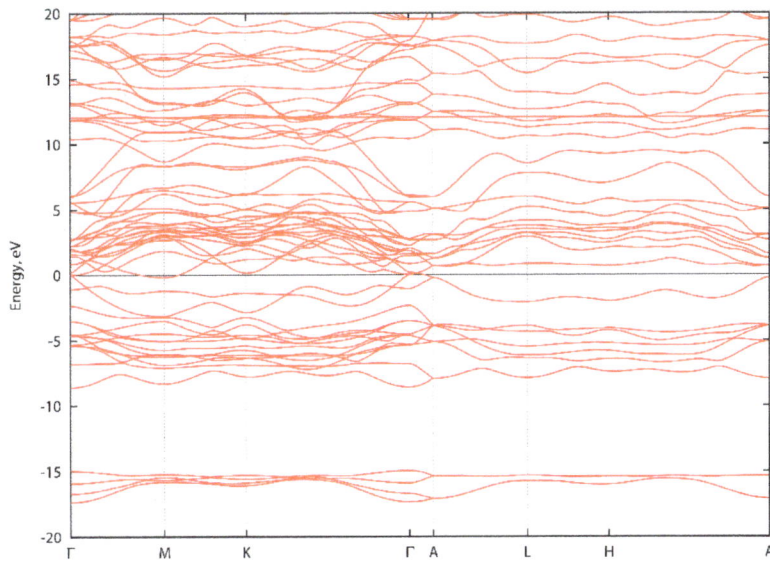

Figure A1. Energy band structure of ScTaN$_2$ along high symmetry directions.

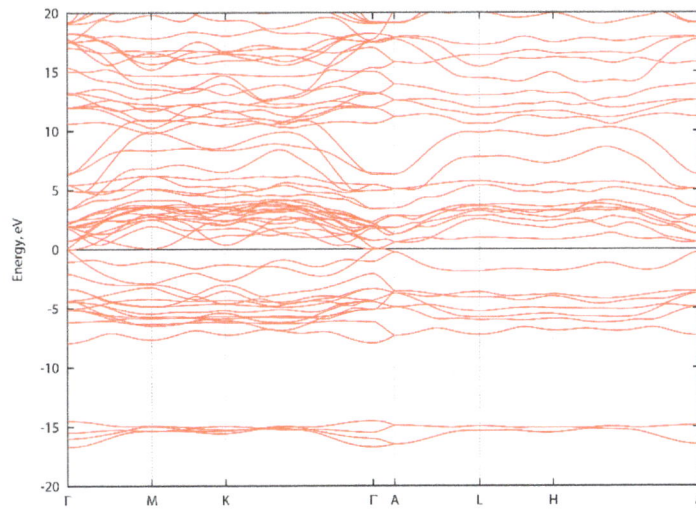

Figure A2. Energy band structure of ScNbN$_2$ along high symmetry directions.

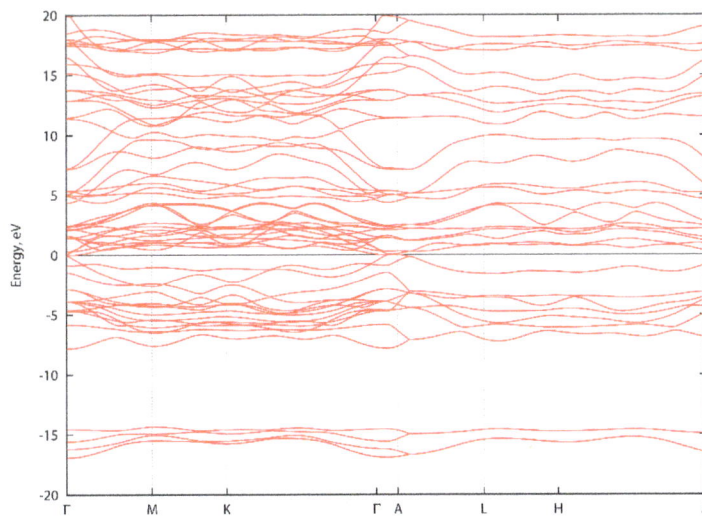

Figure A3. Energy band structure of ScVN$_2$ along high symmetry directions.

References

1. Eklund, P.; Beckers, M.; Jansson, U.; Högberg, H.; Hultman, L. The $M_{n+1}AX_n$ phases: Materials science and thin-film processing. *Thin Solid Films* **2010**, *518*, 1851–1878. [CrossRef]

2. Eklund, P.; Rosén, J.; Persson, P.O.A. Layered ternary $M_{n+1}AX_n$ phases and their 2D derivative MXene: An overview from a thin-film perspective. *J. Phys. D* **2017**, *50*, 113001. [CrossRef]

3. Magnuson, M.; Mettesini, M. Chemical Bonding and Electronic-Structure in MAX phases as Viewed by X-ray Spectroscopy and Density Functional Theory. *Thin Solid Films* **2017**, *621*, 108–130. [CrossRef]

4. Radovic, M.; Barsoum, M.W. MAX phases: Bridging the gap between metals and ceramics. *Am. Ceram. Soc. Bull.* **2013**, *92*, 20–27.

5. Sun, Z.M. Progress in research and development on MAX phases: a family of layered ternary compounds. *Int. Mater. Rev.* **2013**, *56*, 143–166. [CrossRef]

6. Wang, X.H.; Zhou, Y.C. Layered Materials and Electrically Conductive $T_{i2}AlC$ and Ti_3AlC_2 Ceramics: A Review. *J. Mater. Sci. Technol.* **2010**, *26*, 385–416. [CrossRef]

7. Barsoum, M.W. The $M_{N+1}AX_N$ Phases: A New Class of Solids; Thermodynamically Stable Nanolaminates. *Prog. Solid State* **2000**, *28*, 201–281. [CrossRef]

8. Barsoum, M.W. The MAX Phases: Unique New Carbide and Nitride Materials. *Am. Sci.* **2001**, *89*, 334–343. [CrossRef]

9. Lengauer, W. The Crystal Structure of $ScNbN_{1-x}$ and Comparisons with Related Nitride and Carbide Structures. *J. Solid State Chem.* **1989**, *82*, 186–191. [CrossRef]

10. Lengauer, W.; Ettmayer, P. THE CRYSTAL STRUCTURE OF $ScTaN_{1-x}$. *J. Less Common Met.* **1988**, *141*, 157–162. [CrossRef]

11. Niewa, R.; Zherebtsov, D.A.; Schnelle, W.; Wagner, F.R. Metal-Metal Bonding in $ScTaN_2$. A New Compound in the System ScN-TaN. *Inorg. Chem.* **2004**, *43*, 6188–6194. [PubMed]

12. Kerdsongpanya, S.; Alling, B.; Eklund, P. Phase stability of ScN-based solid solutions for thermoelectric applications from first-principles calculations. *J. Appl. Phys.* **2013**, *114*, 073512. [CrossRef]

13. Burmistrova, P.V.; Maassen, J.; Favaloro, T.; Saha, B.; Salamat, S.; Rui Koh, Y.; Lundstrom, M.S.; Shakouri, A.; Sands, T.D. Thermoelectric properties of epitaxial ScN films deposited by reactive magnetron sputtering onto MgO(001) substrates. *Appl. Phys.* **2013**, *113*, 153704. [CrossRef]

14. Burmistrova, P.V.; Zakharov, D.N.; Favaloro, T.; Mohammed, A.; Stach, E.A.; Shakouri, A.; Sands, T.D. Effect of deposition pressure on the microstructure and thermoelectric properties of epitaxial ScN(001) thin films sputtered onto Mgo(001) substrates. *J. Mater. Res.* **2015**, *30*, 626–634. [CrossRef]

15. Gall, D.; Petrov, I.; Hellgren, N.; Hultman, L.; Sundgren, J.E.; Greene, J.E. Growth of poly- and single-crystal ScN on MgO(001): Role of low-energy N^+_2 irradiation in determining texture, microstructure evolution, and mechanical properties. *J. Appl. Phys.* **1998**, *84*, 6034–6041. [CrossRef]

16. Kerdsongpanya, S.; Alling, B.; Eklund, P. Effect of point defects on the electronic density of states of ScN studied by first-principles calculations and implications for thermoelectric properties. *Phys. Rev. B* **2012**, *86*, 195140. [CrossRef]

17. Le Febvrier, A.; Tureson, N.; Stilkerich, N.; Greczynski, G.; Eklund, P. Effect of impurities on morphology, growth mode, and thermoelectric properties of (111) and (001) epitaxial-like ScN films. *J. Phys. D Appl. Phys.* **2019**, *52*, 035302. [CrossRef]

18. Kerdsongpanya, S.; Hellman, O.; Sun, B.; Koh, Y.K.; Lu, J.; Van Nong, N.; Simak, S.I.; Alling, B.; Eklund, P. Phonon thermal conductivity of scandium nitride for thermoelectrics from first-principles calculations and thin-film growth. *Phys. Rev. B* **2017**, *96*, 195417. [CrossRef]

19. Kerdsongpanya, S.; Sun, B.; Eriksson, F.; Jensen, J.; Lu, J.; Koh, Y.K.; Nong, N.V.; Balke, B.; Alling, B.; Eklund, P. Experimental and theoretical investigation of $Cr_{1-x}Sc_xN$ solid solutions for thermoelectrics. *J. Appl. Phys.* **2016**, *120*, 215103. [CrossRef]

20. Saha, B.; Garbrecht, M.; Perez-Taborda, J.A.; Fawey, M.H.; Koh, Y.R.; Shakouri, A.; Martin-Gonzalez, M.; Hultman, L.; Sands, T.D. Compensation of native donor doping in ScN: Carrier concentration control and p-type ScN. *Appl. Phys. Lett.* **2017**, *110*, 252104. [CrossRef]

21. Tureson, N.; Van Nong, N.; Fournier, D.; Singh, N.; Acharya, S.; Schmidt, S.; Belliard, L.; Soni, A.; Eklund, P. Reduction of the thermal conductivity of the thermoelectric material ScN by Nb alloying. *J. Appl. Phys.* **2017**, *122*, 025116. [CrossRef]

22. Rawat, V.; Koh, Y.K.; Cahill, D.G.; Sands, T.D. Thermal conductivity of (Zr,W)N/ScN metal/semiconductor multilayers and superlattices. *J. Appl. Phys.* **2009**, *105*, 024909. [CrossRef]

23. Saha, B.; Koh, Y.R.; Comparan, J.; Sadasivam, S.; Schroeder, J.L.; Garbrecht, M.; Mohammed, A.; Birch, J.; Fisher, T.; Shakouri, A.; et al. Cross-plane thermal conductivity of (Ti;W)N/(Al,Sc)N metal. *Phys. Rev. B* **2016**, *93*, 045311. [CrossRef]

24. Schroeder, J.L.; Saha, B.; Garbrecht, M.; Schell, N.; Sands, T.D.; Birch, J. Thermal stability of epitaxial cubic-TiN/(Al,Sc)N metal/semiconductor superlattices. *J. Mater. Sci.* **2015**, *50*, 3200. [CrossRef]

25. Zebarjadi, M.; Bian, Z.; Singh, R.; Shakouri, A.; Wortman, R.; Rawat, V.; Sands, T. Thermoelectric Transport in a ZrN/ScN Superlattice. *J. Elctron Mater.* **2009**, *38*, 960–963. [CrossRef]

26. Gharavi, M.A.; Kerdsongpanya, S.; Schmidt, S.; Eriksson, F.; Nong, N.V.; Lu, J.; Balke, B.; Fournier, D.; Belliard, L.; le Febvrier, A.; et al. Microstructure and thermoelectric properties of CrN and CrN/Cr$_2$N thin films. *J. Phys. D* **2018**, *51*, 355302. [CrossRef]

27. Ravi, C. First-principles study of ground-state properties and phase stability of vanadium nitrides. *CALPHAD* **2009**, *33*, 469–477. [CrossRef]

28. Grumski, M.; Dholabhai, P.P.; Adams, J.B. Ab initio study of the stable phases of 1:1 tantalum nitride. *Acta Mater.* **2013**, *61*, 3799–3807. [CrossRef]

29. Mei, A.B.; Hellman, O.; Wireklint, N.; Schlepuetz, C.M.; Sangiovanmi, D.; Alling, B.; Rockett, A.; Hultman, L.; Petrov, I.; Greene, J.E. Dynamic and structural stability of cubic vanadium nitride. *Phys. Rev. B* **2015**, *91*, 054101. [CrossRef]

30. Kresse, G.; Furthmüller, J. Efficient iterative schemes for *ab initio* total-energy calculations using a plane-wave basis set. *Phys. Rev. B Condens. Matter Mater. Phys.* **1994**, *54*, 11169.

31. Kresse, G.; Furthmüller, J. Efficiency of ab-initio total energy calculations for metals and semiconductors using a plane wave basis set. *Comput. Mater. Sci.* **1996**, *6*, 15–50. [CrossRef]

32. Kresse, G.; Hafner, J. *Ab initio* molecular dynamics for liquid metals. *Phys. Rev. B Condens. Matter Mater. Phys.* **1993**, *47*, 558–561. [CrossRef]

33. Kresse, G.; Hafner, J. *Ab initio* molecular-dynamis simulation of the liquid-metal-amorphous-semiconductor transition in germaniium. *Phys. Rev. B Condens. Matter Mater. Phys.* **1994**, *49*, 14251–14269. [CrossRef]

34. Blöchl, P.E. Projector augmented-wave method. *Phys. Rev. B Condens. Matter Mater. Phys.* **1994**, *50*, 17953–17979. [CrossRef]

35. Perdew, J.P.; Burke, F.; Ernzerhof, M. Generalized Gradient Approxiamtion Made Simple. *Phys. Rev. Lett.* **1996**, *77*, 3865–3868. [CrossRef]

36. Le Page, Y.; Saxe, P. Symmetry-general least-squares extraction of elastic data for strained materials from *ab initio* calculations of stress. *Phys. Rev. B* **2002**, *65*, 104104. [CrossRef]

37. Blöchl, P.E.; Jepsen, O.; Andersen, O.K. Improved tetrahedron method for Brillouin-zone integrations. *Phys. Rev. B* **1994**, *49*, 16223. [CrossRef]

38. Heyd, J.; Scuseria, G.E.; Ernzerhof, M. Hybrid functionals based on a screened Coulomb potential. *J. Chem. Phys.* **2003**, *118*, 8207–8215. [CrossRef]

39. Blaha, P.; Schwarz, K.; Madsen, G.; Kvasnicka, D.; Luitz, J. *WIEN2k, An augmented Plane Wave + Local Orbitals for Calculating Crystal Properteis*; Techn. Universität Wien: Vienna, Austria, 2001.

40. Madsen, G.K.H.; Singh, D.J. BoltzTraP. A code for calculating band-structure dependent quantities. *Comput. Phys. Commun.* **2006**, *175*, 67–71. [CrossRef]

41. Thore, A.; Dahlqvist, M.; Alling, B.; Rosén, J. Temperature dependent phase stability of nanolaminated ternaries from first-principles calculations. *Comput. Mater. Sci.* **2014**, *91*, 251–257. [CrossRef]

42. Mouhat, F.; Coudert, F.-X. Necessary and sufficient elastic stability conditions in various crystal systems. *Phys. Rev. B* **2014**, *90*, 224104. [CrossRef]

43. Haas, P.; Tran, F.; Blaha, P. Calculation of the lattice constant of solids with semilocal functionals. *Phys. Rev. B* **2009**, *79*, 085109. [CrossRef]

44. Schimka, L.; Harl, J.; Kresse, G. Improved hybrid functional for solids: the HSEsol functional. *J. Chem. Phys.* **2011**, *134*, 024116. [CrossRef]

45. Hill, R. The Elastic Behaviour of a Crystalline Aggregate. *Proc. Phys. Soc. A* **1952**, *65*, 349–354. [CrossRef]

46. Reuss, A. Berechnung der Fließgrenze von Mischkristrallen auf der Grund der Plastizitätsbedingung für Einkristalle. *Z. Angew. Math. Mech.* **1929**, *9*, 55. [CrossRef]

47. Pugh, S.F. Relations between the elastic moduli and the plastic properties of polycrystalline pure metals. *Philos. Mag. Ser.* **1954**, *45*, 823–843. [CrossRef]

48. Barsoum, M.W. Mechanical properties of the MAX Phases. *Annu. Rev. Mater. Res.* **2011**, *41*, 195–227. [CrossRef]

49. Bagayoko, D. Understanding density functional theory (DFT) and completing it in practice. *AIP Adv.* **2014**, *4*, 127104. [CrossRef]

50. Kerdsongpanya, S.; Nong, N.V.; Pryds, N.; Žukauskaitė, A.; Jensen, J.; Birch, J.; Lu, J.; Hultman, L.; Wingqvist, G.; Eklund, P. Anomalously high thermoelectric power factor in epitaxial ScN thin films. *Appl. Phys. Lett.* **2011**, *99*, 132113. [CrossRef]

51. Lambrecht, W.R.L. Electronic structure and optical spectra of the semimetal ScAs and of the indirect-band-gap semiconductors ScN and GdN. *Phys. Rev. B* **2000**, *62*, 13538. [CrossRef]

52. Tran, F.; Blaha, P. Accurate Band Gaps of Semiconductors and Insulators with a Semilocal Exchange-Correlation Potential. *Phys. Rev. Lett.* **2009**, *102*, 226401. [CrossRef]

53. Chaput, L.; Hug, G.; Pécher, P.; Scherrer, H. Thermopower of the 312 MAX phases $Ti_3 Si C_2$, $Ti3 Ge C_2$ and $Ti_3 Al C_2$. *Phys. Rev. B* **2007**, *75*, 035107. [CrossRef]

54. Chaput, L.; Hug, G.; Pécher, P.; Scherrer, H. Anisotropy and thermopower in Ti_3SiC_2. *Phys. Rev. B* **2005**, *71*, 121104. [CrossRef]

55. Yoo, H.-I.; Barsoum, M.W.; El-Raghy, T. Ti_3SiC_2 has negligible thermopower. *Nature* **2000**, *407*, 581–582. [CrossRef]

56. Saha, B.; Perez-Taborda, J.A.; Bahk, J.-H.; Koh, Y.R.; Shakouri, A.; Martin-Gonzalez, M.; Sands, T.D. Temperature-dependent thermal and thermoelectric properties of n-type and p-type $Sc_{1-x}Mg_xN$. *Phys. Rev. B* **2018**, *97*, 085301. [CrossRef]

11

Electronic Structure of Boron Flat Holeless Sheet

Levan Chkhartishvili [1,2,3,*], **Ivane Murusidze** [4] **and Rick Becker** [3]

1   Engineering Physics Department, Georgian Technical University, 77 Kostava Ave., Tbilisi 0175, Georgia
2   Boron and Composite Materials Laboratory, Ferdinand Tavadze Institute of Metallurgy and Materials Science, 10 Mindeli Str., Tbilisi 0186, Georgia
3   Boron Metamaterials, Cluster Sciences Research Institute, 39 Topsfield Rd., Ipswich, MA 01938, USA; rbecker@clustersciences.com
4   Institute of Applied Physics, Ilia State University, 3/5 Cholokashvili Ave., Tbilisi 0162, Georgia; miv@iliauni.edu.ge
*   Correspondence: levanchkhartishvili@gtu.ge or chkharti2003@yahoo.com

Abstract: The electronic band structure, namely energy band surfaces and densities-of-states (DoS), of a hypothetical flat and ideally perfect, i.e., without any type of holes, boron sheet with a triangular network is calculated within a quasi-classical approach. It is shown to have metallic properties as is expected for most of the possible structural modifications of boron sheets. The Fermi curve of the boron flat sheet is found to be consisted of 6 parts of 3 closed curves, which can be approximated by ellipses representing the quadric energy-dispersion of the conduction electrons. The effective mass of electrons at the Fermi level in a boron flat sheet is found to be too small compared with the free electron mass m_0 and to be highly anisotropic. Its values distinctly differ in directions Γ–K and Γ–M: $m_{\Gamma-K}/m_0 \approx 0.480$ and $m_{\Gamma-M}/m_0 \approx 0.052$, respectively. The low effective mass of conduction electrons, $m_\sigma/m_0 \approx 0.094$, indicates their high mobility and, hence, high conductivity of the boron sheet. The effects of buckling/puckering and the presence of hexagonal or other type of holes expected in real boron sheets can be considered as perturbations of the obtained electronic structure and theoretically taken into account as effects of higher order.

Keywords: boron sheet; electronic structure; density-of-states (DoS); Fermi curve; effective mass

1. Introduction

Borophene—a one-atom-thick sheet of boron—is one of the most promising nanomaterials [1]. Most all-boron quasi-planar clusters and nanosurfaces, including nanosheets, constructed of triangular atomic rings, which Ihsan Boustani et al. predicted theoretically in 1994 and the following years (quasi-planar clusters [2], nanotubes [2–5], nanotube-to-sheet transitions [6], and sheets [5,7]), are now experimentally confirmed (quasi-planar clusters [8–10], nanotubes [11–13], nanotube-to-sheet transitions [14,15], and sheets [16,17]).

1.1. Why should Boron Sheets be Formed?

There are different reasons in favor of the formation of stable 2-D all-boron structures. They can be divided into several groups. Let's consider them separately.

1.1.1. 3-D All-Boron Structures

Boron, the fifth element of the Periodic Table, is located at the intersection of semiconductors and metals. Due to a small covalent radius (only 0.84 Å) and number (only 3) of valence electrons, boron does not form simple three-dimensional structures but crystals with icosahedral clusters with many atoms in the unit cell. At least three all-boron allotropes are known—α- and β-rhombohedral and high-pressure γ-orthorhombic phases—for the experimental phase diagram of boron see Reference [18]. In addition, α- and β-tetragonal and a number of other boron structures, probably stabilized by the presence of impurities/defects, were reported. Theoretical studies of the five boron crystal structures (α, dhcp, sc, fcc, and bcc) were carried out using the LAPW (linearized plane wave) method in Reference [19]. The current state of research on the phase diagram of boron from a theoretical point of view is given in Reference [20]. It should be noted that, in the last decade, several new structures of boron allotropes were discovered and some have been disproved. Currently, even the number of allotropes of boron is uncertain. The reason for this is that there are many such structures, all of them complex, and some of them are minimally different from others. A pseudo-cubic tetragonal boron recently discovered under high-pressure and high-temperature conditions may also be another form of boron allotropes; however, its structure, studied in Reference [21] using a DFT (density functional theory) calculation, is abnormal compared to other allotropes of boron in many ways.

The almost regular icosahedron B_{12} with B-atoms at the vertices (Figure 1) serves as the main structural motif not only of boron allotropes but also of all known boron-rich compounds. In the boron icosahedron, each atom is surrounded by 5 neighboring atoms and, as usual, with one more atom from the rest of the crystal. For this reason, the average coordination number of a boron-rich lattice ranges from 5 to 5+1=6.

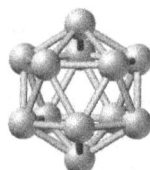

Figure 1. A regular icosahedron B_{12} with B-atoms at the vertices.

However, an isolated regular boron icosahedron is an electron-deficient structure—the total number of valence electrons of 12 boron atoms is not sufficient to fill all the covalent bonding orbitals corresponding to such a cage-molecule. Thus, if it were a stable structure, then intra-icosahedral bonds would be only partially covalent but also to some extent metallic. As for boron icosahedra constituting real crystals, it was clearly demonstrated, for example, for β-rhombohedral boron [22–27], that they are stabilized by the presence of point structural defects—vacancies and interstitials, in other words, both partially filled regular or irregular boron sites—at very high concentrations. For example, in the case of β-rhombohedral boron, the total effect of such a stabilization is to increase the average number of boron atoms inside the unit cell from the ideal value 105 (Figure 2) to 106.7 [28], which leads to the saturation of the electron-deficient orbitals and a 5- or 6-coordination number for the majority of constituent boron atoms.

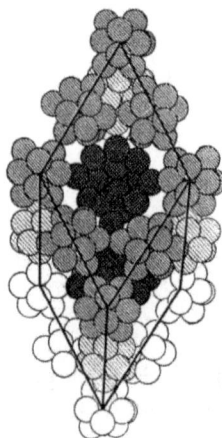

Figure 2. An idealized unit cell of a β-rhombohedral boron crystal.

Thus, all-boron 5- and 6-coordinated regular 3-D lattices cannot exist, but one can naturally imagine 2-D flat or buckled/puckered structures with a triangular arrangement of atoms with and without periodically spaced hexagonal (rarely quadric, pentagonal, or heptagonal) holes. Obviously, most of them are expected to be (semi)metallic. At the moment, a number of different atomic geometries for quasi-planar boron sheets are theoretically proposed [4–7,29–51].

Using the ab initio evolutionary structure prediction approach, a novel reconstruction of the α-boron (111) surface with the lowest energy was discovered [52]. In this reconstruction, all single interstitial boron atoms bridge neighboring icosahedra by polar covalent bonds, and this satisfies the electron counting rule, leading to the reconstruction-induced semiconductor–metal transition. The new stable boron sheet, called H-borophene, proposed in Reference [53] and constructed by tiling 7-membered rings side by side, should be especially noted.

As for the irregularly distributed holes, they have to be considered as defects. The research [54] is focused on the formation of local vacancy defects and pinholes in a 2-D boron structure—the so-called γ_3-type boron monolayer.

1.1.2. Boron Quasi-Planar Clusters

Indirectly, the reality of boron sheets can be proved by the presence of various quasi-planar boron clusters, i.e., finite fragments of sheets, in gaseous state and also boron nanotubes, which are the fragments of boron sheets wrapped into cylinders (see for example, the review from Reference [55] and the references therein). Experimental and theoretical evidences that small boron clusters prefer planar structures were reported in Reference [8].

In addition, recently, a highly stable quasi-planar boron cluster B_{36} of hexagonal shape with a central hexagonal hole [9], which is viewed as a potential basis for an extended 2-D boron sheet, and boron fullerene B_{40} [10], which can be imagined as the fragment of a boron sheet wrapped into the sphere, were discovered experimentally. Photoelectron spectroscopy in combination with ab initio calculations have been carried out to probe the structure and chemical bonding of the $B_{27}{}^-$ cluster [56]. A comparison between the experimental spectrum and the theoretical results reveals a 2-D global minimum with a triangular lattice containing a tetragonal defect and two low-lying 2-D isomers, each with a hexagonal vacancy.

1.1.3. Liquid Boron Structure

There are also evidences [57–59] that liquid boron does not consist of icosahedra but mainly of quasi-planar clusters. Ab initio MD (molecular-dynamics) simulations of the liquid boron structure yields that at short length scales, B_{12} icosahedra, a main structural motif of boron crystals and boron-rich solid compounds, are destroyed upon melting. Although atoms form an open packing, they maintain the 6-coordination.

According to measurements of the structure factor and the pair distribution function, the melting process is associated with relatively small changes in both the volume and the short-range order of the system. Results of a comprehensive study of liquid boron with X-ray measurements of the atomic structure and dynamics coupled with ab initio MD simulations also show that there is no evidence of survival of the icosahedral arrangements into the liquid, but many atoms appear to adopt a geometry corresponding to the quasi-planar pyramids.

1.1.4. Growing of Boron Sheets

Currently, some of the 2-D materials beyond graphene also are used [60]. But for non-layer structured 3-D materials such as boron, it is a real challenge to fabricate the corresponding 2-D nanosheets due to the absence of the driving force of anisotropic growth. There are rare examples of some 2-D metal nanosheets; see for example, the recent report [61] on single-crystalline Rh nanosheets with a 3–5 atomic layers thickness. Boron sheets are expected to be metallic as well. Thus, this should increase the chances of their actual formation.

In this regard, we have to mention the recent report [62] in which large-scale single-crystalline ultrathin boron nanosheets have been fabricated via the thermal decomposition of diborane.

It is obvious that an infinite boron sheet does not exist in nature and that its finite pieces are not stable compared to bulk and/or nanotubular structures of boron. To grow boron sheets, one needs a substrate which binds boron atoms strongly to avoid bulk phases while, at the same time, provides sufficient mobility of boron atoms on the substrate. Possible candidates for substrates are surfaces of (close-packed transition) metals. The feasibility of different synthetic methods for 2-D boron sheets was assessed [47,63–66] using ab initio calculations, i.e., "synthesis in theory" approach. A large-scale boron monolayer has been predicted with mixed hexagonal-triangular geometry obtained via either depositing boron atoms directly on the surface or soft landing of small planar B-clusters.

Recently, a series of planar boron allotropes with honeycomb topology has been proposed [67]. Although the free-standing honeycomb B allotropes are higher in energy than α-sheets, these calculations show that a metal substrate can greatly stabilize these new allotropes.

The atomically thin, crystalline 2-D boron sheets, i.e., borophene, were actually synthesized [16] on silver surfaces under ultrahigh-vacuum conditions (Figure 3). An atomic-scale characterization, supported by theoretical calculations, revealed structures reminiscent of fused boron clusters with multiple scales of anisotropic, out-of-plane buckling. Unlike bulk boron allotropes, borophene shows metallic characteristics that are consistent with predictions of a highly anisotropic 2-D metal.

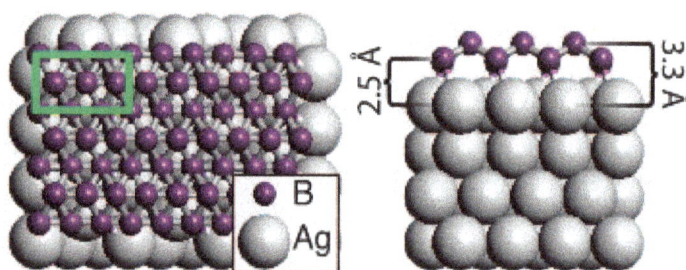

Figure 3. The borophene structure on a silver substrate: the top and side views of the monolayer structure (unit cell indicated by the box) [16].

The experimental work in Reference [17] shows that 2-D boron sheets can be grown epitaxially on a Ag(111) substrate. Two types of boron sheets, β_{12} and χ_3, both exhibiting a triangular lattice but with different arrangements of periodic holes, were observed by scanning tunneling microscopy. DFT simulations indicate that both sheets are planar without obvious vertical undulations.

According to the ab initio calculations [68], the periodic nanoscale 1-D undulations can be preferred in borophenes on concertedly reconstructed Ag(111). This "wavy" configuration is more stable than its planar form on flat Ag(111) due to an anisotropic high bending flexibility of borophene. An atomic-scale ultrahigh vacuum scanning tunneling microscopy characterization of a borophene grown on Ag(111) reveals such undulations, which agree with the theory. Although the lattice is coherent within a borophene island, the undulations nucleated from different sides of the island form a distinctive domain boundary when they are laterally misaligned.

Recently, borophene synthesis monitored in situ by low-energy electron microscopy, diffraction, and scanning tunneling microscopy and modeled using ab initio theories has been reported in Reference [69]. By resolving the crystal structure and phase diagram of borophene on Ag(111), the domains are found to remain nanoscale for all growth conditions. However, by growing borophene on Cu(111) surfaces, large single-crystal domains (up to 100 μm) are obtained. The crystal structure is a novel triangular network with a concentration of hexagonal vacancies of $\eta = 1/5$. These experimental data together with ab initio calculations indicate a charge-transfer coupling to the substrate without significant covalent bonding.

Boron on a Pb(110) surface was simulated [70] by using ab initio evolutionary methodology and found that 2-D *Pmmn* structures can be formed because of a good lattice matching. By increasing the thickness of 2-D boron, the three-bonded graphene-like $P2_1/a$ boron was revealed to possess lower molar energy, indicating the more stable 2-D boron.

The influence of the excess negative charge on the stability of borophenes—2-D boron crystals—was examined in Reference [71] using an analysis of the decomposition of the binding energy of a given boron layer into contributions coming from boron atoms that have different coordination numbers to understand how the local neighborhood of an atom influences the overall stability of the monolayer structure. The decomposition is done for the α-sheet related family of structures. It was found a preference for 2-D boron crystals with very small or very high charges per atom. Structures with intermediate charges are not energetically favorable. It has been also found a clear preference in terms of binding energy for the experimentally observed γ-sheet and δ-sheet structures that is almost independent on the considered excess of negative charge of the structures.

Two-dimensional boron monolayers have been extensively investigated using ab initio calculations [72]. A series of boron bilayer sheets with pillars and hexagonal holes have been constructed. Many of them have a lower formation energy than an α-sheet boron monolayer. However, the distribution and arrangement of hexagonal holes can cause a negligible effect on the stability of these structures.

Recently, an ab initio study [73] of the effect of electron doping on the bonding character and stability of borophene for the neutral system has revealed previously unknown stable 2-D structures: ε-B and ω-B. The chemical bonding characteristic in this and other boron structures is found to be strongly affected by an extra charge. Beyond a critical degree of electron doping, the most stable allotrope changes from ε-B to a buckled honeycomb structure. Additional electron doping, mimicking a transformation of boron to carbon, causes a gradual decrease in the degree of buckling of the honeycomb lattice.

1.2. Applications

In general, the formation of a boron sheet would have wide applications in techniques because the boronizing of metal surfaces is known as an effective method of formation of protective coatings [74]. In particular, quasi-planar bare boron surfaces can serve as lightweight protective armor.

Boron sheets are expected to be very good conductors with potential applications in nanoelectronics, e.g., in high-temperature nanodevices. Boron sheets could have potential as metallic interconnects and wiring in electronic devices and IC (integrated circuits) [41].

A theoretical investigation [32] of both the molecular physisorption and dissociative atomic chemisorption of hydrogen by boron sheets predicts physisorption as the leading mechanism at moderate temperatures and pressures. Further calculations performed on hydrogen-storage properties showed that the decoration of pristine sheets with the right metal elements provide additional absorption sites for hydrogen [47]. Thus, boron sheets can serve for good nanoreservoirs of fuel hydrogen used in green-energy production.

Due to the high neutron-capture cross section of [10]B nuclei, solid-state boron allotropies, as well as boron-rich compounds and composites, are good candidates to be used as neutron-protectors. Boron sheets will be especially useful as an absorbing component in composite neutron shields [75]. Materials with the high bulk concentration of B-atoms usually are nonmetals and, therefore, not suitable for electromagnetic shielding purposes. However, frequently, the simultaneous protection against both the neutron irradiation and electromagnetic waves is needed, in particular, because neutron absorption by [10]B nuclei is accompanied by a gamma-radiation. For this reason, in the boron-containing nanocomposites designed for neutron-protection, it is necessary to introduce some foreign components with metallic conductivity. Utilizing of the metallic boron sheet as a component may resolve this problem [76].

Recently, the mechanical properties of 2-D boron—borophene—have been studied by ab initio calculations [77]. The borophene with a 1/6 concentration of hollow hexagons is shown to have the Föppl–von Kármán number per unit area over twofold higher than graphene's value. The record high flexibility combined with excellent elasticity in boron sheets can be utilized for designing advanced composites and flexible devices. The transfer of undulated borophene onto an elastomeric substrate would allow for high levels of stretchability and compressibility with potential applications for emerging stretchable and foldable devices [68].

The boron sheets are quite inert to oxidization and interact only weakly with their substrate. For this reason, they may find applications in electronic devices in the future [17].

In large-scale single-crystalline ultrathin boron nanosheets fabricated [62] via the thermal decomposition of diborane, the strong combination performances of low turn-on field-of-field emissions, favorable electron transport characteristics, high sensitivity, and fast response time to illumination reveal that the nanosheets have high potential as applications in field-emitters, interconnects, IC, and optoelectronic devices.

Some other applications of borophene are described in recent reviews [78,79].

1.3. Available Electron Structure Calculations

Because boron sheets are of great academic and practical interests, their electronic structure is studied intensively. Most of them are found to be metallic.

Let us note that there are some indirect evidences for metallic conduction in boron sheets. The absence of icosahedra in liquid boron affects its properties including electrical conductivity [57,59], and it behaves like a metal.

The very stable quasi-planar clusters of boron B_n for n up to 46, considered to be fragments of bare boron quasi-planar surfaces, have to possess a singly occupied bonding orbital [29]. Assuming that conduction band of the infinite surface is generated from the HOMO (highest-occupied-molecular-orbital) of a finite fragment, it means the partial filling of the conduction band, i.e., the metallic mechanism of conductance.

Diamond-like, metallic boron crystal structures were predicted in Reference [80] employing so-called decoration schemes of calculations, in which the normal and hexagonal diamond-like frameworks are decorated with extra atoms across the basal plane. They should have an overly high DoS near the Fermi level. This result may provide a plausible explanation for not only the anomalous superconductivity of boron under high pressure but also the nonmetal–metal transition in boron structures.

In Reference [45], it was presented the results of a theoretical study of the phase diagram of elemental boron showing that, at high pressures, boron crystallizes in quasi-layered bulk phases characterized by in-plane multicenter bonds and out-of-plane bonds. All these structures are metallic.

Usually, direct ab initio calculations performed for boron sheets of different structures reveal the pronounced metallic-like total DoSs [4,5,7,30,31,35,38,40,41,44,46,64], and therefore, boron sheets show a strongly conducting character.

Band structures of a series of planar boron allotropes with honeycomb topologies recently proposed in Reference [67], exhibit Dirac cones at the K-point, the same as in graphene. In particular, the Dirac point of honeycomb boron sheet locates precisely on the Fermi level, rendering it as a topologically equivalent material to graphene. Its Fermi velocity is of $6 \cdot 10^5$ m/s, close to that of graphene. However, in H-borophene [53] constructed by tiling 7-membered rings side by side, a Dirac point appeared at about 0.33 eV below the Fermi level.

According to some theoretical results [36,37,47,48], boron sheets can be not only metal but in some cases also an almost zero band-gap semiconductor depending on its atomistic configuration. Probably, the semiconducting character is related to the nonzero thickness of buckled/puckered 2-D boron sheets or double-layered structures.

Some of borophenes can be magnetic. Based on a tight-binding model of 8-*Pmmn* borophene developed in Reference [81], it is confirmed that the crystal hosts massless Dirac fermions and the Dirac points are protected by symmetry. Strain is introduced into the model, and it is shown to induce a pseudomagnetic field vector potential and a scalar potential. The 2-D antiferromagnetic boron, designated as M-boron, has been predicted [82] using an ab initio evolutionary methodology. M-boron is entirely composed of B_{20} clusters in a hexagonal arrangement. Most strikingly, the highest valence band of M-boron is isolated, strongly localized, and quite flat, which induces spin polarization on either cap of the B_{20} cluster. This flat band originates from the unpaired electrons of the capping atoms and is responsible for magnetism. M-boron is thermodynamically metastable.

Boron sheets grown on metal surfaces are predicted [63] to be strongly doped with electrons from the substrate, showing that a boron sheet is an electron-deficient material. As mentioned, by simulating [70] boron on Pb(110) surface using ab initio evolutionary methodology, it was found that the 2-D Dirac $Pmmn$ boron can be formed. Unexpectedly, by increasing the thickness of 2-D boron, the three-bonded graphene-like structure $P2_1/a$ was revealed to possess double anisotropic Dirac cones. It is the most stable 2-D boron with particular Dirac cones. The puckered structure of $P2_1/a$ boron results in the peculiar Dirac cones.

This present work aims to provide more detailed calculations on the electronic structure of boron sheet including not only DoS but also band structure, electron effective mass, Fermi curve, etc.

2. Theoretical Approach

We use an original theoretical method of the quasi-classical type [83] based on the proof that the electronic system of any substance is a quasi-classical system; that is, its exact and quasi-classical energy spectra are close to each other.

This approach successfully was applied to all-boron structures to determine their structural parameters, binding energy, vibrational frequencies, and isotopic effects as well; see for example, References [84–88].

As for the determination of the materials' electron structure, the quasi-classical method is reduced to the LCAO (linear combination of atomic orbitals) method with a basis set of quasi-classical atomic orbitals. Within the initial quasi-classical approximation, the solution of the corresponding mathematical problem consists of two main stages:

1. The construction of matrix elements for secular equation, which, within the initial quasi-classical approximation, reduces to a geometric task of determining the volume of the intersection of three spheres [89], and
2. The solving of the secular equation, which determines the crystalline electronic energy spectrum [90].

This method has been successfully applied to electronic structure calculations performed for various modifications of boron nitride, BN, one of the most important boron compounds [91–94], as well as metal-doped β-rhombohedral boron [95].

The maximal relative error of a quasi-classical calculation itself, i.e., without errors arisen from input data, is estimated as approx. 4%.

As for the input data, the quasi-classical method of band structure calculations requires them to be in the form of quasi-classical parameters of constituent atoms: the inner and outer radii of the classical turning points for electron states in atoms, the radii of layers of the quasi-classical averaging of potential in atoms, and the averaged values of the potential within corresponding radial layers of atoms. These quantities for an isolated boron atom (as well as for other atoms) in the ground state were pre-calculated in Reference [96] on the basis of ab initio theoretical, namely Hartree–Fock (HF), values of electron levels [97]. Thus, in our case, the accuracy of the quasi-classical parameters is determined by that of the HF approach.

As is known, the electronic structure of any atomic system is influenced by its geometric structure and vice versa. Often, one starts with the question of how to find the most stable idealized atomic configuration. Despite this, here, we will directly begin with the electron band structure of a flat triangular boron sheet, neglecting the buckling/puckering effects and hexagonal holes (see references above), assuming that in real sheets (e.g., grown on metal surfaces), their buckled/puckered or vacant parts would not be arranged in a periodic manner, and thus, they should be regarded as perturbations which can be taken into account within a higher-order perturbation theory.

Multi-layered (buckled) boron sheets can be imagined by substituting metal Me atoms again with boron B atoms in the layered structure of a metal boride MeB_2. Analyses of the isolated layer instead of multilayered structure also seems to be quite sufficient for the initial approximation because, in such structures, only intra-layer conductivity is metallic, while interlayer conductivity is nonmetallic due to larger interlayer bond lengths if compared with those in layers.

The 2-D unit cell of the perfectly flat boron sheet without hexagonal holes (Figure 4) is a rhomb with an acute angle of $\beta = \pi/3$, i.e., with a single lattice constant a (Figure 5). Let $\vec{a}_1$ and $\vec{a}_2$ be lattice vectors, $a_1 = a_2 = a$. Then, the radius-vectors of the lattice sites are $n_1\vec{a}_1 + n_2\vec{a}_2$ with $n_1, n_2 = \ldots, -3, -2, -1, 0, +1, +2, +3, \ldots$. Vectors $\vec{k}_1$ and $\vec{k}_2$ of a reciprocal lattice (Figure 6) have to be determined from the relations $\vec{k}_1 \perp \vec{a}_2$, $\vec{k}_2 \perp \vec{a}_1$, and $k_1 = k_2 = 4\pi/\sqrt{3}a$. There is a number of different values for the lattice constant of a boron sheet suggested theoretically. For the self-consistency of calculations, we use $a = 3.37$ a.u. of length, i.e., 1.78 Å, which corresponds to the B–B pair interatomic potential in the same quasi-classical approximation [91].

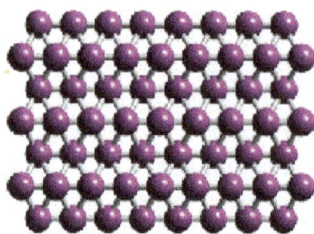

Figure 4. A boron perfect flat sheet without hexagonal holes.

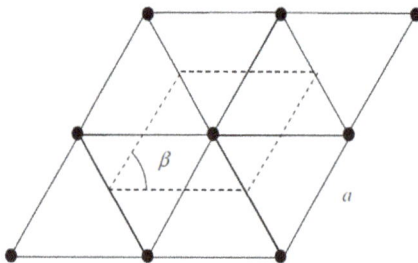

Figure 5. A 2-D rhombic unit cell of a boron flat sheet.

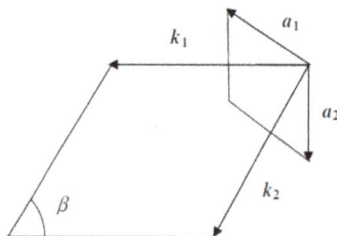

Figure 6. The vectors of a reciprocal lattice of a boron flat sheet.

Into the basis set of a simple LCAO formalism, it has been included core ($1s$), fully ($2s$), and partially filled ($2p$) valence and empty excited ($2p$) atomic orbitals.

Experimentally, there are 10 detected different existing states in the boron atom. To minimize the calculation errors related to the approximation of the crystalline potential by the superposition of atomic potentials, we choose orbitals with the same symmetry as the partially filled valence orbital, i.e., $2p$, with the closest energy level and, consequently, with the closest classical turning point radii of electrons.

Taking into account the degeneracy of atomic energy levels by magnetic and spin quantum numbers of 2, 2, 6, and 6, respectively, we can state that this set of 4 orientation-averaged orbitals replaces 16 angularly dependent atomic orbitals.

The secular equation takes the form

$$\det(H(\alpha_1, \alpha_2) - E(\alpha_1, \alpha_2)S(\alpha_1, \alpha_2)) = 0 \tag{1}$$

where $S(\alpha_1, \alpha_2)$ and $H(\alpha_1, \alpha_2)$ are 16×16 matrices of overlapping integrals and single-electron Hamiltonian, respectively, reducible to 4×4 matrices. $E(\alpha_1, \alpha_2)$ is a required electron energy band. About the parameters α_1 and α_2, see below.

This equation has 4 different real and negative roots $E_m(\alpha_1, \alpha_2)$, $m = 1, 2, 3, 4$. It can be demonstrated that they exhibit all the different solutions of the corresponding secular equation with 16×16 matrices. Within the initial quasi-classical approximation, these matrix elements can be found from the relations shown in the Appendix A.

Formally, these expressions contain infinite series. However, within the initial quasi-classical approximation, due to the finiteness of quasi-classical atomic radii, only a finite number of summands differs from zero. Thus, the series are terminated unambiguously.

The input data in a.u. in the form of quasi-classical parameters of boron atoms are shown in Tables 1 and 2. As it was mentioned above, the parameters of electron states fully or partially filled with electrons in the ground state were calculated on the basis of the theoretical, namely HF, values of electron levels, while for the excited state, we use the experimental value [98], which, however, is modulated by the multiplier of order of 1, 0.984151, leading to the coincidence between experimental and HF-theoretical first ionization potentials for an isolated boron atom: 0.304945 and 0.309856 a.u., respectively. Note that for the ground state, the $1s^2 2s^2 2p$ configuration is considered, not the $1s^2 2s 2p^2$ configuration, from which the ground state and first excited states of some boron-like ions arise [99].

Table 1. The inner and outer classical turning point radii r'_i and r''_i of electrons in boron atom.

Orbital	State	$-E_i$ (a.u.)	r'_i (a.u.)	r''_i (a.u.)
1	$1s$	7.695335	0	0.509802
2	$2s$	0.494706	0	4.021346
3	$2p$	0.309856	0.744122	4.337060
4	$2p$	0.214595	0.894159	5.211538

Table 2. The radii r_λ of radial layers of quasi-classical averaging of potential in boron atoms and the averaged values of potential φ_λ.

λ	r_λ (a.u.)	φ_λ (a.u.)
0	0	–
1	0.027585	210.5468
2	0.509802	8.882329
3	0.744122	3.65292
4	4.021346	0.206072
5	4.33706	0.000614

All the matrix elements and electron energies are calculated in points $\alpha_1 \vec{k}_1 + \alpha_2 \vec{k}_2$ of the reciprocal space with parameters $-1/2 \leq \alpha_1, \alpha_2 \leq +1/2$, i.e., within a rhombic unit cell of the reciprocal lattice (Figure 7). The first Brillouin zone for a boron flat sheet has a hexagonal shape. Of course, the areas of hexagonal and rhombic unit cells are equal (Figure 8).

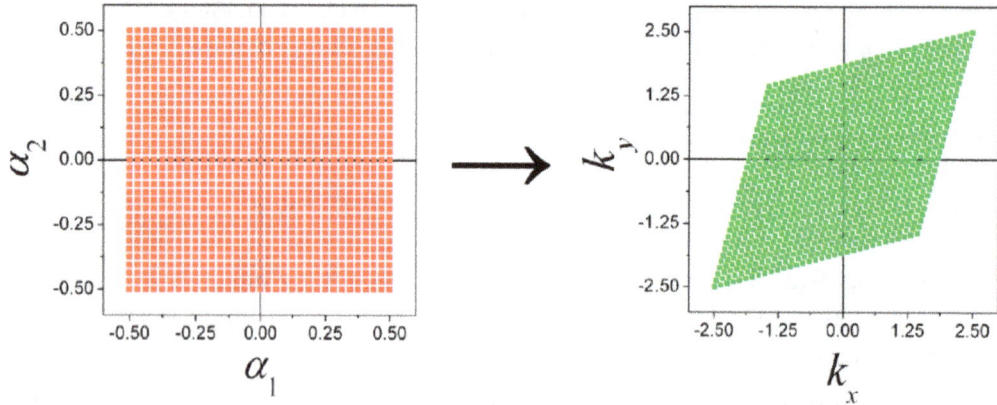

Figure 7. The transform from the (α_1, α_2) domain to the (k_x, k_y) domain of reciprocal space.

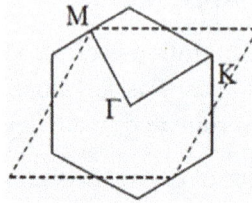

Figure 8. The hexagonal (first Brillouin zone) and rhombic unit cells of a reciprocal lattice of a flat boron sheet.

The unit cell is covered evenly by 1,002,001 points, at which the energy is found as a solution to the generalized eigenvalue problem.

The calculation has been performed in atomic units, a.u. Then, the results have been converted according to the relations: 1 a.u. of energy = 27.212 eV and 1 a.u. of length = 0.52918 Å.

Based on the resulting data set, we have constructed the electron band surfaces, the distribution of DoS in the bands, and the Fermi curve, emphasizing that, instead of the Fermi surface, the characteristics of 3-D crystals, 2-D crystals are characterized by Fermi curves.

3. Results and Discussion

In the quasi-classically calculated electronic structure of the flat boron sheet, we resolve four bands of energy. We have to emphasize that for simplicity, the band surfaces below are shown over a rhombic (not hexagonal) domain.

The lowest energy band E_1 surface is found to be almost a plane placed at the level of $E_{1min} = E_{1max} = -276.21$ eV. Thus, the chemical shift against the core $1s$ atomic level $E_{1s} = -209.41$ eV equals to $\delta E_1 = E_{1s} - E_1 = 66.80$ eV. Dispute the shift of the B $1s$ atomic level, it retains an order of magnitude after transforming in an electronic band of the boron flat sheet. The lowest-lying band E_1 is fully filled with electrons.

The band E_2 is the highest fully filled band (Figure 9) with bottom at $E_{2min} = -37.21$ eV and top at $E_{2max} = -19.85$ eV, i.e., with a width of $\Delta E_2 = E_{2max} - E_{2min} = 17.36$ eV. Note that, this range of energies is comparable in order of magnitude with a valence $2s$ atomic level of $E_{2s} = -13.46$ eV.

The band E_3 (Figure 10) is partially filled, i.e., partially empty, with a bottom at $E_{3min} = -23.08$ eV and top at $E_{3max} = -17.16$ eV, i.e., with a width of $\Delta E_3 = E_{3max} - E_{3min} = 5.92$ eV. Note that this range of energies is comparable in order of magnitude with a valence $2p$ atomic level $E_{2p} = -8.43$ eV.

The band E_4 (Figure 11) is empty, with a bottom at $E_{4min} = -17.65$ eV and top at $E_{4max} = -8.08$ eV, i.e., with a width of $\Delta E_4 = E_{4max} - E_{4min} = 9.57$ eV. Note that this range of energies is comparable in order of magnitude with the modulated value of the excited $2p$ level $E'_{2p} = -5.84$ eV.

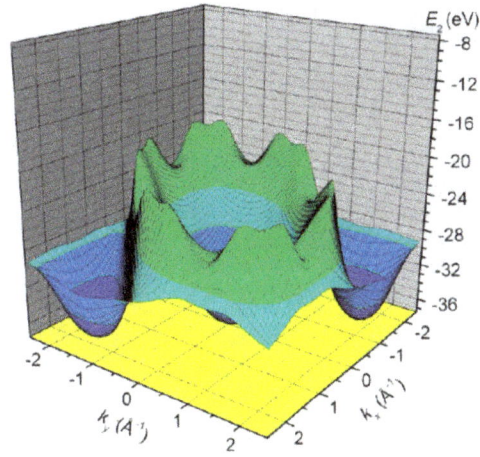

(a)

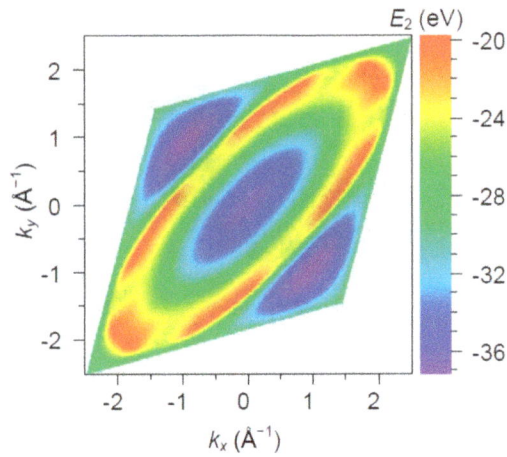

(b)

Figure 9. The band E_2 energy surface (**a**) and contour plots (**b**) over a rhombic unit cell.

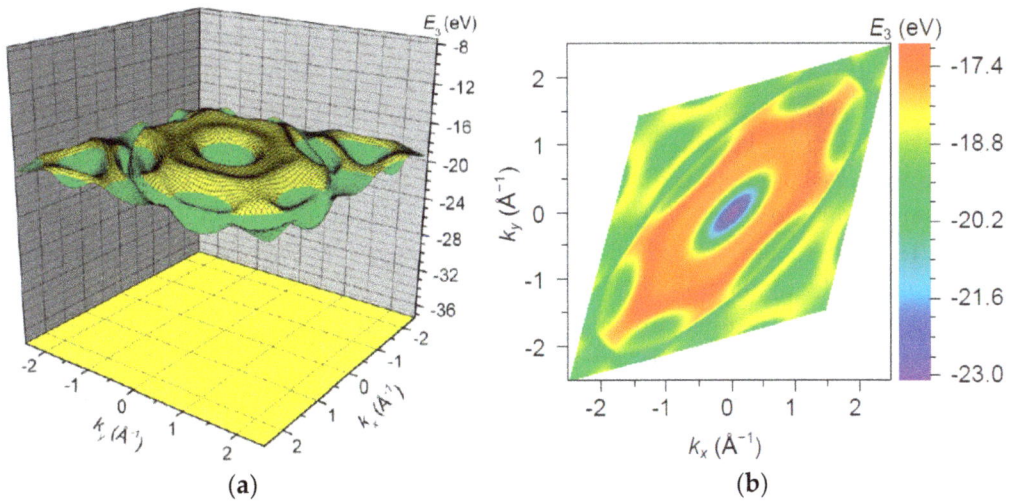

(a)　　　　　　　(b)

Figure 10. The band E_3 energy surface (**a**) and contour plots (**b**) over a rhombic unit cell.

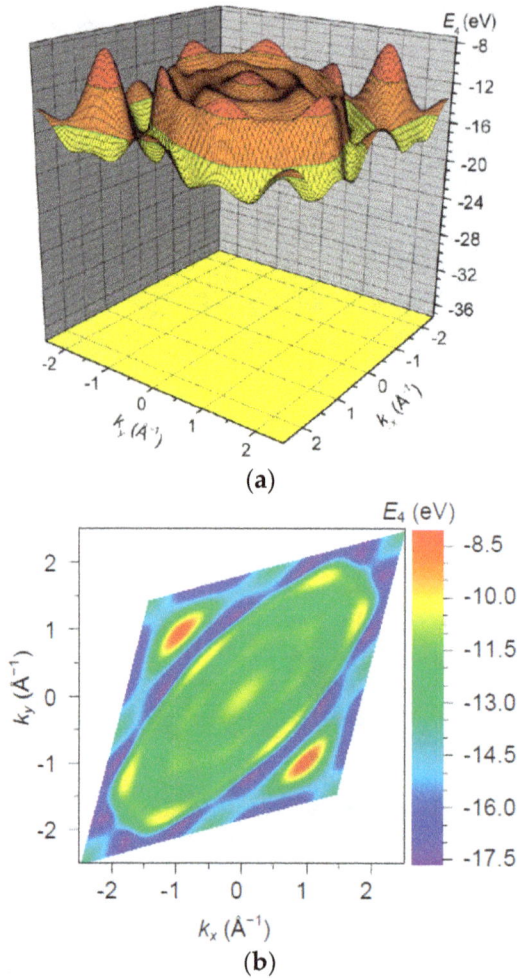

Figure 11. The band E_4 energy surface (**a**) and contour plots (**b**) over a rhombic unit cell.

Between bands E_1 and E_2, there is a very wide energy gap of $\Delta E_{12} = E_{2\min} - E_{1\max} = 239.00$ eV, while pairs of bands E_2 and E_3, and E_3 and E_4 overlap each with other, i.e. there are obtained pseudo-gaps of $\Delta E_{23} = E_{3\min} - E_{2\max} = -3.23$ eV and $\Delta E_{34} = E_{4\min} - E_{3\max} = -0.49$ eV.

The Fermi level is found at $E_{\text{Fermi}} = -19.42$ eV, within the part of the band E_3 without overlapping with other bands. This result confirms the metallicity of the boron sheet.

Thus, all the electron energies are found to be negative. It means that all electrons, including conduction electrons at the Fermi level, are bounded inside the 2-D crystal. This result once more evidences the correctness of the calculations performed in this work. The total width of valence and conduction bands equal to $\Delta E_V = E_{\text{Fermi}} - E_{2\min} = 17.79$ eV and $\Delta E_C = E_{4\max} - E_{\text{Fermi}} = 11.34$ eV, respectively. The upper valence band width is $\Delta E_{\text{VU}} = E_{\text{Fermi}} - E_{2\max} = 0.43$ eV. As expected, it is negligible if compared with that of a conduction band.

To compare easily our results with the literature data, in addition to the presentation of the band structure using the contour plots of the whole Brillouin zone in Figures 12 and 13, we plot the band energies (as well as their second derivatives and corresponding parabolic approximations) along the main lines of symmetry.

Our quasi-classical calculation of the crystalline band structure, like any other approach also utilizing HF parameters of constituting atoms, cannot determine the absolute values of energy parameters with a high accuracy. By this reason, the above mentioned value E_{Fermi} cannot be used directly to determine the electron work function of the boron sheet. This goal can be achieved only after corrections are made to include the electron-correlation and to exclude the electron-self-interacting effects, which have to allow an accurate determination of the position for the vacuum level of energy

$E = 0$. However, the shifting of the reference point at the energy axis does not affect the energy differences, which are credible as are determined with quite an acceptable accuracy. They are collected in Table 3.

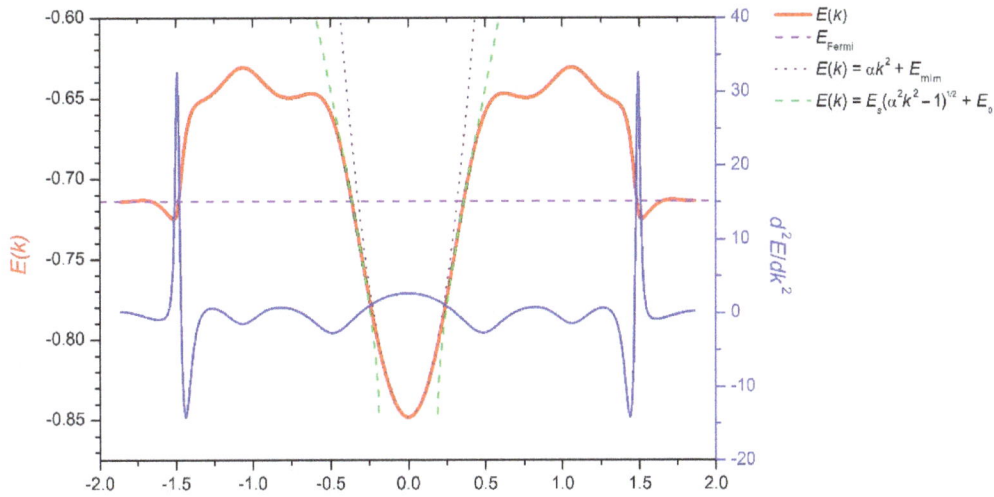

Figure 12. The section of the conduction band surface along the main diagonal of a rhombic unit cell (direction Γ–K) of reciprocal space (in atomic units).

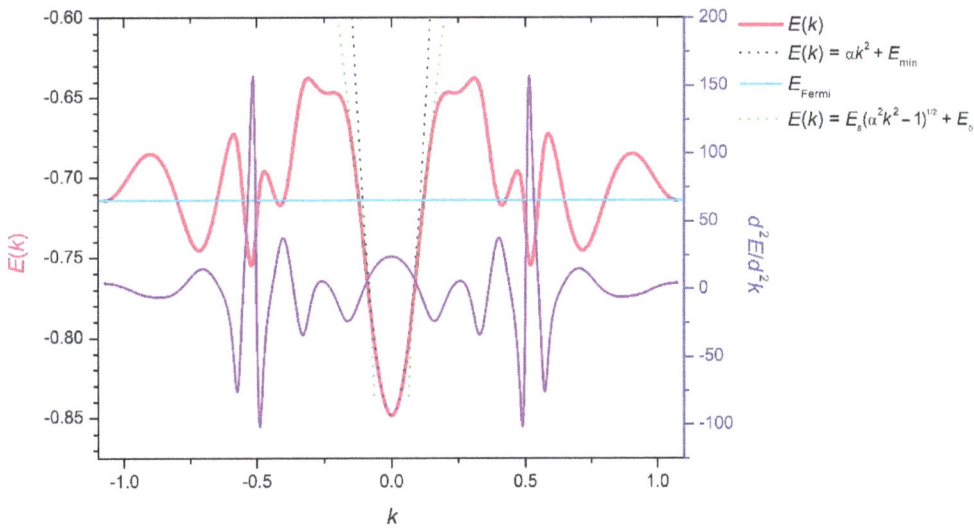

Figure 13. The section of the conduction band surface along a small diagonal of a rhombic unit cell (direction Γ–M) of reciprocal space (in atomic units).

Table 3. The band widths and (pseudo)gaps between bands.

Band	ΔE_i, eV	ΔE_{ij}, eV
1	0	
2	17.36	239
3	5.92	−3.23
4	9.57	−0.49

The Fermi curve of a boron flat sheet is found to be consisting of parts of a number of closed curves including concentric ones, the centre of which can be approximated by ellipse with long and short axes along the Γ–K and Γ–M directions, respectively (Figure 14).

Figure 14. Curves of the intersection of the band surface with the Fermi plane in neighboring rhombic unit cells of reciprocal space.

As it is known, for semiconductors, the effective mass conception referring to band zone curvatures is used to approximate the wave-vector dependence of electron energies near the band gap. As for metals, the Fermi surface curvature can be used to estimate the effective mass of conduction electrons and hence their mobility.

The ellipse representing a branch of intersection between the E_3-band surface with the E_{Fermi}-plane can be described by the equation

$$\frac{\hbar^2 k_{\Gamma-K}^2}{2m_{\Gamma-K}} + \frac{\hbar^2 k_{\Gamma-M}^2}{2m_{\Gamma-M}} = F \tag{2}$$

where $k_{\Gamma-K}$ and $k_{\Gamma-M}$ are wave-number components along perpendicular axes Γ–K and Γ–M and $F = E_{Fermi} - E_{3min} = 3.66$ eV is the Fermi energy. The effective masses $m_{\Gamma-K}$ and $m_{\Gamma-M}$ can be estimated from this equation if it is rewritten in the form of a normalized ellipse equation

$$\frac{k_{\Gamma-K}^2}{k_{\Gamma-K\ 0}^2} + \frac{k_{\Gamma-M}^2}{k_{\Gamma-M\ 0}^2} = 1 \tag{3}$$

where $k_{\Gamma-K\ 0}$ and $k_{\Gamma-M\ 0}$ are half-axes in the directions Γ–K and Γ–M, respectively:

$$k_{\Gamma-K\ 0} = \frac{\sqrt{2m_{\Gamma-K}F}}{\hbar} \approx 0.679/A$$

and

$$k_{\Gamma-M\ 0} = \frac{\sqrt{2m_{\Gamma-M}F}}{\hbar} \approx 0.224/A.$$

Then, one can calculate the effective mass of the conduction electrons m_σ, i.e., electrons placed at the Fermi level, from the relation

$$\frac{2}{m_\sigma} = \frac{1}{m_{\Gamma-K}} + \frac{1}{m_{\Gamma-M}} \tag{4}$$

The effective electron mass at the Fermi level reveals a significant anisotropy. For the central ellipse, the effective masses are $m_{\Gamma-K}/m_0 \approx 0.480$ and $m_{\Gamma-M}/m_0 \approx 0.052$, with $m_\sigma/m_0 \approx 0.094$, where m_0 is the free electron mass.

The Fermi curve of a boron flat sheet is found to consist of 6 parts of 3 ellipses representing the quadric energy-dispersion of the conduction electrons; see Figure 15.

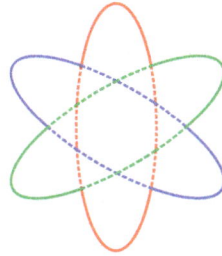

Figure 15. The Fermi curve of a boron flat sheet.

DoS within the bands E_2, E_3, and E_4 against electron energy renormalized to the Fermi level $E \rightarrow E - E_{\text{Fermi}}$ are presented in Figure 16 in two different scales for the convenient consideration. As for the band E_1, DoS within this band is proportional to the Dirac function, $\sim \delta(E + 256.42\,[\text{eV}])$, with the accordingly renormalized argument -276.21 eV $\rightarrow -276.21$ eV $- (-19.42$ eV$) = 256.79$ eV.

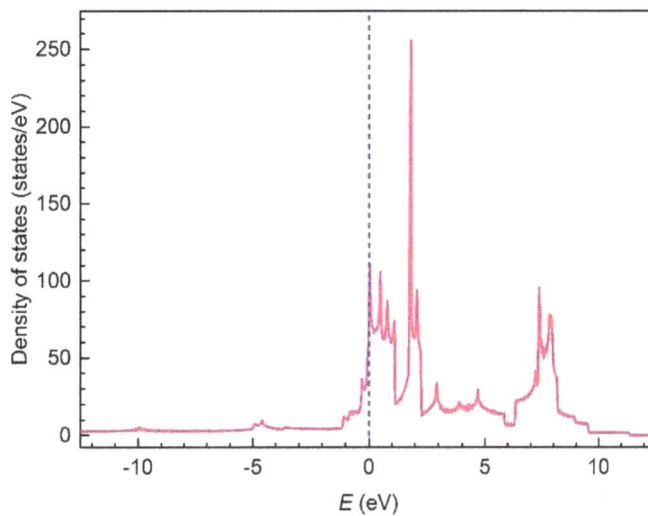

(a)

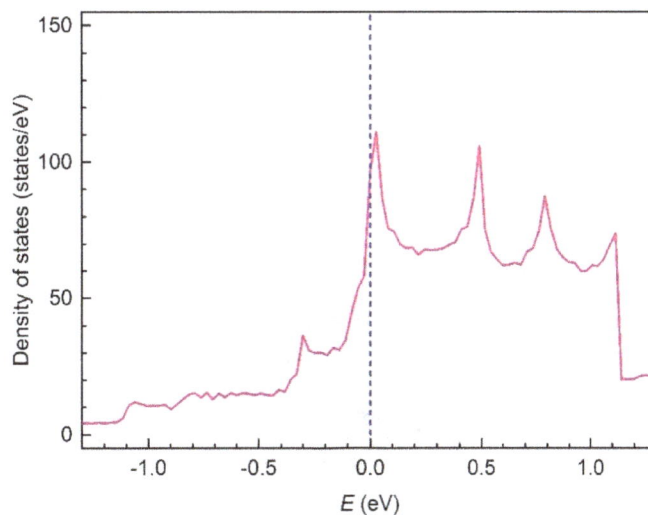

(b)

Figure 16. The density-of-electron-states renormalized to the Fermi level in a valence band and the lower and upper conduction bands of a boron flat sheet in two different scales: general view (**a**) and in Fermi level vicinity (**b**).

The overall shapes of DoSs obtained by us and previously by others, especially, in References [31,33,47] are rather similar but with some differences. It is understandable as these structures are buckled/puckered or flat variants of the same triangular lattice with or without hexagonal holes (Figure 17). The discrepancies may be attributed with the perturbations related to the mentioned structural changes and differences between the computing methods utilized, as well as the difference between projected onto in-plane or out-of-plane orbitals' densities-of-states (PDoSs) from the total DoS of the sheet.

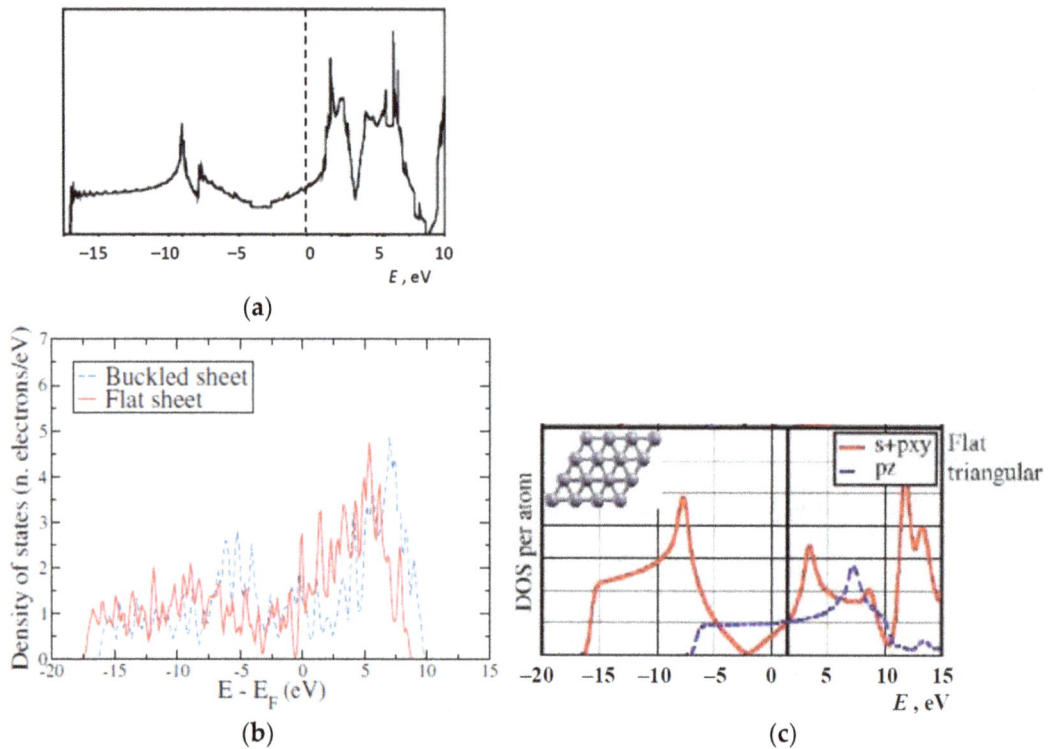

Figure 17. The densities-of-electron-states of boron sheets calculated by different methods: (**a**) modified from Reference [31], (**b**) from Reference [33], and (**c**) modified from Reference [37].

The Fermi curve of the monolayer flat boron sheet approximated by parts of concentric closed ellipse-like curves could be considered as a certain kind of topological analog of the Fermi surface (Figure 18) in the form of a half-torus and distorted cylinder of magnesium diboride MgB_2 [100], which is believed to be a structural analog of the hypothetical multilayered boron sheet, where metal Me atoms in a metal diboride MeB_2 structure are replaced by B-atoms themselves.

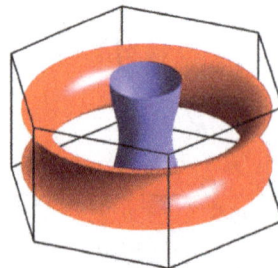

Figure 18. The Fermi surface of magnesium diboride MgB_2 [100].

The low effective mass of conduction electrons at the Fermi level indicates a high mobility of electrons and, hence, a high conductivity of the flat boron sheet.

4. Conclusions

In summary, we can conclude that the electronic band structure of a boron flat triangular sheet has been calculated within a quasi-classical approach for the quasi-classical structural parameter (B–B bond length) of $a = 1.78$Å. It is shown to have metallic properties like most of other modifications of boron sheets.

There are resolved four electronic bands: E_1, E_2, E_3, and E_4. The bands' widths are $\Delta E_1 = 0.00$, $\Delta E_2 = 17.36$, $\Delta E_3 = 5.92$, and $\Delta E_4 = 9.57$ eV, respectively. The (pseudo)gaps between the bands are $\Delta E_{12} = 239.00$, $\Delta E_{23} = -3.23$, and $\Delta E_{34} = -0.49$ eV. The Fermi level E_{Fermi} is located within the filled band E_3, confirming the metallicity of the boron flat sheet.

The Fermi curve of a boron flat sheet consists of parts of 3 ellipses with semimajor and semiminor axes along the Γ–K and Γ–M directions, respectively. The effective electron mass at the Fermi level reveals a distinct anisotropy: $m_{\Gamma-K}/m_0 \approx 0.480$ and $m_{\Gamma-M}/m_0 \approx 0.052$, with $m_\sigma/m_0 \approx 0.094$ for conduction mass. The low effective mass of conduction electrons indicates a high mobility of electrons and, hence, a high conductivity of flat boron sheets.

The shapes of density-of-states obtained here for flat holeless boron sheets and previously calculated ones are rather similar, which is understandable as these structures are buckled/puckered or flat but with hexagonal holes, variants of the same triangular lattice. The remaining discrepancies may be attributed to the perturbations associated with the mentioned structural changes and differences in the used models.

Author Contributions: Conceptualization, L.C. and R.B.; methodology, L.C.; software, I.M.; validation, L.C., I.M., and R.B.; formal analysis, I.M.; investigation, L.C., I.M., and R.B.; resources, L.C. and R.B.; data curation, I.M.; writing—original draft preparation, L.C.; writing—review and editing, I.M.; visualization, L.C. and I.M.; project administration, L.C.

Funding: This research was funded by Shota Rustaveli National Science Foundation of Georgia (SRNSFG), Grant AR–18–1045: "Obtaining of boron carbide-based nanostructured heterophase ceramic materials and products with improved performance characteristics".

Conflicts of Interest: The authors declare no conflict of interest. The funder had no role in the design of the study; in the collection, analyses, or interpretation of data; in the writing of the manuscript; or in the decision to publish the results.

Appendix A

Within the initial quasi-classical approximation, the matrix elements of overlapping integrals $S(\alpha_1, \alpha_2)$ and single-electron Hamiltonian $H(\alpha_1, \alpha_2)$ can be respectively found from following explicit relations:

$$S_{jl}(\alpha_1, \alpha_2) = \frac{3}{4\pi} \sum_{n_1=-\infty}^{n_1=+\infty} \sum_{n_2=-\infty}^{n_2=+\infty} \frac{V_{jl}(n_1,n_2)}{\sqrt{(r''^3_j - r'^3_j)(r''^3_l - r'^3_l)}} \cos 2\pi(n_1\alpha_1 + n_2\alpha_2)$$

$$V_{jl}(n_1, n_2) =$$
$$= V\left(r''_j, r''_l, a\sqrt{n_1^2 + n_1 n_2 + n_2^2}\right) +$$
$$+ V\left(r'_j, r'_l, a\sqrt{n_1^2 + n_1 n_2 + n_2^2}\right) -$$
$$- V\left(r''_j, r'_l, a\sqrt{n_1^2 + n_1 n_2 + n_2^2}\right) -$$
$$- V\left(r'_j, r''_l, a\sqrt{n_1^2 + n_1 n_2 + n_2^2}\right)$$

and

$$H_{jl}(\alpha_1, \alpha_2) = \frac{3}{4\pi} \sum_{n_1=-\infty}^{n_1=+\infty} \sum_{n_2=-\infty}^{n_2=+\infty} \sum_{v_1=-\infty}^{v_1=+\infty} \sum_{v_2=-\infty}^{v_2=+\infty} \sum_{\lambda=1}^{\lambda=5} \frac{\varphi_\lambda V_{jl\lambda}(n_1,n_2,v_1,v_2)}{\sqrt{(r''_j^3 - r'_j^3)(r''_l^3 - r'_l^3)}} \cos 2\pi(n_1\alpha_1 + n_2\alpha_2)$$

$$V_{jl\lambda}(n_1, n_2, v_1, v_2) =$$

$$= V(r''_j, r''_l, r_\lambda, a\sqrt{n_1^2 + n_1 n_2 + n_2^2}, a\sqrt{v_1^2 + v_1 v_2 + v_2^2}, a\sqrt{(n_1-v_1)^2 + (n_1-v_1)(n_2-v_2) + (n_2-v_2)^2}) +$$

$$+ V(r''_j, r'_l, r_{\lambda-1}, a\sqrt{n_1^2 + n_1 n_2 + n_2^2}, a\sqrt{v_1^2 + v_1 v_2 + v_2^2}, a\sqrt{(n_1-v_1)^2 + (n_1-v_1)(n_2-v_2) + (n_2-v_2)^2}) +$$

$$+ V(r'_j, r''_l, r_{\lambda-1}, a\sqrt{n_1^2 + n_1 n_2 + n_2^2}, a\sqrt{v_1^2 + v_1 v_2 + v_2^2}, a\sqrt{(n_1-v_1)^2 + (n_1-v_1)(n_2-v_2) + (n_2-v_2)^2}) +$$

$$+ V(r'_j, r'_l, r_\lambda, a\sqrt{n_1^2 + n_1 n_2 + n_2^2}, a\sqrt{v_1^2 + v_1 v_2 + v_2^2}, a\sqrt{(n_1-v_1)^2 + (n_1-v_1)(n_2-v_2) + (n_2-v_2)^2}) -$$

$$- V(r'_j, r''_l, r_\lambda, a\sqrt{n_1^2 + n_1 n_2 + n_2^2}, a\sqrt{v_1^2 + v_1 v_2 + v_2^2}, a\sqrt{(n_1-v_1)^2 + (n_1-v_1)(n_2-v_2) + (n_2-v_2)^2}) -$$

$$- V(r''_j, r'_l, r_\lambda, a\sqrt{n_1^2 + n_1 n_2 + n_2^2}, a\sqrt{v_1^2 + v_1 v_2 + v_2^2}, a\sqrt{(n_1-v_1)^2 + (n_1-v_1)(n_2-v_2) + (n_2-v_2)^2}) -$$

$$- V(r''_j, r''_l, r_{\lambda-1}, a\sqrt{n_1^2 + n_1 n_2 + n_2^2}, a\sqrt{v_1^2 + v_1 v_2 + v_2^2}, a\sqrt{(n_1-v_1)^2 + (n_1-v_1)(n_2-v_2) + (n_2-v_2)^2}) -$$

$$- V(r'_j, r'_l, r_{\lambda-1}, a\sqrt{n_1^2 + n_1 n_2 + n_2^2}, a\sqrt{v_1^2 + v_1 v_2 + v_2^2}, a\sqrt{(n_1-v_1)^2 + (n_1-v_1)(n_2-v_2) + (n_2-v_2)^2}),$$

where $V(R_1, R_2, D_{12})$ and $V(R_1, R_2, R_3, D_{12}, D_{13}, D_{23})$ are intersection volumes of two and three spheres, respectively, with radii R_1, R_2, and R_3 and intercentral distances D_{12}, D_{13}, and D_{23}.

References

1. Becker, R.; Chkhartishvili, L.; Martin, P. Boron, the new graphene? *Vac. Technol. Coat.* **2015**, *16*, 38–44.

2. Boustani, I. Systematic ab initio investigation of bare boron clusters: Determination of the geometry and electronic structures of B_n (n=12–14). *Phys. Rev. B* **1997**, *55*, 16426–16438. [CrossRef]

3. Boustani, I.; Rubio, A.; Alonso, J.A. Ab initio study of B_{32} clusters: Competition between spherical, quasiplanar and tubular isomers. *Chem. Phys. Lett.* **1999**, *311*, 21–28. [CrossRef]

4. Boustani, I.; Quandt, A.; Hernandez, E.; Rubio, A. New boron based nanostructured materials. *J. Chem. Phys.* **1999**, *110*, 3176–3185. [CrossRef]

5. Boustani, I.; Quandt, A.; Rubio, A. Boron quasicrystals and boron nanotubes: Ab initio study of various B_{96} isomers. *J. Solid State Chem.* **2000**, *154*, 269–274. [CrossRef]

6. Mukhopadhyay, S.; He, H.; Pandey, R.; Yap, Y.K.; Boustani, I. Novel spherical boron clusters and structural transition from 2D quasi-planar structures to 3D double-rings. *J. Phys. Conf. Ser.* **2009**, *176*, 012028. [CrossRef]

7. Ozdogan, C.; Mukhopadhyay, S.; Hayami, W.; Guvenc, Z.B.; Pandey, R.; Boustani, I. The unusually stable B_{100} fullerene, structural transitions in boron nanostructures, and a comparative study of α- and γ-boron and sheets. *J. Phys. Chem. C* **2010**, *114*, 4362–4375. [CrossRef]

8. Zhai, H.-J.; Kiran, B.; Li, J.; Wang, L.-S. Hydrocarbon analogues of boron clusters – Planarity, aromaticity and antiaromaticity. *Nat. Mater.* **2003**, *2*, 827–833. [CrossRef] [PubMed]

9. Piazza, Z.A.; Hu, H.-S.; Li, W.-L.; Zhao, Y.-F.; Li, J.; Wang, L.-S. Planar hexagonal B_{36} as a potential basis for extended single-atom layer boron sheets. *Nat. Commun.* **2014**, *5*, 3113–3118. [CrossRef] [PubMed]

10. Zhai, H.-J.; Zhao, Y.-F.; Li, W.-L.; Chen, Q.; Bai, H.; Hu, H.-S.; Piazza, Z.A.; Tian, W.-J.; Lu, H.-G.; Wu, Y.-B.; et al. Observation of an all-boron fullerene. *Nat. Chem.* **2014**, *6*, 727–731. [CrossRef] [PubMed]

11. Ciuparu, D.; Klie, R.F.; Zhu, Y.; Pfefferle, L. Synthesis of pure boron single-wall nanotubes. *J. Phys. Chem. B* **2004**, *108*, 3967–3969. [CrossRef]

12. Kiran, B.; Bulusu, S.; Zhai, H.-J.; Yoo, S.; Zeng, X.C.; Wang, L.-S. Planar-to-tubular structural transition in boron clusters: B_{20} as the embryo of single-walled boron nanotubes. *Proc. Natl. Acad. Sci. USA* **2005**, *102*, 961–964. [CrossRef] [PubMed]

13. Liu, F.; Shen, C.; Su, Z.; Ding, X.; Deng, S.; Chen, J.; Xu, N.; Gao, H. Metal-like single crystalline boron nanotubes: Synthesis and in situ study on electric transport and field emission properties. *J. Mater. Chem.* **2010**, *20*, 2197–2205. [CrossRef]

14. Oger, E.; Crawford, N.R.M.; Kelting, R.; Weis, P.; Kappes, M.M.; Ahlrichs, R. Boron cluster cations: Transition from planar to cylindrical structures. *Angew. Chem. Int. Ed.* **2007**, *46*, 8503–8506. [CrossRef] [PubMed]

15. Oger, E. Strukturaufklärung durch Mobilitätsmessungen an massenselektierten Clusterionen in der Gasphase. Ph.D. Thesis, University of Karlsruhe—Karlsruher Institute of Technology, Karlsruher, Germany, 2010.

16. Mannix, A.J.; Zhou, X.-F.; Kiraly, B.; Wood, J.D.; Alducin, D.; Myers, B.D.; Liu, X.; Fisher, B.L.; Santiago, U.; Guest, J.R.; et al. Synthesis of borophenes: Anisotropic, two-dimensional boron polymorphs. *Science* **2015**, *350*, 1513–1516. [CrossRef] [PubMed]

17. Feng, B.; Zhang, J.; Zhong, Q.; Li, W.; Li, S.; Li, H.; Cheng, P.; Meng, S.; Chen, L.; Wu, K. Experimental realization of two-dimensional boron sheets. *Nat. Chem.* **2016**, *8*, 563–568. [CrossRef] [PubMed]

18. Parakhonskiy, G.; Dubrovinskaia, N.; Bykova, E.; Wirth, R.; Dubrovinsky, L. Experimental pressure-temperature phase diagram of boron: Resolving the long-standing enigma. *Sci. Rep.* **2011**, *1*, 96. [CrossRef] [PubMed]

19. McGrady, J.W.; Papaconstantopoulos, D.A.; Mehl, M.J. Tight-binding study of boron structures. *J. Phys. Chem. Solids* **2014**, *75*, 1106–1112. [CrossRef]

20. Shirai, K. Phase diagram of boron crystals. *Jpn. J. Appl. Phys.* **2017**, *56*, 05FA06. [CrossRef]

21. Shirai, K.; Uemura, N.; Dekura, H. Structure and stability of pseudo-cubic tetragonal boron. *Jpn. J. Appl. Phys.* **2017**, *56*, 05FB05. [CrossRef]

22. Imai, Y.; Mukaida, M.; Ueda, M.; Watanabe, A. Band-calculation of the electronic densities of states and the total energies of boron–silicon system. *J. Alloys Comp.* **2002**, *347*, 244–251. [CrossRef]

23. Masago, A.; Shirai, K.; Katayama–Yoshida, H. Crystal stability of α- and β-boron. *Phys. Rev. B* **2006**, *73*, 104102. [CrossRef]

24. Jemmis, E.D.; Prasad, D.L.V.K. Icosahedral B_{12}, macropolyhedral boranes, β-rhombohedral boron and boron-rich solids. *J. Solid State Chem.* **2006**, *179*, 2768–2774. [CrossRef]

25. Van Setten, M.J.; Uijttewaal, M.A.; DeWijs, G.A.; DeGroot, R.A. Thermodynamic stability of boron: The role of defects and zero point motion. *J. Am. Chem. Soc.* **2007**, *129*, 2458–2465. [CrossRef] [PubMed]

26. Ogitsu, T.; Gygi, F.; Reed, J.; Motome, Y.; Schwegler, E.; Galli, G. Imperfect crystal and unusual semiconductor: Boron, a frustrated element. *J. Am. Chem. Soc.* **2009**, *131*, 1903–1909. [CrossRef] [PubMed]

27. Widom, M.; Mihalkovic, M. Relative stability of α and β boron. *J. Phys. Conf. Ser.* **2009**, *176*, 012024. [CrossRef]

28. Slack, G.A.; Hejna, C.I.; Garbauskas, M.F.; Kasper, J.S. The crystal structure and density of β-rhombohedral boron. *J. Solid State Chem.* **1988**, *76*, 52–63. [CrossRef]

29. Boustani, I. New quasi-planar surfaces o fbare boron. *Surf. Sci.* **1997**, *377*, 355–363. [CrossRef]

30. Boustani, I.; Quandt, A. Boronin ab initio calculations. *Comput. Mater. Sci.* **1998**, *11*, 132–137. [CrossRef]

31. Evans, M.H.; Joannopoulos, J.D.; Pantelides, S.T. Electronic and mechanical properties of planar and tubular boron structures. *Phys. Rev. B* **2005**, *72*, 045434. [CrossRef]

32. Cabria, I.; Lopez, M.J.; Alonso, J.A. Density functional calculations of hydrogen adsorption on boron nanotubes and boron sheets. *Nanotechnology* **2006**, *17*, 778–785. [CrossRef]

33. Cabria, I.; Alonso, J.A.; Lopez, M.J. Buckling in boron sheets and nanotubes. *Phys. Status Solidi A* **2006**, *203*, 1105–1110. [CrossRef]

34. Kunstmann, J.; Quandt, A. Broad boron sheets and boron nanotubes: An ab initio study of structural, electronic, and mechanical properties. *Phys. Rev. B* **2006**, *74*, 035413. [CrossRef]

35. Lau, K.C.; Pati, R.; Pandey, R.; Pineda, A.C. First-principles study of the stability and electronic properties of sheets and nanotubes of elemental boron. *Chem. Phys. Lett.* **2006**, *418*, 549–554. [CrossRef]

36. Lau, K.C.; Pandey, R. Stability and electronic properties of atomistically-engineered 2D boron sheets. *J. Phys. Chem. C* **2007**, *111*, 2906–2912. [CrossRef]

37. Tang, H.; Ismail–Beigi, S. Novel precursors for boron nanotubes: The competition of two-center and three-center bonding in boron sheets. *Phys. Rev. Lett.* **2007**, *99*, 115501. [CrossRef] [PubMed]

38. Yang, X.; Ding, Y.; Ni, J. Ab initio prediction of stable boron sheets and boron nanotubes: Structure, stability, and electronic properties. *Phys. Rev. B* **2008**, *77*, 041402(R). [CrossRef]

39. Sebetci, A.; Mete, E.; Boustani, I. Freestanding double walled boron nanotubes. *J. Phys. Chem. Solids* **2008**, *69*, 2004–2012. [CrossRef]

40. Singh, A.K.; Sadrzadeh, A.; Yakobson, B.I. Probing properties of boron α-tubes by ab initio calculations. *Nano Lett.* **2008**, *8*, 1314–1317. [CrossRef] [PubMed]

41. Wang, J.; Liu, Y.; Li, Y.-C. A new class of boron nanotubes. *Chem. Phys. Chem.* **2009**, *10*, 3119–3121. [CrossRef] [PubMed]

42. Tang, H.; Ismail–Beigi, S. First-principles study of boron sheets and nanotubes. *Phys. Rev. B* **2010**, *82*, 115412. [CrossRef]

43. Zope, R.R.; Baruah, T. Snub boro nnanostructures: Chiral fullerenes, nanotubes and planar sheet. *Chem. Phys. Lett.* **2011**, *501*, 193–196. [CrossRef]

44. Simsek, M.; Aydın, S. First-principles calculations of two dimensional boron sheets. In *Abstracts of the 17th International Symposium on Boron, Borides and Related Materials, Istanbul, Turkey, 11–17 September 2011*; Yucel, O., Ed.; BKM: Ankara, Turkey, 2011; p. 86.

45. Kunstmann, J.; Boeri, L.; Kortus, J. Bonding in boron: Building high-pressure phases from boron sheets. In *Abstracts of the 17th International Symposium on Boron, Borides and Related Materials, Istanbul, Turkey, 11–17 September 2011*; Yucel, O., Ed.; BKM: Ankara, Turkey, 2011; p. 289.

46. Bezugly, V.; Kunstmann, J.; Grundkotter-Stock, B.; Frauenheim, T.; Niehaus, T.; Cuniberti, G. Highly conductive boron nanotubes: Transport properties, work functions, and structural stabilities. *ACS Nano* **2011**, *5*, 4997–5005. [CrossRef] [PubMed]

47. Tang, H. First-Principles Investigation on Boron Nanostructures. Ph.D. Thesis, Yale University, New Haven, CT, USA, 2011.

48. Zhou, X.-F.; Dong, X.; Oganov, A.R.; Zhu, Q.; Tian, Y.; Wang, H.-T. Semimetallic two-dimensional boron allotrope with massless Dirac fermions. *Phys. Rev. Lett.* **2014**, *112*, 085502. [CrossRef]

49. Wu, X.; Dai, J.; Zhao, Y.; Zhuo, Z.; Yang, J.; Zeng, X. Two-dimensional boron monolayer sheets. *ACS Nano* **2012**, *6*, 7443–7453. [CrossRef] [PubMed]

50. Lu, H.; Mu, Y.; Li, S.-D. Comment on "Two-dimensional boron monolayer sheets". *ACS Nano* **2013**, *7*, 879. [CrossRef] [PubMed]

51. Wu, X.; Dai, J.; Zhao, Y.; Zhuo, Z.; Yang, J.; Zeng, X.C. Reply to "Comment on 'Two-dimensional boron monolayer sheets'". *ACS Nano* **2013**, *7*, 880–881. [CrossRef] [PubMed]

52. Zhou, X.-F.; Oganov, A.R.; Shao, X.; Zhu, Q.; Wang, H.-T. Unexpected reconstruction of the α-boron (111) surface. *Phys. Rev. Lett.* **2014**, *113*, 176101. [CrossRef] [PubMed]

53. Mu, Y.; Chen, Q.; Chen, N.; Lu, H.; Li, S.-D. A novel borophene featuring heptagonal holes: Common precursor of borospherenes. *Phys. Chem. Chem. Phys.* **2017**, *30*, 19890–19895. [CrossRef] [PubMed]

54. Boroznina, E.V.; Davletova, O.A.; Zaporotskova, I.V. Boron monolayer γ_3-type. Formation of the vacancy defect and pinhole. *J. Nano-Electron. Phys.* **2016**, *8*, 04054. [CrossRef]

55. Chkhartishvili, L. All-boron nanostructures. In *CRC Concise Encyclopedia of Nanotechnology*; Kharisov, B.I., Kharissova, O.V., Ortiz–Mendez, U., Eds.; CRC Press: Boca Raton, FL, USA, 2016; pp. 53–69.

56. Li, W.-L.; Pal, R.; Piazza, Z.A.; Zeng, X.C.; Wang, L.-S. B_{27}^-: Appearance of the smallest planar boron cluster containing a hexagonal vacancy. *J. Chem. Phys.* **2015**, *142*, 204305. [CrossRef] [PubMed]

57. Vast, N.; Bernard, S.; Zerah, G. Structural and electronic properties of liquid boron from a molecular-dynamics simulation. *Phys. Rev. B* **1995**, *52*, 4123–4130. [CrossRef]

58. Krishnan, S.; Ansell, S.; Felten, J.J.; Volin, K.J.; Price, D.L. Structure of liquid boron. *Phys. Rev. Lett.* **1998**, *81*, 586–589. [CrossRef]

59. Price, D.L.; Alatas, A.; Hennet, L.; Jakse, N.; Krishnan, S.; Pasturel, A.; Pozdnyakova, I.; Saboungi, M.-L.; Said, A.; Scheunemann, R.; et al. Liquid boron: X-ray measurements and ab initio molecular dynamics simulations. *Phys. Rev. B* **2009**, *79*, 134201. [CrossRef]

60. Butler, S.Z.; Hollen, S.M.; Cao, L.; Cui, Y.; Gupta, J.A.; Gutierrez, H.R.; Heinz, T.F.; Hong, S.S.; Huang, J.; Ismach, A.F.; et al. Progress, challenges, and opportunities in two-dimensional materials beyond graphene. *ACS Nano* **2013**, *7*, 2898–2926. [CrossRef] [PubMed]

61. Zhao, L.; Xu, C.; Su, H.; Liang, J.; Lin, S.; Gu, L.; Wang, X.; Chen, M.; Zheng, N. Single-crystalline rhodium nanosheets with atomic thickness. *Adv. Sci.* **2015**, *2*, 1500100. [CrossRef] [PubMed]

62. Xu, J.; Chang, Y.; Gan, L.; Ma, Y.; Zhai, T. Ultra thin nanosheets: Ultrathin single-crystalline boron nanosheets for enhanced electro-optical performances. *Adv. Sci.* **2015**, *2*, 1500023. [CrossRef] [PubMed]

63. Zhang, L.Z.; Yan, Q.B.; Du, S.X.; Su, G.; Gao, H.-J. Boron sheet adsorbed on metal surfaces: Structures and electronic properties. *J. Phys. Chem. C* **2012**, *116*, 18202–18206. [CrossRef]

64. Penev, E.S.; Bhowmick, S.; Sadrzadeh, A.; Yakobson, B.I. Polymorphism of two-dimensional boron. *Nano Lett.* **2012**, *12*, 2441–2445. [CrossRef] [PubMed]

65. Liu, Y.; Penev, E.S.; Yakobson, B.I. Probing the synthesis of two-dimensional boron by first-principles computations. *Angew. Chem. Int. Ed.* **2013**, *52*, 3156–3159. [CrossRef] [PubMed]

66. Liu, H.; Gao, J.; Zhao, J. Fromboronclustertotwo-dimensionalboronsheetonCu (111) surface: Growth mechanism and hole formation. *Sci. Rep.* **2013**, *3*, 3238.

67. Yi, W.-C.; Liu, W.; Botana, J.; Zhao, L.; Liu, Z.; Liu, J.-Y.; Miao, M.-S. Honeycomb boron allotropes with Dirac cones: A true analogue to graphene. *J. Phys. Chem. Lett.* **2017**, *8*, 2647–2653. [CrossRef] [PubMed]

68. Zhang, Z.; Mannix, A.J.; Hu, Z.; Kiraly, B.; Guisinger, N.P.; Hersam, M.C.; Yakobson, B.I. Substrate-induced nanoscale undulations of borophene on silver. *Nano Lett.* **2016**, *16*, 6622–6627. [CrossRef] [PubMed]

69. Wu, R.; Drozdov, I.K.; Eltinge, S.; Zahl, P.; Ismail–Beigi, S.; Bozovic, I.; Gozar, A. Large-area single-crystal sheets of borophene on Cu(111) surfaces. *Nat. Nanotechnol.* **2019**, *14*, 44–49. [CrossRef] [PubMed]

70. He, X.-L.; Weng, X.-J.; Zhang, Y.; Zhao, Z.; Wang, Z.; Xu, B.; Oganov, A.R.; Tian, Y.; Zhou, X.-F.; Wang, H.-T. Two-dimensional boron on Pb(110) surface. *Flat Chem.* **2018**, *7*, 34–41.

71. Tarkowski, T.; Majewski, J.A.; Gonzalez Szwacki, N. Energy decomposition analysis of neutral and negatively charged borophenes. *Flat Chem.* **2018**, *7*, 42–47. [CrossRef]

72. Gao, N.; Wu, X.; Jiang, X.; Bai, Y.; Zhao, J. Structure and stability of bilayer borophene: The roles of hexagonal holes and interlayer bonding. *Flat Chem.* **2018**, *7*, 48–54. [CrossRef]

73. Liu, D.; Tomanek, D. Effect of net charge on the relative stability of different 2D boron allotropes (ContributedTalk). In Proceedings of the APS March Meeting 2019, Boston, MA, USA, March 4–8 2019. Abstract: X13.00003.

74. Vangaveti, R. Boron Induced Surface Modification of Transition Metals. MSc Thesis, New Jersey's Science & Technology University, Newark, NJ, USA, 2006.

75. Chkhartishvili, L. Interaction between neutron-radiation and boron-containing materials. In *Radiation Synthesis of Materials and Compounds*; Kharisov, B.I., Kharissova, O.V., Ortiz–Méndez, U., Eds.; CRC Press—Taylor & Francis Group: Boca Raton, FL, USA, 2013; pp. 43–80.

76. Chkhartishvili, L.; Murusidze, I. Band structure of all-boron 2D metallic crystals as a prospective electromagnetic shielding material. In Proceedings of the International Conference on Fundamental & Applied Nano Electro Magnetics, Minsk, Belarus, 22–25 May 2012; Belarusian State University: Minsk, Belarus, 2012; p. 11.

77. Zhang, Z.; Yang, Y.; Penev, E.S.; Yakobson, B.I. Elasticity, flexibility, and ideal strength of borophenes. *Adv. Funct. Mater.* **2017**, *27*, 1605059. [CrossRef]

78. Martin, P.M. Active thin films: Graphene-related materials graphene oxide and borophene. *Vac. Technol. Coat.* **2018**, *19*, 6–13.

79. Martin, P.M. Active thin films: Applications for graphene and related materials. *Vac. Technol. Coat.* **2018**, *19*, 6–14.

80. Zhao, Y.; Xu, Q.; Simpson, L.J.; Dillon, A.C. Prediction of diamond-like, metallic boron structures. *Chem. Phys. Lett.* **2010**, *496*, 280–283. [CrossRef]

81. Zabolotsky, A.D.; Lozovik, Y.E. Strain-induced pseudomagnetic field in Dirac semimetal borophene. *arXiv* **2016**, arXiv:1607.02530v2. [CrossRef]

82. Zhou, X.-F.; Oganov, A.R.; Wang, Z.; Popov, I.A.; Boldyrev, A.I.; Wang, H.-T. Two-Dimensional magnetic boron. *Phys. Rev. B* **2016**, *93*, 085406. [CrossRef]

83. Chkhartishvili, L. *Quasi-Classical Theory of Substance Ground State*; Technical University Press: Tbilisi, Georgia, 2004.

84. Chkhartishvili, L.; Mamisashvili, N.; Maisuradze, N. Single-Parameter model for multi-walled geometry of nanotubular boron. *Solid State Sci.* **2015**, *47*, 61–67. [CrossRef]

85. Chkhartishvili, L. Boron quasi-planar clusters. A mini-review on diatomic approach. In Proceedings of the IEEE 7th International Conference on Nanomaterials: Applications and Properties, Part 4, Track: Nanomaterials for Electronics, Spintronics and Photonics, Zatoka, Ukraine, 10–15 September 2017; Pogrebnjak, A.D., Ed.; Sumy State University: Sumy, Ukraine, 2017; pp. 1–5.

86. Chkhartishvili, L. Relative stability of planar clusters B_{11}, B_{12}, and B_{13} in neutral- and charged states. *Char. Appl. Nanomater.* **2018**, *1*, 3. [CrossRef]

87. Chkhartishvili, L.; Murusidze, I. Frequencies of vibrations localized on interstitial metal impurities in beta-rhombohedral boron based materials. *Am. J. Mater. Sci.* **2014**, *4*, 103–110.

88. Chkhartishvili, L.; Tsagareishvili, O.; Gabunia, D. Isotopic expansion of boron. *J. Metall. Eng.* **2014**, *3*, 97–103. [CrossRef]

89. Chkhartishvili, L.S. Volume of the intersection of three spheres. *Math. Notes* **2001**, *69*, 421–428. [CrossRef]

90. Chkhartishvili, L.S. Iterative solution of the secular equation. *Math. Notes* **2005**, *77*, 273–279. [CrossRef]

91. Chkhartishvili, L.; Lezhava, D.; Tsagareishvili, O. Quasi-classical determination of electronic energies and vibration frequencies in boron compounds. *J. Solid State Chem.* **2000**, *154*, 148–152. [CrossRef]

92. Chkhartishvili, L. Quasi-classical approach: Electronic structure of cubic boron nitride crystals. *J. Solid State Chem.* **2004**, *177*, 395–399. [CrossRef]

93. Chkhartishvili, L.S. Quasi-classical estimates of the lattice constant and bandgap of a crystal: Two-dimensional boron nitride. *Phys. Solid State* **2004**, *46*, 2126–2133. [CrossRef]

94. Chkhartishvili, L. Density of electron states in wurtzite-like boron nitride: A quasi-classical calculation. *Mater. Sci. Ind. J.* **2006**, *2*, 18–23.

95. Chkhartishvili, L.; Murusidze, I.; Darchiashvili, M.; Tsagareishvili, O.; Gabunia, D. Metal impurities in crystallographic voids of beta-rhombohedral boron lattice: Binding energies and electron levels. *Solid State Sci.* **2012**, *14*, 1673–1682. [CrossRef]

96. Chkhartishvili, L.; Berberashvili, T. Intra-atomic electric field radial potentials in step-like presentation. *J. Electr. Magn. Anal. Appl.* **2010**, *2*, 205–243. [CrossRef]

97. Froese–Fischer, C. *Th eHartree–Fock Method for Atoms. A Numerical Approach*; Wiley: New York, NY, USA, 1977.

98. Radtsig, A.A.; Smirnov, B.M. *Parameters of Atoms and Atomic Ions. Reference Book*; Energoatomizdat: Moscow, Russia, 1986.

99. Galvez, F.J.; Buendia, E.; Sarsa, A. Excited states of boron isoelectronic series from explicitly correlated wave functions. *J. Chem. Phys.* **2005**, *122*, 154307. [CrossRef] [PubMed]

100. Dahm, T.; Schopohl, N. Fermi surface topology and the upper critical field in two-band superconductors—Application to MgB_2. *Phys. Rev. Lett.* **2003**, *91*, 017001. [CrossRef] [PubMed]

12

Classifying Induced Superconductivity in Atomically Thin Dirac-Cone Materials

Evgueni F. Talantsev [1,2] (iD)

[1] M. N. Miheev Institute of Metal Physics, Ural Branch, Russian Academy of Sciences, 18, S. Kovalevskoy St., Ekaterinburg 620108, Russia; evgeny.talantsev@imp.uran.ru; Tel.: +7-912-676-0374

[2] NANOTECH Centre, Ural Federal University, 19 Mira St., Ekaterinburg 620002, Russia

Abstract: Recently, Kayyalha et al. (*Phys. Rev. Lett.*, **2019**, *122*, 047003) reported on the anomalous enhancement of the self-field critical currents (I_c(sf,T)) at low temperatures in Nb/BiSbTeSe$_2$-nanoribbon/Nb Josephson junctions. The enhancement was attributed to the low-energy Andreev-bound states arising from the winding of the electronic wave function around the circumference of the topological insulator BiSbTeSe$_2$ nanoribbon. It should be noted that identical enhancement in I_c(sf,T) and in the upper critical field ($B_{c2}(T)$) in approximately the same reduced temperatures, were reported by several research groups in atomically thin junctions based on a variety of Dirac-cone materials (DCM) earlier. The analysis shows that in all these S/DCM/S systems, the enhancement is due to a new superconducting band opening. Taking into account that several intrinsic superconductors also exhibit the effect of new superconducting band(s) opening when sample thickness becomes thinner than the out-of-plane coherence length ($\xi_c(0)$), we reaffirm our previous proposal that there is a new phenomenon of additional superconducting band(s) opening in atomically thin films.

Keywords: superconductivity enhancement in atomically thin films; Dirac-cone materials; single layer graphene; Josephson junctions; multiple-band superconductivity

---

1. Introduction

Intrinsic superconductors can be grouped into 32 classes under "conventional", "possibly unconventional", and "unconventional" categories, according to the mechanism believed to give rise to superconductivity [1]. One of the most widely used concepts to represent all 32 classes of superconductors was proposed by Uemura et al. [2,3]. The concept of the Uemura plot is based on the utilization of two fundamental temperatures of superconductors: one is the Fermi temperature (T_F) (X-axis), and the superconducting transition temperature (T_c) (Y-axis). In the most recently updated Uemura plot (Figure 1), it can be seen that elemental superconductors are located for wide range of $T_c/T_F \leq 0.001$, while all unconventional superconductors, including both nearly-room-temperature superconductors of H$_3$S [4] and LaH$_{10}$ [5] (for which experimentally measured upper critical field data [6,7] was analyzed in Refs. [8,9]), are located within a narrow band of $0.01 \leq T_c/T_F \leq 0.05$.

It should be mentioned that Hardy et al. [10] in 1993 (seven years after the discovery of high-temperature superconductivity in cuprates by Bednoltz and Mueller [11]) were the first to experimentally find that YBa$_2$Cu$_3$O$_{7-x}$ has nodal superconducting gap. This experimental result was used to propose *d*-wave superconducting gap symmetry in HTS cuprates by Won and Maki [12]. It should be noted that several researchers and research groups (over last 33 years) have proposed different mechanisms for high-temperature superconductivity in cuprates, the first two-band BCS superconductor (MgB$_2$), pnictides, and hydrogen-rich superconductors, for which we refer the reader to original papers and comprehensive reviews [13–31].

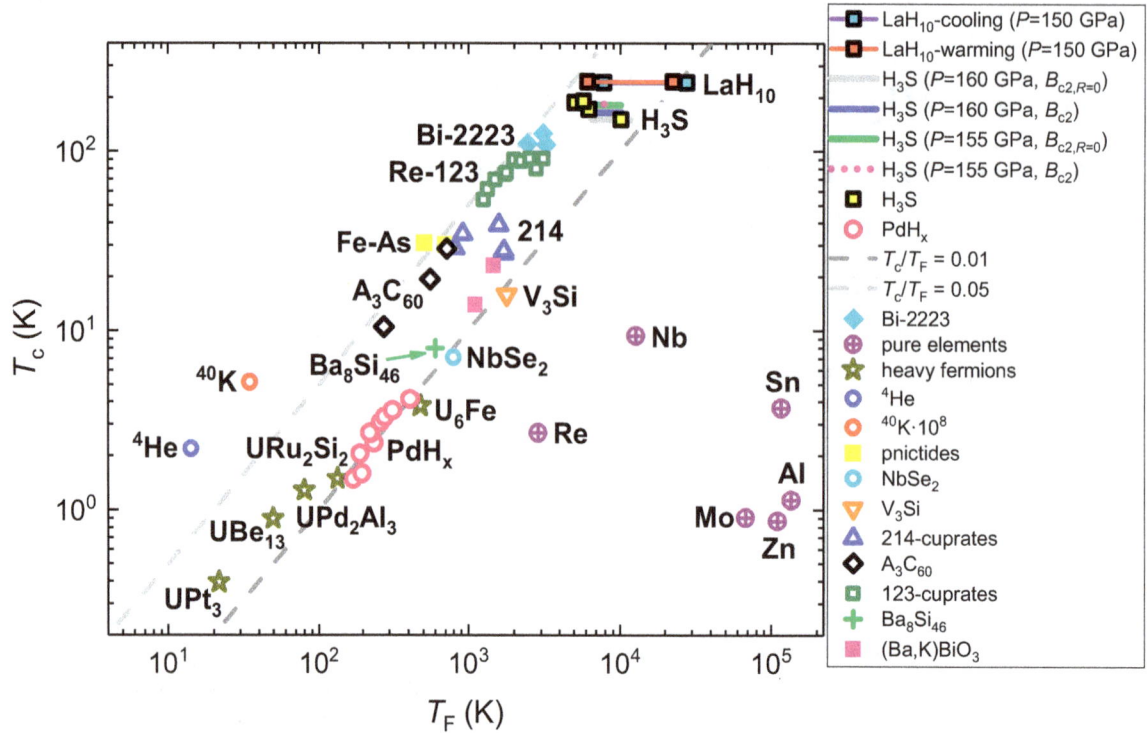

Figure 1. A plot of superconducting transition temperature (T_c) versus Fermi temperature (T_F) obtained for most representative superconducting families. Data was taken from [3,8,9,32–35].

Despite some differences, all intrinsic superconductors can induce superconducting state in non-superconducting materials via the Holm-Meissner effect [36] (also designated as the proximity effect [37,38]). As direct consequence of this, non-dissipative transport current can flow throw the non-superconducting material at superconductor/non-superconductor/superconductor (S/N/S) junctions. The amplitude of this non-dissipative transport current at self-field conditions (when no external magnetic field is applied) (I_c (sf,T)) was given by Ambegaokar and Baratoff (AB) [39,40]:

$$I_c(sf, T) = \frac{\pi \Delta(T)}{2eR_n} tanh\left(\frac{\Delta(T)}{2k_B T}\right), \tag{1}$$

where $\Delta(T)$ is the temperature-dependent superconducting gap, e is the electron charge, normal-state tunneling resistance (R_n) is the normal-state tunneling resistance in the junction, and k_B is the Boltzmann constant.

Many interesting physical effects are expected if the non-superconducting part of the S/N/S junction is made of single-layer graphene (SLG) [41]; multiple-layer graphene (MLG) [42]; graphene-like materials [43]; and many other new 2D- and nano-DCMs, which are under on-going discovery/invention/exploration [44–73]. One interesting class of S/N/S junctions is the non-superconducting part of the device made of topological insulators (TI) [74–81]. Temperature-dependent self-field critical currents (I_c(sf,T)) in this class of junctions were first reported by Veldhorst et al. in Nb/Bi$_2$Te$_3$/Nb [52], and later by Kurter et al. in Nb/Bi$_2$Se$_3$/Nb [53], by Charpentier et al. in Al/Bi$_2$Te$_3$/Al [76], and by other research groups in different systems (extended reference list for studied S/TI/S junctions can be found in Refs. [79,80]).

Recently, Kayyalha et al. [82] have reported on anomalous enhancement of I_c(sf,T) at the Nb/BiSbTeSe$_2$-nanoribon/Nb junction at temperatures of $T \leq 0.25T_c$. They confirmed the effect in all five studied junctions [82], for which TI parts were made of BiSbTeSe$_2$ flakes with thicknesses of $2b$, which varied from 30 nm to 50 nm, and flakes widths of $2a$, which varied from 266 nm to 390 nm. It should be noted that in all these S/TI/S junctions, BiSbTeSe$_2$-nanoribbons thicknesses and widths were smaller than the ground state superconducting coherence length ($2b << 2a < \xi(0) \sim 600$ nm) in these devices [82]. For one junction, made of wider BiSbTeSe$_2$-nanoribon ($2a = 4$ μm (Figure S4 of Supplementary Information of Ref. [82])), measurements were performed only at low temperatures ($T < 2$ K, which is about $T < 0.2T_c$ (taking into account that Nb has $T_c = 8.9$–9.6 K [83])), and thus more experimental studies are required for this 4-μm wide Nb/BiSbTeSe$_2$-nanoribon/Nb junction to see the $I_c(\mathrm{sf},T)$ enhancement.

It needs to be stressed that identical $I_c(\mathrm{sf},T)$ enhancement (or, in another words, $I_c(\mathrm{sf},T)$ upturn [46]) at approximately the same reduced temperature of $T \leq 0.25T_c$ in atomically-thin S/N/S junction was first reported by Calado et al. [46] in MoRe/SLG/MoRe junction in 2015. One year later, less prominent $I_c(\mathrm{sf},T)$ enhancement (which wass, however, still very clearly visible in raw experimental data [84]) in nominally the same MoRe/SLG/MoRe junctions at $T \leq 0.25T_c$ was reported by Borzenets et al. [49]. Based on this, it would be incorrect to attribute the $I_c(\mathrm{sf},T)$ enhancement at low reduced temperatures in Nb/BiSbTeSe$_2$-nanoribon/Nb [82] to unique property of S/TI/S junctions.

In addition, it is important to mention that Kurter et al. [53] were the first to report $I_c(\mathrm{sf},T)$ enhancement at the S/TI/S junction at a reduced temperature of $T \leq 0.25T_c$. At Nb/Bi$_2$Se$_3$/Nb junctions, Bi$_2$Se$_3$ flake had a thickness of $2b = 9$ nm, and, thus, the condition of $2b < \xi_c(0)$ was also satisfied.

Overall, both S/TI/S [53,82], as S/SLG/S [46,49], studied junctions, for which the effect of the low-temperature $I_c(\mathrm{sf},T)$ enhancement was observed to have non-superconducting parts thinner than the ground state out-of-plane coherence lengths, $\xi_c(0)$. SLG thickness is $2b = 0.4$–1.7 nm [84], and thus the condition of $2b << \xi_c(0)$ satisfies any SLG-based junctions.

It should be noted that several intrinsic superconductors exhibit multiple-band superconducting gapping [85,86] and the enhancement of the transition temperature [87–93] when the condition of $2b < \xi_c(0)$ [86] is satisfied. The first discovered material in this class of superconductors is atomically thin FeSe [88–90], in which a 13-fold increase (i.e., 100 K vs 7.5 K) is experimentally registered. Another milestone experimental finding in this field was reported by Liao et al. [43], who observed the effect of new superconducting band opening and T_c enhancement in few layers of stanene (which is the closest counterpart of graphene) by tuning the films' thicknesses. To date, maximal T_c increase due to the effect [86] stands with another single-atomic layer superconductor, T_d-MoTe$_2$, for which Rhodes et al. [93] reported a 30-fold T_c increase when samples were thinneed down to a single atomic layer.

This paper reports the results of an analysis of $I_c(\mathrm{sf},T)$ in Nb/BiSbTeSe$_2$-nanoribbon/Nb [82] and Nb/(Bi$_{0.06}$Sb$_{0.94}$)$_2$Te$_3$/Nb [94] junctions, and of the upper critical field ($B_{c2}(T)$) in Sn/single-layer graphene (SLG)/Sn junctions [95]. In the results, it is shown that a new superconducting band opening phenomenon in atomically thin superconductors, which we proposed earlier [85,86], has further experimental support.

2. Models Description

In [85], it was proposed to substitute $\Delta(T)$ in Equation (1) by analytical expression given by Gross et al. [96], as follows:

$$\Delta(T) = \Delta(0)\tanh\left(\frac{\pi k_B T_c}{\Delta(0)} \sqrt{\eta\left(\frac{\Delta C}{C}\right)\left(\frac{T_c}{T} - 1\right)}\right), \tag{2}$$

where $\Delta(0)$ is the ground-state amplitude of the superconducting band, $\Delta C/C$ is the relative jump in electronic specific heat at the T_c, and $\eta = 2/3$ for s-wave superconductors [96]. In result, T_c, $\Delta C/C$, $\Delta(0)$, and R_n of the S/N/S junction can be deduced by fitting experimental of an $I_c(\mathrm{sf},T)$ dataset to Equation (1) (full expression for Equation (1) is given in Ref. [85]).

In [85], it was shown that S/SLG/S and S/Bi$_2$Se$_3$/S junctions exhibit two-decoupled band superconducting state, for which, for general case of multiple-decoupled bands, I_c(sf,T) can be described by the following equation:

$$I_c(sf, T) = \sum_{i=1}^{N} \frac{\pi\Delta_i(T)}{2eR_{n,i}}\theta(T_{c,i} - T)tanh\left(\frac{\Delta_i(T)}{2k_BT}\right),$$ (3)

where the subscript i refers to the i-band, $\theta(x)$ is the Heaviside step function, and each band has its own independent parameters of $T_{c,i}$, $\Delta C_i/C_i$, $\Delta_i(0)$, and $R_{n,i}$.

It should be noted that multiple-band induced superconductivity in junctions should be detectable by any technique which is sensitive to additional bands crossing the Fermi surface, for instance multiple distinct gaps should be evident in the temperature-dependence of the upper critical field, $B_{c2}(T)$, for which general equation is:

$$B_{c2}(T) = \sum_{i=1}^{N} B_{c2,i}(T)\theta(T_{c,i} - T),$$ (4)

where, within each i-band, the upper critical field can be described by known model. In this paper, four $B_{c2}(T)$ models were used to show that main result is model-independent. For instance, the following was used:

1. Two-fluid Gorter-Casimir (GC) model [97,98], as follows:

$$B_{c2}(T) = \sum_{i=1}^{N}\left[B_{c2,i}(0)\left(1-\left(\frac{T}{T_{c,i}}\right)^2\right)\theta(T_{c,i} - T)\right] = \frac{\phi_0}{2\pi}\sum_{i=1}^{N}\left[\frac{\theta(T_{c,i} - T)}{\xi_i^2(0)}\left(1-\left(\frac{T}{T_{c,i}}\right)^2\right)\right],$$ (5)

where $\varphi_0 = 2.068 \times 10^{-15}$ Wb is flux quantum and ξ_i (0) is the ground state in-plane coherence length of the i–band. This model is widely used for single-band superconductors ranging from 3D near-room-temperature superconducting hydrides [4,7–9,99,100] to 2D superconductors [88,89,95,101].

2. Jones-Hulm-Chandrasekhar (JHC) model [102], as follows:

$$B_{c2}(T) = \frac{\phi_0}{2\pi}\sum_{i=1}^{N}\frac{\theta(T_{c,i} - T)}{\xi_i^2(0)}\left(\frac{1-\left(\frac{T}{T_{c,i}}\right)^2}{1+\left(\frac{T}{T_{c,i}}\right)^2}\right),$$ (6)

3. Werthamer-Helfand-Hohenberg model [103,104], for which we use analytical expression given by Baumgartner et al. [105] (we will designate this model as B-WHH herein), as follows:

$$B_{c2}(T) = \frac{\phi_0}{2\pi}\sum_{i=1}^{N}\frac{\theta(T_{c,i} - T)}{\xi_i^2(0)}\left(\frac{\left(1-\frac{T}{T_{c,i}}\right)-0.153\left(1-\frac{T}{T_{c,i}}\right)^2-0.152\left(1-\frac{T}{T_{c,i}}\right)^4}{0.693}\right),$$ (7)

4. Gor'kov model [106], for which simple analytical expression was given by Jones et al. [102], as follows:

$$B_{c2}(T) = \frac{\phi_0}{2\pi}\sum_{i=1}^{N}\frac{\theta(T_{c,i} - T)}{\xi_i^2(0)}\left(\left(\frac{1.77-0.43\left(\frac{T}{T_{c,i}}\right)^2+0.07\left(\frac{T}{T_{c,i}}\right)^4}{1.77}\right)\left[1-\left(\frac{T}{T_{c,i}}\right)^2\right]\right),$$ (8)

3. Results

3.1. Planar Sn/SLG/Sn Array

Superconductivity in planar graphene junctions varies by the change of the charge carrier density when it moves away from the Dirac point in the dispersion [46,47,51]. This change is usually controlled by the gate voltage (V_g) that applies to the junction. Han et al. [95] reported on a proximity-coupled array of Sn discs with diameter of 400 nm on SLG that were placed in a hexagonal lattice separated by 1 μm between disks centers.

In Figures 2 and 3, we show reported $B_{c2}(T)$ for Sn/SLG/Sn array by Han et al. [95] in their Figures 4 and 5 at gate voltage of $V_g = 30$ V. We defined $B_{c2}(T)$ by two criteria of $R = 0.01$ kΩ (Figure 2) and $R = 0.2$ kΩ (Figure 3). It can be seen that there is an obvious upturn in $B_{c2}(T)$ at $T \leq 0.4T_c$, independent of the upper critical field definition criterion. It should be noted that the upturn occurs at practically the same reduced temperature at which Borzenets et al. [49] observed the $I_c(\mathrm{sf},T)$ enhancement in MoRe/SLG/MoRe junctions.

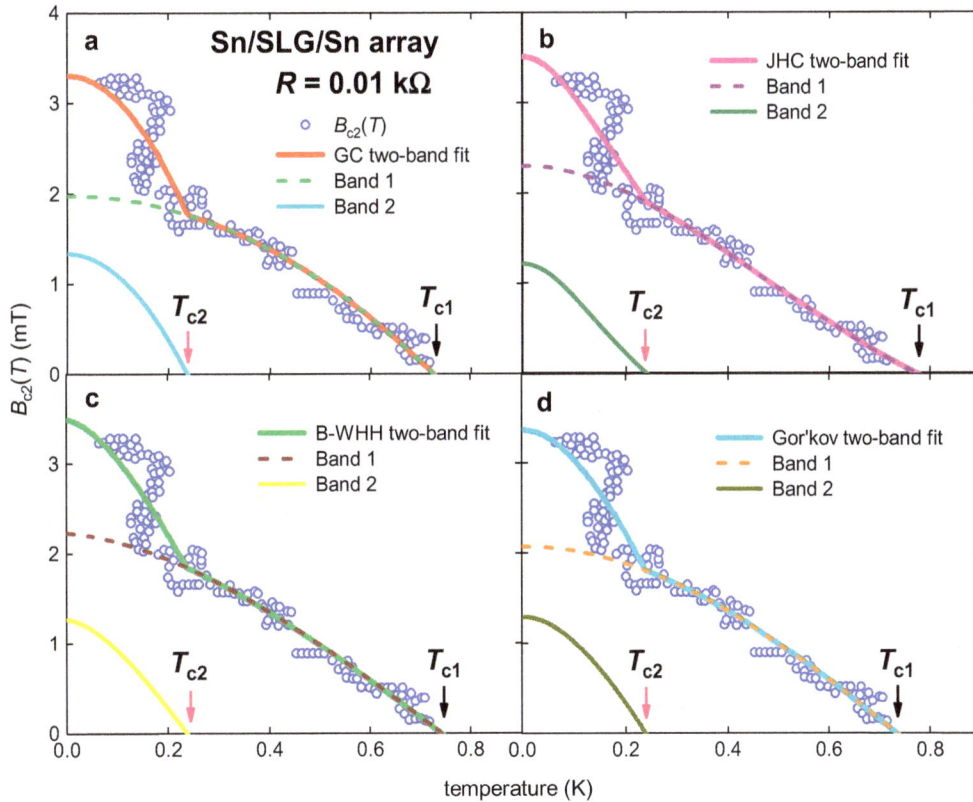

Figure 2. Experimental upper critical field ($B_{c2}(T)$) for Sn/single-layer graphene (SLG)/Sn array at gate voltage of $V_g = 30$ V [95] and data fits to Equations (5)–(8). B_{c2} criterion is $R = 0.01$ kΩ. (**a**) Gorter-Casimir (GC) model. Derived parameters are as follows: $T_{c1} = 0.72 \pm 0.01$ K, $\xi_1(0) = 408 \pm 7$ nm, $T_{c2} = 0.24 \pm 0.01$ K, $\xi_2(0) = 497 \pm 22$ nm, $\frac{T_{c2}}{T_{c1}} = 0.33 \pm 0.02$, and fit quality is $R = 0.9059$; (**b**) Jones-Hulm-Chandrasekhar (JHC) model. Derived parameters are as follows: $T_{c1} = 0.77 \pm 0.02$ K, $\xi_1(0) = 378 \pm 8$ nm, $T_{c2} = 0.24 \pm 0.02$ K, $\xi_2(0) = 521 \pm 34$ nm, $\frac{T_{c2}}{T_{c1}} = 0.31 \pm 0.04$, and fit quality is $R = 0.9101$; (**c**) Werthamer-Helfand-Hohenberg model [103,104], for which we use analytical expression given by Baumgartner et al. [105] (B-WHH) model. Derived parameters: $T_{c1} = 0.74 \pm 0.02$ K, $\xi_1(0) = 385 \pm 7$ nm, $T_{c2} = 0.24 \pm 0.01$ K, $\xi_2(0) = 510 \pm 28$ nm, $\frac{T_{c2}}{T_{c1}} = 0.32 \pm 0.02$, and fit quality is $R = 0.9093$. (**d**) Gor'kov model. Derived parameters: $T_{c1} = 0.74 \pm 0.02$ K, $\xi_1(0) = 398 \pm 7$ nm, $T_{c2} = 0.24 \pm 0.01$ K, $\xi_2(0) = 504 \pm 25$ nm, $\frac{T_{c2}}{T_{c1}} = 0.32 \pm 0.02$, and fit quality is $R = 0.9082$.

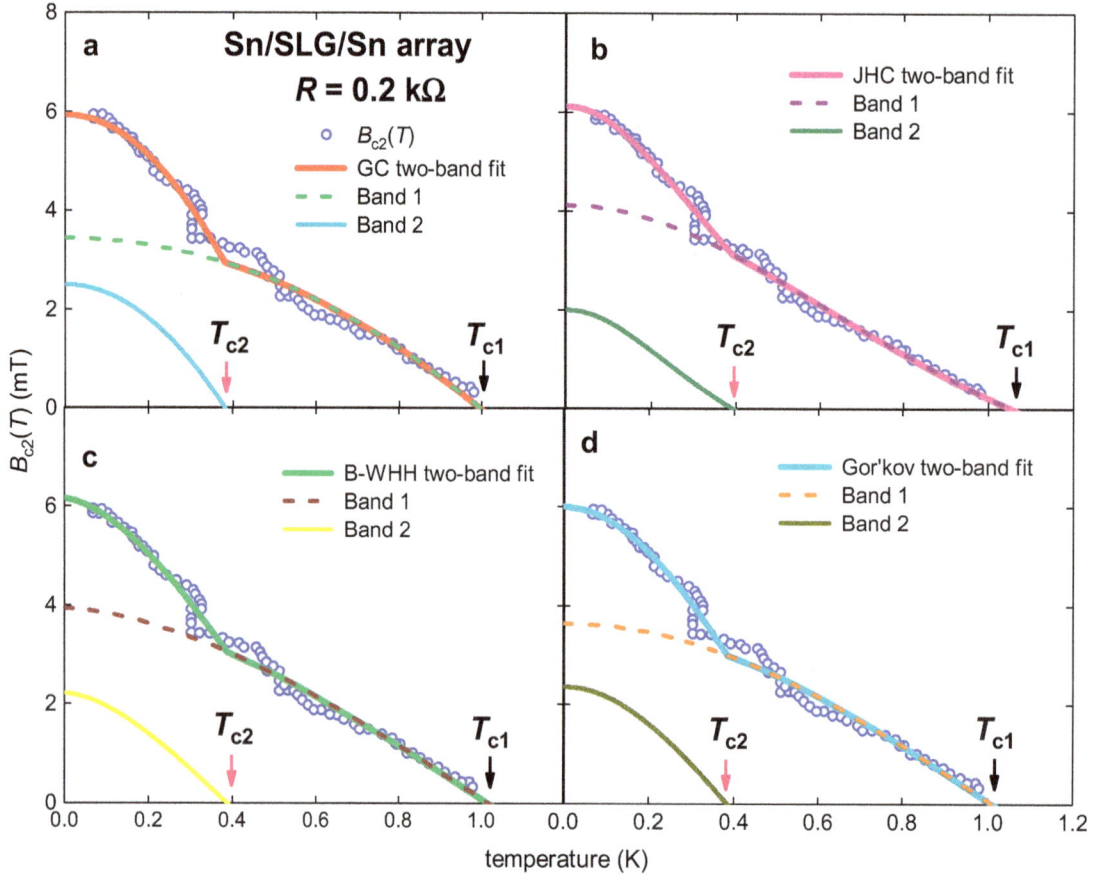

Figure 3. Experimental $B_{c2}(T)$ for Sn/SLG/Sn array at gate voltage of $V_g = 30$ V [95] and data fits to Equations (5)–(8). B_{c2} criterion is $R = 0.2$ kΩ. (**a**) GC model. Derived parameters: $T_{c1} = 1.00 \pm 0.01$ K, $\xi_1(0) = 309 \pm 3$ nm, $T_{c2} = 0.38 \pm 0.01$ K, $\xi_2(0) = 363 \pm 7$ nm, $\frac{T_{c2}}{T_{c1}} = 0.38 \pm 0.01$ and fit quality is $R = 0.9847$; (**b**) JHC model. Derived parameters: $T_{c1} = 1.06 \pm 0.01$ K, $\xi_1(0) = 283 \pm 3$ nm, $T_{c2} = 0.39 \pm 0.01$ K, $\xi_2(0) = 405 \pm 11$ nm, $\frac{T_{c2}}{T_{c1}} = 0.37 \pm 0.01$, and fit quality is $R = 0.9903$; (**c**) B-WHH model. Derived parameters: $T_{c1} = 1.02 \pm 0.01$ K, $\xi_1(0) = 289 \pm 3$ nm, $T_{c2} = 0.39 \pm 0.01$ K, $\xi_2(0) = 385 \pm 9$ nm, $\frac{T_{c2}}{T_{c1}} = 0.38 \pm 0.01$, and fit quality is $R = 0.9885$. (**d**) Gor'kov model. Derived parameters: $T_{c1} = 1.01 \pm 0.01$ K, $\xi_1(0) = 300 \pm 3$ nm, $T_{c2} = 0.38 \pm 0.01$ K, $\xi_2(0) = 374 \pm 7$ nm, $\frac{T_{c2}}{T_{c1}} = 0.38 \pm 0.01$, and fit quality is $R = 0.9873$.

Accordingly, these $B_{c2}(T)$ datasets were fitted to four two-band models (Equations (5)–(8)); they are shown in Figures 2 and 3. Deduced parameters, including the ratio of transition temperatures for two bands, $\frac{T_{c2}}{T_{c1}} = 0.32 \pm 0.02$ for $R = 0.01$ kΩ criterion (Figure 2), and $\frac{T_{c2}}{T_{c1}} = 0.38 \pm 0.01$ for $R = 0.2$ kΩ criterion (Figure 3), agreed with each other despite the fact that experimental $B_{c2}(T)$ data were processed by four different models.

It should be noted that experimental data of Han et al. [95] show that there is a third upturn in $B_{c2}(T)$ that can be seen at lowest experimentally available temperatures of $T < 0.1$ K and applied fields of about $B \sim 4.5$ mT in Figures 4 and 5 [95], if the criterion of $R \sim 0.05$ kΩ (for the $B_{c2}(T)$ definition) are applied.

Despite the fact that authors [61] did not mention the presence of these two upturns in raw experimental $B_{c2}(T)$ data and more detailed measurements of $B_{c2}(T)$ requires to reveal more accurately the position and parameters for the third band, there is already enough experimental evidence that Sn/SLG/Sn array exhibits at least two-superconducting bands gapping, and thus the report of Han et al. [61] supports the idea that atomically thin films exhibit multiple-band superconducting gapping phenomenon [85,86].

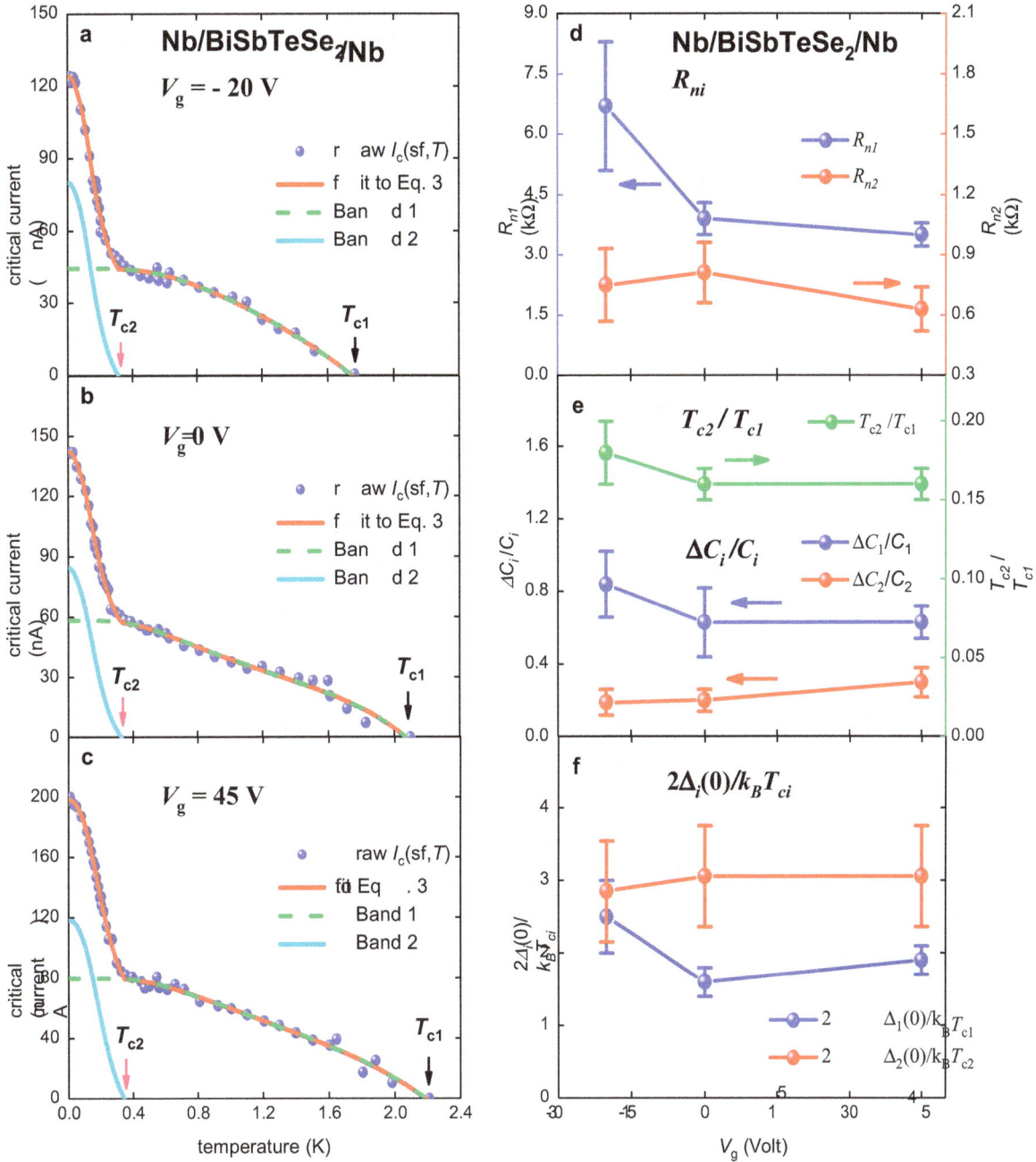

Figure 4. Experimental self-field critical currents ($I_c(\text{sf},T)$) for Nb/BiSbTeSe$_2$-nanoribbon/Nb junction (Sample 1 [82]), data fits to Equation (3), and major deduced parameters. (**a**) Gate voltage $V_g = -20$ V. Derived parameters: $T_{c1} = 1.74 \pm 0.04$ K, $\Delta_1(0) = 190 \pm 40$ μeV, $\Delta C_1/C_1 = 0.84 \pm 0.18$, $2\Delta_1(0)/k_B T_{c1} = 2.5 \pm 0.5$, $R_{n1} = 6.7 \pm 1.6$ kΩ, $T_{c2} = 0.31 \pm 0.02$ K, $\Delta_2(0) = 38.2 \pm 9.7$ μeV, $\Delta C_2/C_2 = 0.19 \pm 0.07$, $2\Delta_2(0)/k_B T_{c2} = 2.85 \pm 0.70$, $R_{n2} = 0.75 \pm 0.18$ kΩ, $\frac{T_{c2}}{T_{c1}} = 0.18 \pm 0.02$, and fit quality is $R = 0.9953$. (**b**) Gate voltage $V_g = 0$ V. Derived parameters: $T_{c1} = 2.07 \pm 0.03$ K, $\Delta_1(0) = 144 \pm 11$ μeV, $\Delta C_1/C_1 = 0.63 \pm 0.19$, $2\Delta_1(0)/k_B T_{c1} = 1.6 \pm 0.2$, $R_{n1} = 3.9 \pm 0.4$ kΩ, $T_{c2} = 0.33 \pm 0.02$ K, $\Delta_2(0) = 43.5 \pm 8.4$ μeV, $\Delta C_2/C_2 = 0.20 \pm 0.06$, $2\Delta_2(0)/k_B T_{c2} = 3.06 \pm 0.70$, $R_{n2} = 0.81 \pm 0.15$ kΩ, $\frac{T_{c2}}{T_{c1}} = 0.16 \pm 0.01$, and fit quality is $R = 0.9965$. (**c**) Gate voltage $V_g = 45$ V. Derived parameters: $T_{c1} = 2.19 \pm 0.03$ K, $\Delta_1(0) = 176 \pm 13$ μeV, $\Delta C_1/C_1 = 0.63 \pm 0.09$, $2\Delta_1(0)/k_B T_{c1} = 1.9 \pm 0.2$, $R_{n1} = 3.5 \pm 0.3$ kΩ, $T_{c2} = 0.34 \pm 0.01$ K, $\Delta_2(0) = 47.6 \pm 8.7$ μeV, $\Delta C_2/C_2 = 0.30 \pm 0.08$, $2\Delta_2(0)/k_B T_{c2} = 3.06 \pm 0.70$, $R_{n2} = 0.63 \pm 0.11$ kΩ, $\frac{T_{c2}}{T_{c1}} = 0.16 \pm 0.01$, and fit quality is $R = 0.9977$. (**d**) Derived R_{ni} as function of gate voltage V_g; (**e**) Derived $\frac{T_{c2}}{T_{c1}}$ and $\Delta C_i/C_i$ as function of gate voltage V_g. (**f**) Derived $2\Delta_i(0)/k_B T_{ci}$ as function of gate voltage V_g.

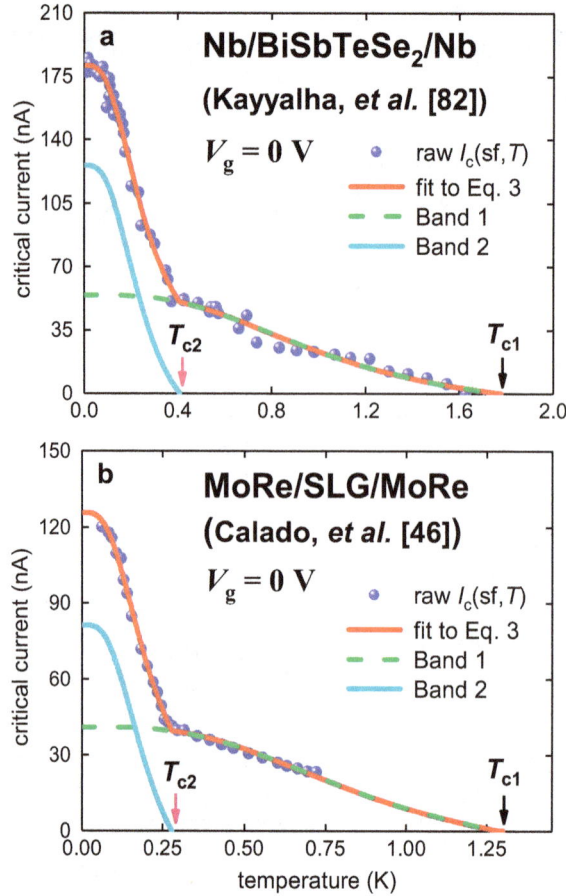

Figure 5. Experimental $I_c(\text{sf},T)$ for two atomically thin Dirac-cone materials (DCM)-based junctions and fits to Equations (3), (9), (10). (**a**) Nb/BiSbTeSe$_2$/Nb (Sample 3 [82]). Derived parameters: $T_{c1} = 1.8 \pm 0.1$ K, $\Delta_1(0) = 179 \pm 51$ μeV, $\Delta C/C = 0.20 \pm 0.04$, $2\Delta(0)/k_B T_c = 2.3 \pm 0.7$, $R_{n1} = 5.2 \pm 1.4$ kΩ, $T_{c2} = 0.41 \pm 0.02$ K, $\Delta_2(0) = 41 \pm 12$ μeV, and $R_{n2} = 0.$ 51 ± 0.15 kΩ, $\frac{T_{c2}}{T_{c1}} = 0.23 \pm 0.02$, $R = 0.9954$; (**b**) MoRe/SLG/MoRe (Sample A [46]). Derived parameters: $T_{c1} = 1.29 \pm 0.07$ K, $\Delta_1(0) = 139 \pm 36$ μeV, $\Delta C/C = 0.30 \pm 0.04$, $2\Delta(0)/k_B T_c = 2.5 \pm 0.7$, $R_{n1} = 5.3 \pm 1.4$ kΩ, $T_{c2} = 0.28 \pm 0.01$ K, $\Delta_2(0) = 30 \pm 8$ μeV, $R_{n2} = 0.56 \pm 0.16$ kΩ, $\frac{T_{c2}}{T_{c1}} = 0.22 \pm 0.01$, and $R = 0.9981$.

3.2. Planar Nb/BiSbTeSe$_2$-Nanoribbon/Nb Junctions

Recently, Kayyalha et al. [82], in their Figure 2 and S1, reported $I_c(\text{sf},T)$ for five Nb/BiSbTeSe$_2$-nanopribbon/Nb junctions at different V_{gs}. The thickness of BiSbTeSe$_2$ flakes varied from $2b = 30$ nm to 50 nm, and based on reported $\xi(0) \sim 600$ nm [82], the condition of $2b < \xi(0)$ [85,86] was satisfied for all junctions.

3.2.1. Nb/BiSbTeSe$_2$-Nanoribbon/Nb Junctions

In Figure 4, we show experimental $I_c(\text{sf},T)$ datasets for Sample 1 [82] reported for three gate voltages: $V_g = -20$ V (Figure 4a), 0 V (Figure 4b), and +45 V (Figure 4c). $I_c(\text{sf},T)$ fits to Equation (3) were performed for all parameters to be free, as experimental raw datasets were rich enough to carry out these sorts of fits.

Deduced R_{ni}, $\frac{T_{c1}}{T_{c2}}$, $\Delta C_i/C_i$, $\Delta_i(0)$, and $\frac{2\Delta_i(0)}{k_B T_{c,i}}$ for both superconducting bands as functions of V_g are shown in Figure 4d–f.

It needs to be stressed that within the range of uncertainties, deduced R_{n1} values are well agree with directly measured values by Kayyalha et al. [82] (these values are reported in Figure 1a of Ref. [82]). Increasingly often, measured raw $I_c(\text{sf},T)$ data, and especially at high reduced temperatures, are required to reduce the uncertainty for R_{n1} values.

Most notable outcome of our analysis is that, within uncertainty ranges, fundamental superconducting parameters for both bands, including the ratio of $\frac{T_{c2}}{T_{c1}}$, remain unchanged vs. gate voltage variation in the range from −20 V to 45 V. This means that two-band superconducting state in Nb/BiSbTeSe$_2$-nanoribbon/Nb junction is very robust and mostly independent from the change in V_g. This is an unexpected result, because there is generally accepted view that because V_g is determined the electronic state in 2D-systems in the normal state, it should also determine the superconducting state. However, performed analysis shows that this is not a case in general. As was already mentioned, there is a need for more frequent measurements of raw $I_c(sf,T)$ data, which will allow one to reduce uncertainties for all deduced parameters.

3.2.2. Nb/BiSbTeSe$_2$-Nanoribbon/Nb Junction

In Figure 5a, we show experimental $I_c(sf,T)$ dataset for Nb/BiSbTeSe$_2$-nanoribbon/Nb (Sample 3) reported by Kayyalha et al. [82].

Raw experimental $I_c(sf,T)$ dataset for this sample was not reach enough at $T \geq 0.6$, and thus we cannot perform the fit to Equation (3) for all parameters to be free. To run the model (Equation (3)), we make the same model restriction, as we did in our previous work [85]:

$$\frac{\Delta C_1}{C_1} = \frac{\Delta C_2}{C_2} = \frac{\Delta C}{C}, \tag{9}$$

$$\frac{2\Delta_1(0)}{k_B T_1} = \frac{2\Delta_2(0)}{k_B T_2} = \frac{2\Delta(0)}{k_B T_c}, \tag{10}$$

i.e., we forced $\Delta C_i/C_i$ and $\frac{2\Delta_i(0)}{k_B T_{c,i}}$ values to be the same for both bands. As a result, deduced R_{ni}, T_{ci}, $\frac{T_{c2}}{T_{c1}} \sim \frac{1}{4}$, $\Delta C/C$, $\Delta_i(0)$ and $\frac{2\Delta(0)}{k_B T_c}$ for this junction are very close to ones deduced for Sample 1 (Figure 4).

3.3. Planar MoRe/SLG/MoRe Junction

To demonstrate that findings in regard of Nb/BiSbTeSe$_2$-nanoribbon/Nb junctions are generic for a much wide range of atomically-thin DCM-based Josephson junctions, in Figure 5b we show raw $I_c(sf,T)$ dataset and fit to the model (Equation (3)) for MoRe/SLG/MoRe reported by Calado et al. [46] for their Device A [46]. For the $I_c(sf,T)$ fit for this device, we used the same parameters restrictions (Equations (9) and (10)), as for Nb/BiSbTeSe$_2$-nanoribbon/Nb Sample 3 [82].

In work [85], this $I_c(sf,T)$ dataset for MoRe/SLG/MoRe Device A [46] was already analyzed. What was found in this paper was that there is remarkable and practically undistinguishable similarity between reduced $I_c(sf,T)$ datasets and fits for Nb/BiSbTeSe$_2$-nanoribbon/Nb [48] and MoRe/SLG/MoRe [46] junctions (Figure 5). In an attempt to further extend atomically-thin S/DCM/S junctions, in next Section we analyze $I_c(sf,T)$ data for Nb/(Bi$_{0.06}$Sb$_{0.94}$)$_2$Te$_3$-nanoribbon/Nb junction [94].

3.4. Planar Nb/(Bi$_{0.06}$Sb$_{0.94}$)$_2$Te$_3$-Nanoribbon/Nb Junction

In Figure 6, $I_c(sf,T)$ in Nb/(Bi$_{0.06}$Sb$_{0.94}$)$_2$Te$_3$-nanoribbon/Nb reported by Schüffelgen et al. [94] are shown. TI nanoribbon has thickness of $2b = 10$ nm, and, thus the condition of $2b < \xi(0)$ [85,86] is satisfied.

Due to the fact that the reported $I_c(sf,T)$ dataset was not rich enough at high reduced temperatures, we restricted the model by utilizing Equations (9) and (10). Overall, fitted curves and all deduced parameters were very close to one reported by Borzenets et al. [49] for MoRe/SLG/MoRe junctions (which we processed and showed in our previous paper [85] in Figure 7).

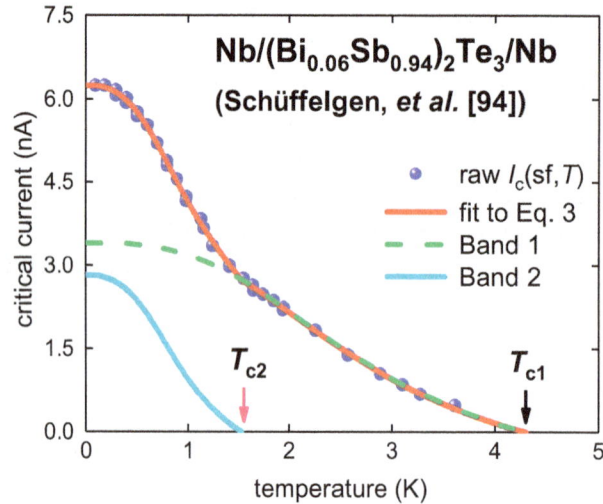

Figure 6. Experimental $I_c(sf,T)$ for atomically thin DCM-based junction Nb/(Bi$_{0.06}$Sb$_{0.94}$)$_2$Te$_3$-nanoribbon/Nb [94] and fit to Equations (3), (9) and (10). Derived parameters: $T_{c1} = 4.30 \pm 0.07$ K, $\Delta_1(0) = 530 \pm 7$ μeV, $\Delta C/C = 0.28 \pm 0.04$, $2\Delta(0)/k_B T_c = 2.87 \pm 0.05$, $R_{n1} = 244 \pm 32$ Ω, $T_{c2} = 1.53 \pm 0.03$ K, $\Delta_2(0) = 189 \pm 3$ μeV, $R_{n2} = 105 \pm 16$ Ω, $\frac{T_{c2}}{T_{c1}} = 0.36 \pm 0.01$ and $R = 0.9995$.

4. Discussion

It should be noted that the idea of multiple-band superconductivity in bulk superconductors was proposed by Suhl et al. [107] in 1959, and it took more than forty years to discover the first two-band BCS superconductor (MgB$_2$) [108] and about fifty years to discover multiple-band iron-based superconductors in 2006 [109]. Interband scattering in these materials have been discussed in details elsewhere [24,110–112].

It needs to be mentioned that Shalnikov discussed the discovery of the T_c increase in thin films [113], who reported the effect for lead and tin thin films more than eighty years ago. Three-fold increase in the transition temperature of thin granular Al films was reported three decades later by Cohen and Abeles [114], and the superconductivity in granular Al films is still active scientific topic [115,116]; the discussion of this effect in intrinsic superconductors, however, is beyond the scope of this paper.

It needs to be stressed that Calado et al. [46] in 2015 emphasized the necessity for a new model to explain the upturn in $I_c(sf,T)$ registered in their MoRe/SLG/MoRe junction (Device A) at $T \sim \frac{1}{4}T_c$ (which we show in Figure 5b), because this $I_c(sf,T)$ enhancement was not possible to explain using either the Eilenberger model (which is used to describe clean S/N/S junctions) [117] or the Usadel model (which describes diffusive S/N/S junctions) [118].

Our explanation for this upturn [85], which is well aligned with the $I_c(sf,T)$ upturn in natural atomically thin superconductors [86], is that this $I_c(sf,T)$ enhancement is due to a new superconducting band opening phenomenon when sample dimensions become smaller than some critical value. For this critical value, we proposed to use [86] the out-of-plane coherence length, $\xi_c(0)$, which is still, after expanding our analysis herein, a good choice for the criterion.

It should be pointed out that this new opening band phenomenon does not necessarily cause the increase in observed transition temperature in comparison with "bulk" material. For instance, in pure Nb films [119], this new "thin film" band has lower transition temperature in comparison with "bulk" band [86]. In these circumstances, the researchers are not able to explore the further creation of devices or films for new superconducting band.

Thus, in many atomically thin films, which in fact exhibit a new band opening phenomenon, this effect has not been registered yet, because there is no guarantee that something important/interesting can be observed at low reduced temperatures, well below "bulk" or observed T_c for given atomically thin film.

It should be noted that the effect of new superconducting band opening [86] in atomically thin films can be detected using any experimental techniques that are sensitive to additional band(s) crossing the Fermi surface. To date, most evident confirmations for the phenomenon are related to the $I_c(sf,T)$ upturn [43,85,86] and $B_{c2}(T)$ upturn [43]; however, other techniques also should be able to detect this.

In this regard, the observation of the $I_c(sf,T)$ upturn reported by Li et al. [78] in their Figure 4a at $T = 2.5$ K in Nb/Cd_3As_2-nanowire/Nb junction should be mentioned. However, raw experimental $I_c(sf,T)$ dataset [78] was limited by measurements at $T < 3.5$ K, and thus we are not able to perform the analysis for this very interesting atomically-narrow S/TI/S junction at the moment.

There are very interesting results reported by Sasaki et al. [120] and by Andersen et al. [121], who found that temperature-dependent $B_{c2}(T)$, in nanostructures of topological insulators, cannot be explained by single-band WHH model [103,104]. However, reported, to date, raw experimental $B_{c2}(T)$ datasets [119,120] are insufficient to perform two-band model fit to reveal the presence of additional band at low reduced temperatures in these structures.

We also need to mention an interesting research field of interfaced superconductivity [71,122–125], where, as was proposed earlier, the enhancement of the superconductivity is also due to new superconducting band opening [86]. However, the discussion of this interesting field is beyond the scope of this paper.

5. Conclusions

In this paper, an analysis of recently reported experimental data on induced superconducting state in atomically thin Dirac-cone films was performed. It was shown that the phenomenon of the new superconducting band opening in atomically thin films [85,86], when the film thickness becomes thinner than the ground state out-of-plane coherence length, $\xi_c(0)$, can be extended to an induced superconducting state in atomically thin DCM, as one was established before for natural superconductors, i.e., pure Nb, exfoliated $2H$-TaS_2, double-atomic layer FeSe, and a few layers of stanene [9].

Funding: This research was funded by the State Assignment of Minobrnauki of Russia, theme "Pressure" No. AAAA-A18-118020190104-3, and by Act 211 Government of the Russian Federation, contract No. 02.A03.21.0006.

Acknowledgments: Author would like to thank Srijit Goswami and Lieven Vandersypen (Kavli Institute of Nanoscience, Delft University of Technology, The Netherlands) for providing raw self-field critical current data for the MoRe/SLG/MoRe devices analyzed in this work.

Conflicts of Interest: The funders had no role in the design of the study; in the collection, analyses, or interpretation of data; in the writing of the manuscript; or in the decision to publish the results.

References

1. Hirsch, J.E.; Maple, M.B.; Marsiglio, F. Superconducting materials classes: Introduction and overview. *Physica C* **2015**, *514*, 1–8. [CrossRef]

2. Uemura, Y.J.; Luke, G.M.; Sternlieb, B.J.; Brewer, J.H.; Carolan, J.F.; Hardy, W.N.; Kadono, R.; Kempton, J.R.; Kiefl, R.F.; Kreitzman, S.R. Universal correlations between T_c and $\frac{n_s}{m^*}$ (carrier density over effective mass) in high-T_c cuprate. *Phys. Rev. Lett.* **1989**, *62*, 2317–2320. [CrossRef] [PubMed]

3. Uemura, Y.J. Condensation, excitation, pairing, and superfluid density in high-T_c superconductors: The magnetic resonance mode as a roton analogue and a possible spin-mediated pairing. *J. Phys. Condens. Matter* **2004**, *16*, S4515–S4540. [CrossRef]

4. Drozdov, A.P.; Eremets, M.I.; Troyan, I.A.; Ksenofontov, V.; Shylin, S.I. Conventional superconductivity at 203 kelvin at high pressures in the sulfur hydride system. *Nature* **2015**, *525*, 73–76. [CrossRef] [PubMed]

5. Somayazulu, M.; Ahart, M.; Mishra, A.K.; Geballe, Z.M.; Baldini, M.; Meng, Y.; Struzhkin, V.V.; Hemley, R.J. Evidence for superconductivity above 260 K in lanthanum superhydride at megabar pressures. *Phys. Rev. Lett.* **2019**, *122*, 027001. [CrossRef] [PubMed]

6. Mozaffari, S.; Sun, D.; Minkov, V.S.; Drozdov, A.P.; Knyazev, D.; Betts, J.B.; Einaga, M.; Shimizu, K.; Eremets, M.I.; Balicas, L. Superconducting phase-diagram of H_3S under high magnetic fields. *Nat. Commun.* **2019**, *10*, 2522. [CrossRef] [PubMed]

7. Drozdov, A.P.; Kong, P.P.; Minkov, V.S.; Besedin, S.P.; Kuzovnikov, M.A.; Mozaffari, S.; Balicas, L.; Balakirev, F.F.; Graf, D.E.; Prakapenka, V.B. Superconductivity at 250 K in lanthanum hydride under high pressures. *Nature* **2019**, *569*, 528–531. [CrossRef] [PubMed]

8. Talantsev, E.F. Classifying superconductivity in compressed H_3S. *Mod. Phys. Lett. B* **2019**, *33*, 1950195. [CrossRef]

9. Talantsev, E.F. Classifying hydrogen-rich superconductors. *Mater. Res. Express* **2019**, *6*, 106002. [CrossRef]

10. Hardy, W.N.; Bonn, D.A.; Morgan, D.C.; Liang, R.; Zhang, K. Precision measurements of the temperature dependence of l in $YBa_2Cu_3O_{6.95}$: Strong evidence for nodes in the gap function. *Phys. Rev. Lett.* **1993**, *70*, 3999–4002. [CrossRef]

11. Bednorz, J.G.; Mueller, K.A. Possible high Tc superconductivity in the Ba-La-Cu-O system. *Z. Phys. B* **1986**, *64*, 189–193. [CrossRef]

12. Won, H.; Maki, K. *d*-wave superconductor as a model of high-T_c superconductors. *Phys. Rev. B* **1994**, *49*, 1397–1402. [CrossRef] [PubMed]

13. Hirsch, J.E. Hole superconductivity. *Phys. Lett. A* **1989**, *134*, 451–455. [CrossRef]

14. Bianconi, A. On the possibility of new high T_c superconductors by producing metal heterostructures as in cuprate perovskites. *Solid State Commun.* **1994**, *89*, 933–936. [CrossRef]

15. Bouquet, F.; Wang, Y.; Fisher, R.A.; Hinks, D.G.; Jorgensen, J.D.; Junod, A.; Phillips, N.E. Phenomenological two-gap model for the specific heat of MgB_2. *Europhys. Lett.* **2001**, *56*, 856–862. [CrossRef]

16. Bauer, E.; Paul, C.H.; Berger, S.T.; Majumdar, S.; Michor, H.; Giovannini, M.; Saccone, A.; Bianconi, A. Thermal conductivity of superconducting MgB_2. *J. Phys. Condens. Matter* **2001**, *13*, L487–L494. [CrossRef]

17. Carrington, A.; Manzano, F. Magnetic penetration depth of MgB_2. *Physica C* **2003**, *385*, 205–214. [CrossRef]

18. Agrestini, S.; Metallo, C.; Filippi, M.; Simonelli, L.; Campi, G.; Sanipoli, C.; Liarokapis, E.; De Negri, S.; Giovannini, M.; Saccone, A. Substitution of Sc for Mg in MgB_2: Effects on transition temperature and Kohn anomaly. *Phys. Rev. B* **2004**, *70*, 134514. [CrossRef]

19. Mazin, I.I.; Singh, D.J.; Johannes, M.D.; Du, M.H. Unconventional superconductivity with a sign reversal in the order parameter of $LaFeAsO_{1-x}F_x$. *Phys. Rev. Lett.* **2008**, *101*, 057003. [CrossRef]

20. Kuroki, K.; Onari, S.; Arita, R.; Usui, H.; Tanaka, Y.; Kontani, H.; Aoki, H. Unconventional pairing originating from the disconnected Fermi surfaces of superconducting $LaFeAsO_{1-x}F_x$. *Phys. Rev. Lett.* **2008**, *101*, 087004. [CrossRef] [PubMed]

21. Innocenti, D.; Caprara, S.; Poccia, N.; Ricci, A.; Valletta, A.; Bianconi, A. Shape resonance for the anisotropic superconducting gaps near a Lifshitz transition: The effect of electron hopping between layers. *Supercond. Sci. Technol.* **2011**, *24*, 015012. [CrossRef]

22. Bianconi, A.; Innocenti, D.; Valletta, A.; Perali, A. Shape resonances in superconducting gaps in a 2DEG at oxide-oxide interface. *J. Phys. Conf. Ser.* **2014**, *529*, 012007. [CrossRef]

23. Hosono, H.; Tanabe, K.; Takayama-Muromachi, E.; Kageyama, H.; Yamanaka, S.; Kumakura, H.; Nohara, M.; Hiramatsu, H.; Fujitsu, S. Exploration of new superconductors and functional materials, and fabrication of superconducting tapes and wires of iron pnictides. *Sci. Technol. Adv. Mater.* **2015**, *16*, 033503. [CrossRef] [PubMed]

24. Hosono, H.; Kuroki, K. Iron-based superconductors: Current status of materials and pairing mechanism. *Physica C* **2015**, *514*, 399–422. [CrossRef]

25. Bianconi, A.; Jarlborg, T. Superconductivity above the lowest Earth temperature in pressurized sulfur hydride. *EPL (Europhys. Lett.)* **2015**, *112*, 37001. [CrossRef]

26. Hirsch, J.E.; Marsiglio, F. Hole superconductivity in H_2S and other sulfides under high pressure. *Physica C* **2015**, *511*, 45–49. [CrossRef]

27. Souza, T.X.R.; Marsiglio, F. Systematic study of the superconducting critical temperature in two- and three-dimensional tight-binding models: A possible scenario for superconducting H_3S. *Phys. Rev. B* **2016**, *94*, 184509. [CrossRef]

28. Harshman, D.R.; Fiory, A.T. Compressed H_3S: Inter-sublattice Coulomb coupling in a high-T_c superconductor. *J. Phys. Condens. Matter* **2017**, *29*, 445702. [CrossRef]

29. Bang, Y.; Stewart, G.R. Superconducting properties of the s±-wave state: Fe-based superconductors. *J. Phys. Condens. Matter* **2017**, *29*, 123003. [CrossRef]

30. Kaplan, D.; Imry, Y. High-temperature superconductivity using a model of hydrogen bonds. *Proc. Natl. Acad. Sci. USA* **2018**, *115*, 5709–5713. [CrossRef]

31. Moskvin, A.S.; Panov, Y.D. Topological structures in unconventional scenario for 2D cuprates. *J. Supercond. Nov. Magn.* **2019**, *32*, 61–84. [CrossRef]

32. Ye, J.T.; Zhang, Y.J.; Akashi, R.; Bahramy, M.S.; Arita, R.; Iwasa, Y. Superconducting dome in a gate-tuned band insulator. *Science* **2012**, *338*, 1193–1196. [CrossRef] [PubMed]

33. Qian, T.; Wang, X.-P.; Jin, W.-C.; Zhang, P.; Richard, P.; Xu, G.; Dai, X.; Fang, Z.; Guo, J.-G.; Chen, X.-L. Absence of a holelike Fermi surface for the iron-based $K_{0.8}Fe_{1.7}Se_2$ superconductor revealed by angle-resolved photoemission spectroscopy. *Phys. Rev. Lett.* **2011**, *106*, 187001. [CrossRef] [PubMed]

34. Hashimoto, K.; Cho, K.; Shibauchi, T.; Kasahara, S.; Mizukami, Y.; Katsumata, R.; Tsuruhara, Y.; Terashima, T.; Ikeda, H.; Tanatar, M.A. A sharp peak of the zero-temperature penetration depth at optimal composition in $BaFe_2(As_{1-x}P_x)_2$. *Science* **2012**, *336*, 1554–1557. [CrossRef] [PubMed]

35. Shang, T.; Philippe, J.; Verezhak, J.A.T.; Guguchia, Z.; Zhao, J.Z.; Chang, L.-J.; Lee, M.K.; Gawryluk, D.J.; Pomjakushina, E.; Shi, M. Nodeless superconductivity and preserved time-reversal symmetry in the noncentrosymmetric Mo_3P superconductor. *Phys. Rev. B* **2019**, *99*, 184513. [CrossRef]

36. Holm, R.; Meissner, W. Messungen mit Hilfe von flüssigem Helium. XIII. Kontaktwiderstand zwischen Supraleitern und Nichtsupraleitern (Measurements using liquid helium. XIII. Contact resistance between superconductors and non-superconductors). *Z. Phys.* **1932**, *74*, 715–735. [CrossRef]

37. Natterer, F.D.; Ha, J.; Baek, H.; Zhang, D.; Cullen, W.G.; Zhitenev, N.B.; Kuk, Y.; Stroscio, J.A. Scanning tunneling spectroscopy of proximity superconductivity in epitaxial multilayer graphene. *Phys. Rev. B* **2016**, *93*, 045406. [CrossRef]

38. Kim, H.; Miyata, Y.; Hasegawa, Y. Superconducting proximity effect on a Rashba-split Pb/Ge(111)-$\sqrt{3} \times \sqrt{3}$ surface. *Supercond. Sci. Technol.* **2016**, *29*, 084006. [CrossRef]

39. Ambegaokar, V.; Baratoff, A. Tunneling between superconductors. *Phys. Rev. Lett.* **1963**, *10*, 486–489. [CrossRef]

40. Ambegaokar, V.; Baratoff, A. Errata: Tunneling between superconductors. *Phys. Rev. Lett.* **1963**, *11*, 104. [CrossRef]

41. Novoselov, K.S.; Geim, A.K.; Morozov, S.V.; Jiang, D.; Zhang, Y.; Dubonos, S.V.; Grigorieva, I.V.; Firsov, A.A. Electric field effect in atomically thin carbon films. *Science* **2004**, *306*, 666–669. [CrossRef] [PubMed]

42. Lee, G.H.; Kim, S.; Jhi, S.-H.; Lee, H.-J. Ultimately short ballistic vertical graphene Josephson junctions. *Nat. Commun.* **2015**, *6*, 6181. [CrossRef] [PubMed]

43. Liao, M.; Zang, Y.; Guan, Z.; Li, H.; Gong, Y.; Zhu, K.; Hu, X.-P.; Zhang, D.; Xu, Y.; Wang, Y.-Y. Superconductivity in few-layer stanene. *Nat. Phys.* **2018**, *14*, 344–348. [CrossRef]

44. Heersche, H.B.; Jarillo-Herrero, P.; Oostinga, J.B.; Vandersypen, L.M.K.; Morpurgo, A.F. Bipolar supercurrent in graphene. *Nature* **2007**, *446*, 56–59. [CrossRef] [PubMed]

45. Du, X.; Skachko, I.; Andrei, E.Y. Josephson current and multiple Andreev reflections in graphene SNS junctions. *Phys. Rev. B* **2008**, *77*, 184507. [CrossRef]

46. Calado, V.E.; Goswami, S.; Nanda, G.; Diez, M.; Akhmerov, A.R.; Watanabe, K.; Taniguchi, T.; Klapwijk, T.M.; Vandersypen, L.M.K. Ballistic Josephson junctions in edge-contacted graphene. *Nat. Nanotechnol.* **2015**, *10*, 761–764. [CrossRef] [PubMed]

47. Ben Shalom, M.; Zhu, M.J.; Fal'ko, V.I.; Mishchenko, A.; Kretinin, A.V.; Novoselov, K.S.; Woods, C.R.; Watanabe, K.; Taniguchi, T.; Geim, A.K. Quantum oscillations of the critical current and high-field superconducting proximity in ballistic graphene. *Nat. Phys.* **2016**, *12*, 318–322. [CrossRef]

48. Amet, F.; Ke, C.T.; Borzenets, I.V.; Wang, J.; Watanabe, K.; Taniguchi, T.; Deacon, R.S.; Yamamoto, M.; Bomze, Y.; Tarucha, S. Supercurrent in the quantum Hall regime. *Science* **2016**, *352*, 966–969. [CrossRef] [PubMed]

49. Borzenets, I.V.; Amet, F.; Ke, C.T.; Draelos, A.W.; Wei, M.T.; Seredinski, A.; Watanabe, K.; Taniguchi, T.; Bomze, Y.; Yamamoto, M. Ballistic graphene Josephson junctions from the short to the long junction regimes. *Phys. Rev. Lett.* **2016**, *117*, 237002. [CrossRef] [PubMed]

50. Island, J.O.; Steele, G.A.; van der Zant, H.S.J.; Castellanos-Gomez, A. Thickness dependent interlayer transport in vertical MoS_2 Josephson. *2D Mater.* **2016**, *3*, 031002. [CrossRef]

51. Zhu, M.J.; Kretinin, A.V.; Thompson, M.D.; Bandurin, D.A.; Hu, S.; Yu, G.L.; Birkbeck, J.; Mishchenko, A.; Vera-Marun, I.J.; Watanabe, K. Edge currents shunt the insulating bulk in gapped graphene. *Nat. Commun.* **2017**, *8*, 14552. [CrossRef] [PubMed]

52. Veldhorst, M.; Snelder, M.; Hoek, M.; Gang, T.; Guduru, V.K.; Wang, X.L.; Zeitler, U.; Van Der Wiel, W.G.; Golubov, A.A.; Hilgenkamp, H. Josephson supercurrent through a topological insulator surface state. *Nat. Mater.* **2012**, *11*, 417–421. [CrossRef] [PubMed]

53. Kurter, C.; Finck, A.D.K.; Hor, Y.S.; Van Harlingen, D.J. Evidence for an anomalous current–phase relation in topological insulator Josephson junctions. *Nat. Commun.* **2015**, *6*, 7130. [CrossRef] [PubMed]

54. Nanda, G.; Aguilera-Servin, J.L.; Rakyta, P.; Kormányos, A.; Kleiner, R.; Koelle, D.; Watanabe, K.; Taniguchi, T.; Vandersypen, L.M.K.; Goswami, S. Current-phase relation of ballistic graphene Josephson junctions. *Nano Lett.* **2017**, *17*, 3396–3401. [CrossRef] [PubMed]

55. Yankowitz, M.; Chen, S.; Polshyn, H.; Watanabe, K.; Taniguchi, T.; Graf, D.; Young, A.F.; Dean, C.R. Tuning superconductivity in twisted bilayer graphene. *Science* **2019**, *363*, 1059–1064. [CrossRef] [PubMed]

56. Lucignano, P.; Alfè, D.; Cataudella, V.; Ninno, D.; Cantele, G. The crucial role of atomic corrugation on the flat bands and energy gaps of twisted bilayer graphene at the "magic angle" $\theta \sim 1.08°$. *Phys. Rev. B* **2019**, *99*, 195419. [CrossRef]

57. Giubileo, F.; Romeo, F.; Di Bartolomeo, A.; Mizuguchi, Y.; Romano, P. Probing unconventional pairing in $LaO_{0.5}F_{0.5}BiS_2$ layered superconductor by point contact spectroscopy. *J. Phys. Chem. Solids* **2018**, *118*, 192–199. [CrossRef]

58. Kizilaslan, O.; Truccato, M.; Simsek, Y.; Aksan, M.A.; Koval, Y.; Müller, P. Interlayer tunneling spectroscopy of mixed-phase BSCCO superconducting whiskers. *Supercond. Sci. Technol.* **2016**, *29*, 065013. [CrossRef]

59. Fête, A.; Rossi, L.; Augieri, A.; Senatore, C. Ionic liquid gating of ultra-thin YBa2Cu3O7-x films. *Appl. Phys. Lett.* **2016**, *109*, 192601. [CrossRef]

60. Mueller, P.; Koval, Y.; Lazareva, Y.; Steiner, C.; Wurmehl, S.; Buechner, B.; Stuerzer, T.; Johrendt, D. C-axis transport of pnictide superconductors. *Phys. Status Solidi B* **2017**, *254*, 1600157. [CrossRef]

61. Fête, A.; Senatore, C. Strong improvement of the transport characteristics of $YBa_2Cu_3O_{7-x}$ grain boundaries using ionic liquid gating. *Sci. Rep.* **2017**, *8*, 17703. [CrossRef] [PubMed]

62. Paradiso, N.; Nguyen, A.-T.; Kloss, K.E.; Strunk, C. Phase slip lines in superconducting few-layer NbSe2 crystals. *2D Mater.* **2019**, *6*, 025039. [CrossRef]

63. Wu, Y.; Xiao, H.; Li, Q.; Li, X.; Li, Z.; Mu, G.; Jiang, D.; Hu, T.; Xie, X.M. The transport properties in graphene/single-unit-cell cuprates van der Waals heterostructure. *Supercond. Sci. Technol.* **2019**, *32*, 085007. [CrossRef]

64. Guo, J.G.; Chen, X.; Jia, X.Y.; Zhang, Q.H.; Liu, N.; Lei, H.C.; Li, S.Y.; Gu, L.; Jin, S.F.; Chen, X.L. Quasi-two-dimensional superconductivity from dimerization of atomically ordered $AuTe_2Se_{4/3}$ cubes. *Nat. Commun.* **2017**, *8*, 871. [CrossRef] [PubMed]

65. Pan, J.; Guo, C.; Song, C.; Lai, X.; Li, H.; Zhao, W.; Zhang, H.; Mu, G.; Bu, K.; Lin, T. Enhanced superconductivity in restacked TaS2 nanosheets. *J. Am. Chem. Soc.* **2017**, *139*, 4623–4626. [CrossRef] [PubMed]

66. Ma, Y.; Pan, J.; Guo, C.; Zhang, X.; Wang, L.; Hu, T.; Mu, G.; Huang, F.; Xie, X. Unusual evolution of B_{c2} and T_c with inclined fields in restacked TaS2 nanosheets. *NPJ Quantum Mater.* **2018**, *3*, 34. [CrossRef]

67. Desrat, W.; Moret, M.; Briot, O.; Ngo, T.-H.; Piot, B.A.; Jabakhanji, B.; Gil, B. Superconducting Ga/GaSe layers grown by van der Waals epitaxy. *Mater. Res. Express* **2018**, *5*, 045901. [CrossRef]

68. Liu, C.; Lian, C.-S.; Liao, M.-H.; Wang, Y.; Zhong, Y.; Ding, C.; Li, W.; Song, C.-L.; He, K.; Ma, X.-C. Two-dimensional superconductivity and topological states in PdTe2 thin films. *Phys. Rev. Mater.* **2018**, *2*, 094001. [CrossRef]

69. Peng, J.; Yu, Z.; Wu, J.; Zhou, Y.; Guo, Y.; Li, Z.; Zhao, J.; Wu, C.; Xie, Y. Disorder enhanced superconductivity toward TaS2 monolayer. *ACS Nano* **2018**, *12*, 9461–9466. [CrossRef]

70. De La Barrera, S.C.; Sinko, M.R.; Gopalan, D.P.; Sivadas, N.; Seyler, K.L.; Watanabe, K.; Taniguchi, T.; Tsen, A.W.; Xu, X.; Xiao, D. Tuning Ising superconductivity with layer and spin-orbit coupling in two-dimensional transition-metal dichalcogenides. *Nat. Commun.* **2018**, *9*, 1427. [CrossRef]

71. Di Castro, D.; Balestrino, G. Superconductivity in interacting interfaces of cuprate-based heterostructures. *Supercond. Sci. Technol.* **2018**, *31*, 073001. [CrossRef]

72. Wu, Y.; He, J.; Liu, J.; Xing, H.; Mao, Z.; Liu, Y. Dimensional reduction and ionic gating induced enhancement of superconductivity in atomically thin crystals of 2H-TaSe2. *Nanotechnology* **2019**, *30*, 035702. [CrossRef]

73. Talantsev, E.F. Angular dependence of the upper critical field in randomly restacked 2D superconducting nanosheets. *Supercond. Sci. Technol.* **2019**, *32*, 015013. [CrossRef]

74. Pankratov, O.A.; Pakhomov, S.V.; Volkov, B.A. Supersymmetry in heterojunctions: Band-inverting contact on the basis of $Pb_{1-x}Sn_xTe$ and $Hg_{1-x}Cd_xTe$. *Solid State Commun.* **1987**, *61*, 93–96. [CrossRef]

75. König, M.; Wiedmann, S.; Brüne, C.; Roth, A.; Buhmann, H.; Molenkamp, L.W.; Qi, X.-L.; Zhang, S.-C. Quantum spin Hall insulator state in HgTe quantum wells. *Science* **2007**, *318*, 766–770. [CrossRef] [PubMed]

76. Charpentier, S.; Galletti, L.; Kunakova, G.; Arpaia, R.; Song, Y.; Baghdadi, R.; Wang, S.M.; Kalaboukhov, A.; Olsson, E.; Tafuri, F. Induced unconventional superconductivity on the surface states of Bi_2Te_3 topological insulator. *Nat. Commun.* **2017**, *8*, 2019. [CrossRef] [PubMed]

77. Qu, D.-X.; Teslich, N.E.; Dai, Z.; Chapline, G.F.; Schenkel, T.; Durham, S.R.; Dubois, J. Onset of a two-dimensional superconducting phase in a topological-insulator—Normal-metal $Bi_{1-x}Sb_x$/Pt junction fabricated by ion-beam techniques. *Phys. Rev. Lett.* **2018**, *121*, 037001. [CrossRef]

78. Li, C.-Z.; Li, C.; Wang, L.-X.; Wang, S.; Liao, Z.-M.; Brinkman, A.; Yu, D.-P. Bulk and surface states carried supercurrent in ballistic Nb-Dirac semimetal Cd_3As_2 nanowire-Nb junctions. *Phys. Rev. B* **2018**, *97*, 115446. [CrossRef]

79. Schüffelgen, P.; Schmitt, T.; Schleenvoigt, M.; Rosenbach, D.; Perla, P.; Jalil, A.R.; Mussler, G.; Lepsa, M.; Schäpers, T.; Grützmacher, D. Exploiting topological matter for Majorana physics and devices. *Solid State Electron.* **2019**, *155*, 99–104. [CrossRef]

80. Kurter, C.; Finck, A.D.K.; Huemiller, E.D.; Medvedeva, J.; Weis, A.; Atkinson, J.M.; Qiu, Y.; Shen, L.; Lee, S.H.; Vojta, T. Conductance spectroscopy of exfoliated thin flakes of $Nb_xBi_2Se_3$. *Nano Lett.* **2019**, *19*, 38–45. [CrossRef] [PubMed]

81. German, R.; Komleva, E.V.; Stein, P.; Mazurenko, V.G.; Wang, Z.; Streltsov, S.V.; Ando, Y.; Van Loosdrecht, P.H.M. Phonon mode calculations and Raman spectroscopy of the bulk-insulating topological insulator $BiSbTeSe_2$. *Phys. Rev. Mater.* **2019**, *3*, 054204. [CrossRef]

82. Kayyalha, M.; Kargarian, M.; Kazakov, A.; Miotkowski, I.; Galitski, V.M.; Yakovenko, V.M.; Rokhinson, L.P.; Chen, Y.P. Anomalous low-temperature enhancement of supercurrent in topological-insulator nanoribbon Josephson junctions: Evidence for low-energy Andreev bound states. *Phys. Rev. Lett.* **2019**, *122*, 047003. [CrossRef] [PubMed]

83. Miyazaki, A.; Delsolaro, W.V. Determination of the Bardeen–Cooper–Schrieffer material parameters of the HIE-ISOLDE superconducting resonator. *Supercond. Sci. Technol.* **2019**, *32*, 025002. [CrossRef]

84. Shearer, C.J.; Slattery, A.D.; Stapleton, A.J.; Shapter, J.G.; Gibson, C.T. Accurate thickness measurement of graphene. *Nanotechnology* **2016**, *27*, 125704. [CrossRef] [PubMed]

85. Talantsev, E.F.; Crump, W.P.; Tallon, J.L. Two-band induced superconductivity in single-layer graphene and topological insulator bismuth selenide. *Supercond. Sci. Technol.* **2018**, *31*, 015011. [CrossRef]

86. Talantsev, E.F.; Crump, W.P.; Island, J.O.; Xing, Y.; Sun, Y.; Wang, J.; Tallon, J.L. On the origin of critical temperature enhancement in atomically thin superconductors. *2D Mater.* **2017**, *4*, 025072. [CrossRef]

87. Wang, Q.-Y.; Li, Z.; Zhang, W.-H.; Zhang, Z.-C.; Zhang, J.-S.; Li, W.; Ding, H.; Ou, Y.-B.; Deng, P.; Chang, K. Interface-induced high-temperature superconductivity in single unit-cell FeSe films on $SrTiO_3$. *Chin. Phys. Lett.* **2012**, *29*, 037402. [CrossRef]

88. Zhang, W.-H.; Sun, Y.; Zhang, J.-S.; Li, F.-S.; Guo, M.-H.; Zhao, Y.-F.; Zhang, H.-M.; Peng, J.-P.; Xing, Y.; Wang, H.-C. Direct observation of high-temperature superconductivity in one-unit-cell FeSe films. *Chin. Phys. Lett.* **2014**, *31*, 017401. [CrossRef]

89. Ge, J.F.; Liu, Z.-L.; Liu, C.; Gao, C.-L.; Qian, D.; Xue, Q.-K.; Liu, Y.; Jia, J.-F. Superconductivity above 100 K in single-layer FeSe films on doped $SrTiO_3$. *Nat. Mater.* **2015**, *14*, 285–289. [CrossRef]

90. Zhang, H.-M.; Sun, Y.; Li, W.; Peng, J.-P.; Song, C.-L.; Xing, Y.; Zhang, Q.; Guan, J.; Li, Z.; Zhao, Y. Detection of a superconducting phase in a two-atom layer of hexagonal Ga film grown on semiconducting GaN(0001). *Phys. Rev. Lett.* **2015**, *114*, 107003. [CrossRef]

91. Xing, Y.; Zhang, H.-M.; Fu, H.-L.; Liu, H.; Sun, Y.; Peng, J.-P.; Wang, F.; Lin, X.; Ma, X.-C.; Xue, Q.-K. Quantum Griffiths singularity of superconductor-metal transition in Ga thin films. *Science* **2015**, *350*, 542–545. [CrossRef] [PubMed]

92. Navarro-Moratalla, E.; Island, J.O.; Manãs-Valero, S.; Pinilla-Cienfuegos, E.; Castellanos-Gomez, A.; Quereda, J.; Rubio-Bollinger, G.; Chirolli, L.; Silva-Guillén, J.A.; Agraït, N. Enhanced superconductivity in atomically thin TaS_2. *Nat. Commun.* **2016**, *7*, 11043. [CrossRef] [PubMed]

93.   Rhodes, D.; Yuan, N.F.; Jung, Y.; Antony, A.; Hua Wang, H.; Kim, B.; Chiu, Y.-C.; Taniguchi, T.; Watanabe, K.; Barmak, K. Enhanced superconductivity in monolayer T_d-MoTe$_2$ with tilted Ising spin texture. *arXiv* **2019**, arXiv:1905.06508.

94.   Schüffelgen, P.; Rosenbach, D.; Li, C.; Schmitt, T.; Schleenvoigt, M.; Jalil, A.R.; Kölzer, J.; Wang, M.; Bennemann, B.; Parlak, U. Boosting transparency in topological Josephson junctions via stencil lithography. *arXiv* **2018**, arXiv:1711.01665.

95.   Han, Z.; Allain, A.; Arjmandi-Tash, H.; Tikhonov, K.; Feigel'man, M.; Sacépé, B.; Bouchiat, V. Collapse of superconductivity in a hybrid tin-graphene Josephson junction array. *Nat. Phys.* **2014**, *10*, 380–386. [CrossRef]

96.   Gross, F.; Chandrasekhar, B.S.; Einzel, D.; Andres, K.; Hirschfeld, P.J.; Ott, H.R.; Beuers, J.; Fisk, Z.; Smith, J.L. Anomalous temperature dependence of the magnetic field penetration depth in superconducting UBe$_{13}$. *Z. Phys. B Condens. Matter* **1986**, *64*, 175–188. [CrossRef]

97.   Gorter, C.J.; Casimir, H. On supraconductivity I. *Physica* **1934**, *1*, 306–320. [CrossRef]

98.   Poole, P.P.; Farach, H.A.; Creswick, R.J.; Prozorov, R. *Superconductivity*, 2nd ed.; Associated Press: London, UK, 2007; pp. 52–55.

99.   Talantsev, E.F.; Crump, W.P.; Storey, J.G.; Tallon, J.L. London penetration depth and thermal fluctuations in the sulphur hydride 203 K superconductor. *Ann. Phys.* **2017**, *529*, 1600390. [CrossRef]

100. Mozaffari, S.; Balicas, L.; Minkov, V.S.; Knyazev, D.; Eremets, M.I.; Einaga, M.; Shimizu, K.; Sun, D.; Balakirev, F.F. *Superconducting Hydride under Extreme Field and Pressure*; LA-UR-18-30460; Los Alamos National Laboratory: Los Alamos, NM, USA, 2019.

101. Pal, B.; Joshi, B.P.; Chakraborti, H.; Jain, A.K.; Barick, B.K.; Ghosh, K.; Bhunia, S.; Laha, A.; Dhar, S.; Gupta, K.D. Experimental evidence of a very thin superconducting layer in epitaxial indium nitride. *Supercond. Sci. Technol.* **2019**, *32*, 015009. [CrossRef]

102. Jones, C.K.; Hulm, J.K.; Chandrasekhar, B.S. Upper critical field of solid solution alloys of the transition elements. *Rev. Mod. Phys.* **1964**, *36*, 74–76. [CrossRef]

103. Helfand, E.; Werthamer, N.R. Temperature and purity dependence of the superconducting critical field, H_{c2}. II. *Phys. Rev.* **1966**, *147*, 288–294. [CrossRef]

104. Werthamer, N.R.; Helfand, E.; Hohenberg, P.C. Temperature and purity dependence of the superconducting critical field, H_{c2}. III. Electron spin and spin-orbit effects. *Phys. Rev.* **1966**, *147*, 295–302. [CrossRef]

105. Baumgartner, T.; Eisterer, M.; Weber, H.W.; Fluekiger, R.; Scheuerlein, C.; Bottura, L. Effects of neutron irradiation on pinning force scaling in state-of-the-art Nb$_3$Sn wires. *Supercond. Sci. Technol.* **2014**, *27*, 015005. [CrossRef]

106. Gor'kov, L.P. The critical supercooling field in superconductivity theory. *Sov. Phys. JETP* **1960**, *10*, 593–599.

107. Suhl, H.; Matthias, B.T.; Walker, L.R. Bardeen-Cooper-Schrieffer theory of superconductivity in the case of overlapping bands. *Phys. Rev. Lett.* **1959**, *3*, 552–554. [CrossRef]

108. Nagamatsu, J.; Nakagawa, N.; Muranaka, T.; Zenitani, Y.; Akimitsu, J. Superconductivity at 39 K in magnesium diboride. *Nature* **2001**, *410*, 63–64. [CrossRef]

109. Kamihara, Y.; Hiramatsu, H.; Hirano, M.; Kawamura, R.; Yanagi, H.; Kamiya, T.; Hosono, H. Iron-based layered superconductor: LaOFeP. *J. Am. Chem. Soc.* **2006**, *128*, 10012–10013. [CrossRef]

110. Buzea, C.; Yamashita, T. Review of the superconducting properties of MgB$_2$. *Supercond. Sci. Technol.* **2001**, *14*, R115–R146. [CrossRef]

111. Zehetmayer, M. A review of two-band superconductivity: Materials and effects on the thermodynamic and reversible mixed-state properties. *Supercond. Sci. Technol.* **2013**, *26*, 043001. [CrossRef]

112. Hänisch, J.; Iida, K.; Hühne, R.; C Tarantini, C. Fe-based superconducting thin films—preparation and tuning of superconducting properties. *Supercond. Sci. Technol.* **2019**, *32*, 093001. [CrossRef]

113. Shalnikov, A. Superconducting thin films. *Nature* **1938**, *132*, 74. [CrossRef]

114. Cohen, R.W.; Abeles, B. Superconductivity in granular aluminum films. *Phys. Rev.* **1968**, *168*, 444–450. [CrossRef]

115. Pracht, U.S.; Bachar, N.; Benfatto, L.; Deutscher, G.; Farber, E.; Dressel, M.; Scheffler, M. Enhanced Cooper pairing versus suppressed phase coherence shaping the superconducting dome in coupled aluminum nanograins. *Phys. Rev. B* **2016**, *93*, 100503(R). [CrossRef]

116. Pracht, U.S.; Cea, T.; Bachar, N.; Deutscher, G.; Farber, E.; Dressel, M.; Scheffler, M.; Castellani, C.; García-García, A.M.; Benfatto, L. Optical signatures of the superconducting Goldstone mode in granular aluminum: Experiments and theory. *Phys. Rev. B* **2017**, *96*, 094514. [CrossRef]

117. Eilenberger, G. Transformation of Gorkov's equation for type II superconductors into transport-like equations. *Z. Phys.* **1968**, *214*, 195–213. [CrossRef]

118. Usadel, K.D. Generalized diffusion equation for superconducting alloys. *Phys. Rev. Lett.* **1970**, *25*, 507–509. [CrossRef]

119. Rusanov, A.Y.; Hesselberth, M.B.S.; Aarts, J. Depairing currents in superconducting films of Nb and amorphous MoGe. *Phys. Rev. B* **2004**, *70*, 024510. [CrossRef]

120. Sasaki, S.; Segawa, K.; Ando, Y. Superconductor derived from a topological insulator heterostructure. *Phys. Rev. B* **2014**, *90*, 220504. [CrossRef]

121. Andersen, L.; Wang, Z.; Lorenz, T.; Ando, Y. Nematic superconductivity in $Cu_{1.5}(PbSe)_5(Bi_2Se_3)_6$. *Phys. Rev. B* **2018**, *98*, 220512(R). [CrossRef]

122. Reyren, N.; Thiel, S.; Caviglia, A.D.; Fitting Kourkoutis, L.; Hammerl, G.; Richter, C.; Schneider, C.W.; Kopp, T.; Rüetschi, A.-S.; Jaccard, D. Superconducting interfaces between insulating oxides. *Science* **2007**, *317*, 1196–1199. [CrossRef]

123. Gozar, A.; Logvenov, G.; Fitting Kourkoutis, L.; Bollinger, A.T.; Giannuzzi, L.A.; Muller, D.A.; Bozovic, I. High-temperature interface superconductivity between metallic and insulating copper oxides. *Nature* **2008**, *455*, 782–785. [CrossRef] [PubMed]

124. Campi, G.; Bianconi, A.; Poccia, N.; Bianconi, G.; Barba, L.; Arrighetti, G.; Innocenti, D.; Karpinski, J.; Zhigadlo, N.D.; Kazakov, S.M. Inhomogeneity of charge-density-wave order and quenched disorder in a high-T_c superconductor. *Nature* **2015**, *525*, 359–362. [CrossRef] [PubMed]

125. Ricci, A.; Poccia, N.; Joseph, B.; Innocenti, D.; Campi, G.; Zozulya, A.; Westermeier, F.; Schavkan, A.; Coneri, F.; Bianconi, A. Direct observation of nanoscale interface phase in the superconducting chalcogenide $K_xFe_{2-y}Se_2$ with intrinsic phase separation. *Phys. Rev. B* **2015**, *91*, 020503. [CrossRef]

13

Superconducting Properties of 3D Low-Density TI-Bipolaron Gas in Magnetic Field

Victor D. Lakhno🆔

Keldysh Institute of Applied Mathematics of Russian Academy of Sciences, 125047 Moscow, Russia; lak@impb.ru

Abstract: Consideration is given to thermodynamical properties of a three-dimensional Bose-condensate of translation-invariant bipolarons (TI-bipolarons) in magnetic field. The critical temperature of transition, critical magnetic fields, energy, heat capacity and the transition heat of TI-bipolaron gas are calculated. Such values as maximum magnetic field, London penetration depth and their temperature dependencies are calculated. The results obtained are used to explain experiments on high-temperature superconductors.

Keywords: Bose-Einstein condensation; BEC; BCS; translation-invariant bipolarons; electron-phonon interaction; energy gap

---

1. Introduction

Before the discovery of high-temperature superconductivity (HTSC) Bardeen-Cooper-Schrieffer theory [1] (BCS) played the role of fundamental microscopic theory of superconductivity with, in fact, no alternative. The discovery of HTSC revealed some problems which arose while trying to describe various properties of high-temperature superconductors within BCS. This gave birth to a great number of alternative theories aimed at resolving the problems. Review of various HTSC theories is presented in numerous papers. The current state-of-the-art can be found in [2–13]. All the approaches, are however, based on the same proposition the phenomenon of bosonization of Fermi-particles, or Cooper effect. This proposition straightforwardly leads to the conclusion that the phenomenon of superconductivity is related to the phenomenon of Bose-Einstein condensation (BEC). Presently the idea that superconductivity is based on BEC is generally recognized.

A great obstacle to the development of the theory which should be based on BEC was a statement made in BCS (see comment in [1], p. 1177 and [14]) about incompatibility of their theory with BEC.

Some evidence that this viewpoint is erroneous was first obtained in paper [15] whose authors, while studying the properties of high-density exciton gas, demonstrated an analogy between BCS theory and BEC. The results of [15] provided the basis for developing the idea of crossover passing on from the BCS theory which is appropriate for the limit of weak electron-phonon interaction (EPI) to BEC which is suitable for the limit of strong EPI [16–22]. It was believed that additional evidence in favor of this approach is Eliashberg theory of strong coupling [23]. According to [24], in the limit of infinitely strong EPI this theory leads to the regime of local pairs, though greatly different from the regime of BEC [25].

The attempts to develop a theory of crossover between BCS and BEC ran, however, into insurmountable obstacles. Thus, for example, an idea was put forward, to construct a theory with the use of T-matrix transition where T-matrix of the initial Fermion system would transform into T-matrix of Boson system as the force of EPI would increase [26–31]. The approach, however, turned out to be unfeasible even in the case of greatly diluted systems. Actually, the point is that when the system consists of only two fermions it is impossible to construct a one-boson state of them. In the EPI theory this problem is known as that of a bipolaron.

The reason the crossover theory failed can be the following: The BCS theory, as the bipolaron theory, proceeds from Froehlich Hamiltonian. For this Hamiltonian, an important theorem of analyticity of the polaron and bipolaron energy on EPI constant is proved [32]. In the BCS theory a very important approximation is, however, made. Namely the actual matrix element of the interaction in Froehlich Hamiltonian is replaced by a model quantity a matrix element truncated near Fermi surface. This procedure is by no means fair. As is shown in [33], in the bipolaron theory it leads to ghost effects existence of a local energy level separated by a gap from the quasicontinuous spectrum (Cooper effect). This solution is isolated and does not possess the property of analyticity on coupling constant. In the BCS theory just this solution provides the basis for constructing the superconductivity theory.

As a result, the theory constructed and its analytic continuation (Eliashberg theory) greatly falsify the reality, in particular, they make impossible development of the superconductivity theory on the basis of BEC. Replacement of the actual matrix element by the model one enables one to make analytical calculations completely. Thus, substitution of local interaction for actual one in BCS enabled the authors of [34] to derive Ginzburg-Landau (GL) phenomenological model which is also a local model. Actually, the power of this approach can hardly be overestimated since it has enabled one to get a lot of statements consistent with the experiment.

This paper is an attempt to develop a HTSC theory on the basis of BEC of translation-invariant bipolarons (TI-bipolarons) [33,35–38], free of approximations made in [1].

If we proceed from the fact that the superconducting mechanism is based on Cooper pairing, then in the case of a strong Froehlich electron-phonon interaction this leads to translation-invariant bipolaron theory of HTSC [39,40]. In distinction from bipolarons with broken symmetry, TI-bipolaron is delocalized in space and polarization potential well is lacking (zero polarization charge). The energy of TI-bipolarons is lower than the energy of bipolarons with broken symmetry, so they are more stable than bipolarons with broken symmetry.

We recall the main results of the theory of TI-polarons and bipolarons obtained in [33,35–38]. Notice that consideration of just electron-phonon interaction is not essential for the theory and can be generalized to any type of interaction, for example the interaction of electrons with the spin subsystem [41].

In what follows, we will deal only with the main points of the theory important for the HTSC theory. The main result of papers [33,35–38] is the construction of delocalized polaron and bipolaron states in the limit of strong electron-phonon interaction. The theory of TI-bipolarons is based on the theory of TI-polarons [42] and retains the validity of basic statements proved for TI-polarons. The chief of them is the theorem of analytic properties of the ground state of a TI-polaron (accordingly TI-bipolaron) depending on the constant of electron-phonon interaction α. The main implication of this statement is the absence of a critical value of the EPI constant α_c below which the bipolaron state becomes impossible since it decays into independent polaron states. In other words, if there exists a value of α_c at which the TI-state becomes energetically disadvantageous with respect to its decay into individual polarons, then nothing occurs at this point but for $\alpha < \alpha_c$ the state becomes metastable. For this reason we can expect that for $\alpha < \alpha_c$ the normal phase occurs rather than the superconducting one.

Another important property of TI-bipolarons is the possibility of changing the correlation length over the whole range of $[0, \infty]$ variation depending on the Hamiltonian parameters [36,38]. Hence, it can be both much larger (as is the case in metals) and much less than the characteristic size between the electrons in an electron gas (as happens with ceramics).

An outstandingly important property of TI-polarons and bipolarons is the availability of an energy gap between their ground and excited states (Section 3).

The above-indicated characteristics can be used to develop a microscopic HTSC theory on the basis of TI-bipolarons.

The paper is arranged as follows. In Section 2 we take Pekar-Froehlich Hamiltonian for a bipolaron as an initial Hamiltonian. The results of three canonical transformations, such as Heisenberg transformation, Lee-Low-Pines transformation and that of Bogolyubov-Tyablikov are briefly outlined. Equations determining the TI-bipolaron spectrum are derived. In Section 3 we analyze solutions of the equations for the TI-bipolaron spectrum. It is shown that the spectrum has a gap separating the ground state of a TI-bipolaron from its excited states which form a quasicontinuous spectrum. The concept of an ideal gas of TI-bipolarons is substantiated.

With the use of the spectrum obtained, in Section 4 we reproduce thermodynamic characteristics of an ideal gas of TI-bipolarons in the absence of a magnetic field, considered earlier in [39,40].

In Section 5 we deal with the case when the external magnetic field differs from zero. It is shown that the current state in the system under discussion is caused by the existence of a constant quantity the total momentum of the electron-phonon system in the magnetic field. Comparison of the value of the total momentum with that obtained within phenomenological approach enables us to determine the London penetration depth which is a very important characteristic. The results of the initial isotropic model are generalized to anisotropic case.

In Section 6 we investigate thermodynamic characteristics of an ideal TI-bipolaron gas in the presence of a magnetic field. It is shown that the availability of an energy gap in the TI-bipolaron spectrum makes possible their Bose-condensation in a magnetic field. A notion of a maximum value of the magnetic field intensity for which homogeneous Bose-condensation is possible is introduced. The temperature dependence of the value of the critical magnetic field and the dependence of the critical temperature on the magnetic field are found. It is shown that the phase transition of an ideal TI-bipolaron gas can be either of the 1-st kind or of infinite kind, depending on the magnetic field value. The theory is generalized to the case of anisotropic superconductor. This generalization enables us to compare the results obtained with experimental data (Section 7).

In Section 8 we consider scaling relations in superconductors. Alexandrov's formula and Homes's law are derived.

In Section 9 the results obtained are summed up.

2. Pekar-Froehlich Hamiltonian. Canonical Transformations

Following [33,35–37], in describing bipolarons we will proceed from Pekar-Froehlich Hamiltonian with non zero magnetic field:

$$H = \frac{1}{2m^*}\left(\hat{p}_1 - \frac{e}{c}\vec{A}_1\right)^2 + \frac{1}{2m^*}\left(\hat{p}_2 - \frac{e}{c}\vec{A}_2\right)^2 + \sum_k \hbar\omega_0(k)a_k^+ a_k + \tag{1}$$

$$\sum_k \left(V_k e^{i\vec{k}\vec{r}_1}a_k + V_k e^{i\vec{k}\vec{r}_2}a_k + H.c.\right) + U\left(|\vec{r}_1 - \vec{r}_2|\right),$$

$$U\left(|\vec{r}_1 - \vec{r}_2|\right) = \frac{e^2}{\epsilon_\infty |\vec{r}_1 - \vec{r}_2|},$$

where $\hat{p}_1, \vec{r}_1, \hat{p}_2, \vec{r}_2$ are momenta and coordinates of the first and second electrons, respectively; a_k^+, a_k are operators of the creation and annihilation of the field quanta with energy $\hbar\omega_0(k)$; m^* is the electron effective mass; the quantity U describes Coulomb repulsion between the electrons; V_k is the function of the wave vector k. We write (1) in general form. For a special case of ion crystals which is typical for HTSC we consider the Pekar-Froehlich Hamiltonian in the form:

$$V_k = \frac{e}{k}\sqrt{\frac{2\pi\hbar\omega_0}{\tilde{\epsilon}V}} = \frac{\hbar\omega_0}{ku^{1/2}}\left(\frac{4\pi\alpha}{V}\right)^{1/2}, \quad u = \left(\frac{2m^*\omega_0}{\hbar}\right)^{1/2}, \quad \alpha = \frac{1}{2}\frac{e^2 u}{\hbar\omega_0\tilde{\epsilon}}, \tag{2}$$

$$\tilde{\epsilon}^{-1} = \epsilon_\infty^{-1} - \epsilon_0^{-1},$$

where e is the electron charge; ϵ_∞ and ϵ_0 are high-frequency and static dielectric permittivities; α is the constant of electron-phonon interaction; V is the system's volume, ω_0 is the optical phonon frequency.

The axis z is chosen along the direction of the magnetic field induction $\vec{B}$ and use is made of symmetrical gauge:

$$\vec{A}_j = \frac{1}{2}\vec{B} \times \vec{r}_j,$$

for $j = 1.2$. For the bipolaron singlet state discussed below, the contribution of the spin term is equal to zero.

In the system of the center of mass Hamiltonian (1) takes the form:

$$H = \frac{1}{2M_e}\left(\hat{p}_R - \frac{2e}{c}\vec{A}_R\right)^2 + \frac{1}{2\mu_e}\left(\hat{p}_r - \frac{e}{2c}\vec{A}_r\right)^2 + \sum_k \hbar\omega_0(k)a_k^+ a_k + \tag{3}$$

$$\sum_k 2V_k\cos\frac{\vec{k}\vec{r}}{2}\left(a_k e^{i\vec{k}\vec{R}} + H.c.\right) + U(|\vec{r}|),$$

$$\vec{R} = (\vec{r}_1 + \vec{r}_2)/2, \ \vec{r} = \vec{r}_1 - \vec{r}_2, \ M_e = 2m^*, \ \mu_e = m^*/2,$$

$$\vec{A}_r = \frac{1}{2}B(-y, x, 0), \ \vec{A}_R = \frac{1}{2}B(-Y, X, 0),$$

$$\hat{p}_R = \hat{p}_1 + \hat{p}_2 = -i\hbar\nabla_{\vec{R}}, \ \hat{p}_r = (\hat{p}_1 - \hat{p}_2)/2 = -i\hbar\nabla_r,$$

where x, y, and X, Y are components of vectors $\vec{r}$, $\vec{R}$ respectively.

Let us subject Hamiltonian H to Heisenberg canonical transformation [43,44]:

$$S_1 = \exp i\left(\vec{G} - \sum_k \vec{k}a_k^+ a_k\right)\vec{R}. \tag{4}$$

$$\vec{G} = \hat{\vec{P}}_R + \frac{2e}{c}\vec{A}_R, \ \hat{\vec{P}}_R = \hat{p}_R + \sum_k \hbar\vec{k}a_k^+ a_k, \tag{5}$$

where $\vec{G}$ is the quantity commuting with the Hamiltonian, thereby being a constant, i.e., c-number, $\hat{\vec{P}}_R$ is the total momentum.

Action of S_1 on the field operator yields:

$$S_1^{-1}a_k S_1 = a_k e^{-i\vec{k}\vec{R}}, \ S_1^{-1}a_k^+ S_1 = a_k^+ e^{-i\vec{k}\vec{R}}. \tag{6}$$

Accordingly, the transformed Hamiltonian $\tilde{H} = S_1^{-1}HS_1$ takes on the form:

$$\tilde{H} = \frac{1}{2M_e}\left(\vec{G} - \sum_k \hbar\vec{k}a_k^+ a_k - \frac{2e}{c}\vec{A}_R\right)^2 + \frac{1}{2\mu_e}\left(\hat{p}_r - \frac{e}{2c}\vec{A}_r\right)^2 + \sum_k \hbar\omega_0(k)a_k^+ a_k + \tag{7}$$

$$\sum_k 2V_k\cos\frac{\vec{k}\vec{r}}{2}\left(a_k + a_k^+\right) + U(|\vec{r}|).$$

In what follows we will believe:

$$\vec{G} = 0. \tag{8}$$

The physical meaning of (8) is the absence of the total momentum (current) in the bulk of superconductor. This fact follows from the Meissner effect which states that the current in the volume of superconductor needs to be zero. We use this fact in Section 5 to obtain the value of London penetration depth λ. We will seek the solution of the stationary Schroedinger equation corresponding to Hamiltonian (7) in the form:

$$\Psi_H(r, R, \{a_k\}) = \phi(R)\Psi_{H=0}(r, R, \{a_k\}) \tag{9}$$

$$\phi(R) = \exp\left(-i\frac{2e}{\hbar c}\int_0^{\vec{R}} \vec{A}_{R'} d\vec{R}'\right)$$

$$\Psi_{H=0}(r, R, \{a_k\}) = \psi(r)\Theta(R, \{a_k\}),$$

where $\Psi_{H=0}(r, R, \{a_k\})$ is bipolaron wave function in the absence of magnetic field. The explicit form of functions $\psi(r)$ and $\Theta(R, \{a_k\})$ is given in [35,38].

Averaging of $\bar{H}$ over the wave function $\phi(R)$ and $\psi(r)$ yields:

$$\bar{H} = \frac{1}{2M_e}\left(\sum_k \hbar\vec{k}a_k^+ a_k\right)^2 + \sum_k \hbar\tilde{\omega}_k a_k^+ a_k + \sum_k \bar{V}_k\left(a_k + a_k^+\right) + \bar{T} + \bar{U} + \bar{\Pi}, \tag{10}$$

where:

$$\bar{T} = \frac{1}{2\mu_e}\left\langle\psi\left|(\hat{p}_r - \frac{e}{2c}\vec{A}_r)^2\right|\psi\right\rangle, \quad \bar{U} = \langle\psi|U(r)|\psi\rangle, \quad \bar{V}_k = 2V_k\left\langle\psi\left|\cos\frac{\vec{k}\vec{r}}{2}\right|\psi\right\rangle, \tag{11}$$

$$\bar{\Pi} = \frac{2e^2}{M_e c^2}\left\langle\phi\left|A_R^2\right|\phi\right\rangle, \quad \hbar\tilde{\omega}_k = \hbar\omega_0(k) + \frac{2\hbar e}{M_e c}\left\langle\phi\left|\vec{k}\vec{A}_R\right|\phi\right\rangle.$$

In what follows in this section we will assume $\hbar = 1$, $\omega_0(k) = \omega_0 = 1$, $M_e = 1$. Equation (10) suggests that the bipolaron Hamiltonian differs from the polaron one in that in the latter the quantity V_k is replaced by $\bar{V}_k$ and $\bar{T}, \bar{U}, \bar{\Pi}$ are added to the polaron Hamiltonian.

With the use of Lee-Low-Pines canonical transformation [45]

$$S_2 = \exp\left\{\sum_k f_k(a_k^+ - a_k)\right\},$$

where f_k are variational parameters with the sense of the distance by which the field oscillators are displaced from their equilibrium positions:

$$S_2^{-1}a_k S_2 = a_k + f_k, \quad S_2^{-1}a_k^+ S_2 = a_k^+ + f_k,$$

for Hamiltonian $\tilde{\tilde{H}}$:

$$\tilde{\tilde{H}} = S_2^{-1}\bar{H}S_2, \tag{12}$$

$$\tilde{\tilde{H}} = H_0 + H_1,$$

we get:

$$H_0 = 2\sum_k \bar{V}_k f_k + \sum_k f_k^2 \tilde{\omega}_k + \frac{1}{2}\left(\sum_k \vec{k}f_k^2\right)^2 + \mathcal{H}_0 + \bar{T} + \bar{U} + \bar{\Pi},$$

$$\mathcal{H}_0 = \sum_k \omega_k a_k^+ a_k + \frac{1}{2}\sum_{k,k'} \vec{k}\vec{k}' f_k f_{k'}(a_k a_{k'} + a_k^+ a_{k'}^+ + a_k^+ a_{k'} + a_{k'}^+ a_k), \tag{13}$$

where:

$$\omega_k = \tilde{\omega}_k + \frac{k^2}{2} + \vec{k}\sum_{k'} \vec{k}' f_{k'}^2. \tag{14}$$

Hamiltonian H_1 contains terms linear, threefold and fourfold in the creation and annihilation operators. Its explicit form is given in [38,42].

Then, as is shown in [38,42], the use of Bogolyubov-Tyablikov canonical transformation [46] for passing on from operators a_k^+, a_k to new operators α_k^+, α_k:

$$a_k = \sum_{k'} M_{1kk'}\alpha_{k'} + \sum_{k'} M_{2kk'}^*\alpha_{k'}^+$$

$$a_k^+ = \sum_{k'} M_{1kk'}^*\alpha_{k'}^+ + \sum_{k'} M_{2kk'}\alpha_{k'}, \tag{15}$$

(in which $\mathcal{H}_0$ is a diagonal operator) makes mathematical expectation of H_1 equal to zero in the absence of magnetic field (see Appendix). The contribution of H_1 to the spectrum of transformed Hamiltonian when magnetic field is non-zero is discussed in Section 3.

In the new operators α_k^+, α_k Hamiltonian (13) takes on the form

$$\tilde{\tilde{H}} = E_{bp} + \sum_k \nu_k \alpha_k^+ \alpha_k,$$

$$E_{bp} = \Delta E_r + 2\sum_k \bar{V}_k f_k + \sum_k \tilde{\omega}_k f_k^2 + \bar{T} + \bar{U} + \bar{\Pi}, \tag{16}$$

where ΔE_r is the so-called recoil energy. The general expression for $\Delta E_r = \Delta E_r\{f_k\}$ was obtained in [42]. Actually, calculation of the ground state energy E_{bp} for the case of Froehlich Hamiltonian was performed in [38,42] by minimization of (16) in f_k and in Ψ in the absence of a magnetic field.

Notice that in the theory of a polaron with broken symmetry a diagonalized electron-phonon Hamiltonian has the form of (16) [47]. This Hamiltonian can be treated as a Hamiltonian of a polaron and a system of its associated renormalized real phonons or as a Hamiltonian whose quasiparticle excitations spectrum is determined by (16) [48]. In the latter case excited states of a polaron are Fermi quasiparticles.

In the case of a bipolaron the situation is qualitatively different since a bipolaron is a boson quasiparticle whose spectrum is determined by (16). Obviously, a gas of such quasiparticles can experience Bose-Einstein condensation (BEC). Treatment of (16) as a bipolaron and its associated renormalized phonons does not prevent their BEC since maintenance of the number of particles required in this case takes place automatically due to commutation of the total number of renormalized phonons with Hamiltonian (16).

Renormalized frequencies ν_k involved in (16), according to [38,42,49] are determined by the equation for s:

$$1 = \frac{2}{3}\sum_k \frac{k^2 f_k^2 \omega_k}{s - \omega_k^2}, \tag{17}$$

solutions of which yield the spectrum of $s = \{\nu_k^2\}$ values. This equation has a general form and can be applied for arbitrary dependence of V_k and ω_k on k.

3. Energy Spectrum of a TI-Bipolaron

Hamiltonian (16) is conveniently presented in the form:

$$\tilde{\tilde{H}} = \sum_{n=0,1,2,\ldots} E_n \alpha_n^+ \alpha_n, \tag{18}$$

$$E_n = \begin{cases} E_{bp}, & n = 0; \\ \nu_n = E_{bp} + \omega_{k_n}, & n \neq 0; \end{cases} \tag{19}$$

where in the case of a three-dimensional ionic crystal $\vec{k}_n$ is a vector with the components:

$$k_{n_i} = \pm \frac{2\pi(n_i - 1)}{N_{a_i}}, \quad n_i = 1, 2, ..., \frac{N_{a_i}}{2} + 1, \ i = x, y, z,$$

N_{ai} is the number of atoms along the i-th crystallographic axis.

Let us prove the validity of the expression for the spectrum (18), (19). Since operators α_n^+, α_n obey Bose commutation relations:

$$\left[\alpha_n, \alpha_{n'}^+\right] = \alpha_n \alpha_{n'}^+ - \alpha_{n'}^+ \alpha_n = \delta_{n,n'},$$

they can be considered to be operators of creation and annihilation of TI-bipolarons. The energy spectrum of TI-bipolarons, according to (17), is determined by the equation:

$$F(s) = 1, \tag{20}$$

where:

$$F(s) = \frac{2}{3} \sum_n \frac{k_n^2 f_{k_n}^2 \omega_{k_n}^2}{s - \omega_{k_n}^2}.$$

It is convenient to solve Equation (20) graphically (Figure 1).

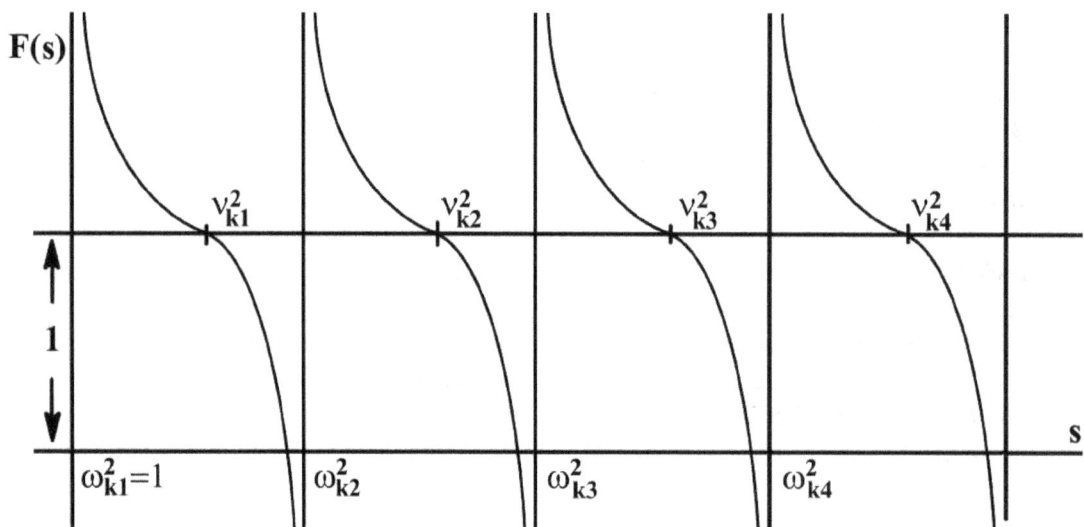

Figure 1. Graphical solution of Equation (20).

Figure 1 suggests that the frequencies ν_{k_n} lie between the frequencies ω_{k_n} and $\omega_{k_{n+1}}$. Hence, the spectrum ν_{k_n} as well as the spectrum ω_{k_n} are quasicontinuous: $\nu_{k_n} - \omega_{k_n} = O(N^{-1})$ which just proves the validity of (18) and (19).

It follows that the spectrum of a TI-bipolaron has a gap between the ground state E_{bp} and the quasicontinuous spectrum, equal to ω_0.

In the absence of an external magnetic field, functions f_k involved in expression for ω_k (14) are independent of the direction of the wave vector $\vec{k}$. When an external magnetic field is applied, f_k cannot be considered to be an isotropic quantity, accordingly, we cannot put the last term in equation (14) for ω_k equal to zero. Besides, the angular dependence involved in the spectrum ω_k in the magnetic field is also contained in the term $\tilde{\omega}_k$ involved in the quantity ω_k. Since in the isotropic system discussed there is only one preferred direction determined by vector $\vec{B}$, for ω_k from (14) we will get:

$$\omega_{k_n} = \omega_0 + \frac{\hbar k_n^2}{2M_e} + \frac{\eta}{M_e} \left(\vec{B}\vec{k}_n \right), \tag{21}$$

where η is a scalar quantity. Notice that the contribution of H_1 to spectrum (21) will lead to the dependence of η from $|\vec{k}|$ and $(\vec{k}\vec{B})$. For weak magnetic field in longitudinal limit (when Froehlich Hamiltonian is valid) we will neglect such dependence and consider η as a constant value.

For a magnetic field directed along the axis z, expression (21) can be written as:

$$\omega_{k_n} = \omega_0 + \frac{\hbar^2}{2M_e} \left(k_{zn} + k_z^0 \right)^2 + \frac{\hbar^2}{2M_e} \left(k_{xn}^2 + k_{yn}^2 \right)^2 - \frac{\eta^2 B^2}{2\hbar^2 M_e}. \tag{22}$$

Note that formula (22) can be generalized to the anisotropic case when in the directions k_x and k_y: $M_{ex} = M_{ey} = M_{||}$, and in the direction k_z: $M_{ez} = M_\perp$(Section 5). Formula (22) in this case takes on the form:

$$\omega_{k_n} = \omega_0 + \frac{\hbar^2}{2M_\perp} \left(k_{zn} + k_z^0 \right)^2 + \frac{\hbar^2}{2M_{||}} \left(k_{xn}^2 + k_{yn}^2 \right) - \frac{\eta^2 B^2}{2\hbar^2 M_\perp}, \tag{22'}$$

if the magnetic field is directed along the axis z and:

$$\omega_{k_n} = \omega_0 + \frac{\hbar^2}{2M_\perp} k_{zn}^2 + \frac{\hbar^2}{2M_{||}} \left(k_{xn} + k_{xn}^0 \right)^2 + \frac{\hbar^2}{2M_{||}} k_{yn}^2 - \frac{\eta^2 B^2}{2\hbar^2 M_{||}}, \tag{22''}$$

if the magnetic field is directed along the axis x.

Below we will consider the case of low concentration of TI-bipolarons in a crystal. Then they can adequately be considered to be an ideal Bose gas, whose properties are determined by Hamiltonian (18).

4. Statistical Thermodynamics of Low-Density TI Bipolarons without Magnetic Field

Let us consider an ideal Bose-gas of TI-bipolarons which represents a system of N particles occurring in some volume V. Let us write N_0 for the number of particles in the lower one-particle state and N' for the number of particles in higher states. Then:

$$N = \sum_{n=0,1,2,\ldots} \bar{m}_n = \sum_n \frac{1}{e^{(E_n - \mu)/T} - 1}, \tag{23}$$

or:

$$N = N_0 + N', \quad N_0 = \frac{1}{e^{(E_{bp} - \mu)/T} - 1}, \quad N' = \sum_{n \neq 0} \frac{1}{e^{(E_n - \mu)/T} - 1}. \tag{24}$$

In expression N' (24) we will perform integration over quasicontinuous spectrum (instead of summation) (18), (19) and (22) and assume $\mu = E_{bp}$. As a result, from (23) and (24) we get an equation for determining the critical temperature of Bose-condensation T_c:

$$C_{bp} = f_{\tilde{\omega}_H} \left(\tilde{T}_c \right), \tag{25}$$

$$f_{\tilde{\omega}_H} \left(\tilde{T}_c \right) = \tilde{T}_c^{3/2} F_{3/2} \left(\tilde{\omega}_H / \tilde{T}_c \right), \quad F_{3/2}(\alpha) = \frac{2}{\sqrt{\pi}} \int_0^\infty \frac{x^{1/2} dx}{e^{x+\alpha} - 1},$$

$$C_{bp} = \left(\frac{n^{2/3} 2\pi \hbar^2}{M_e \omega^*} \right)^{3/2}, \quad \tilde{\omega}_H = \frac{\omega_0 - \eta^2 H^2 / 2M_e}{\omega^*}, \quad \tilde{T}_c = \frac{T_c}{\omega^*},$$

where $n = N/V$. In this section we will deal with the case when the magnetic field is lacking: $H = 0$. Figure 2 shows a graphical solution of Equation (25) for the values of parameters $M_e = 2m^* = 2m_0$, where m_0 is the mass of a free electron in vacuum, $\omega^* = 5$ meV (≈ 58 K), $n = 10^{21}$ cm^{-3} and the values: $\tilde{\omega}_1 = 0.2$; $\tilde{\omega}_2 = 1$; $\tilde{\omega}_3 = 2$; $\tilde{\omega}_4 = 10$; $\tilde{\omega}_5 = 15$; $\tilde{\omega}_6 = 20$; $\tilde{\omega}_H = \tilde{\omega} = \omega_0/\omega^*$.

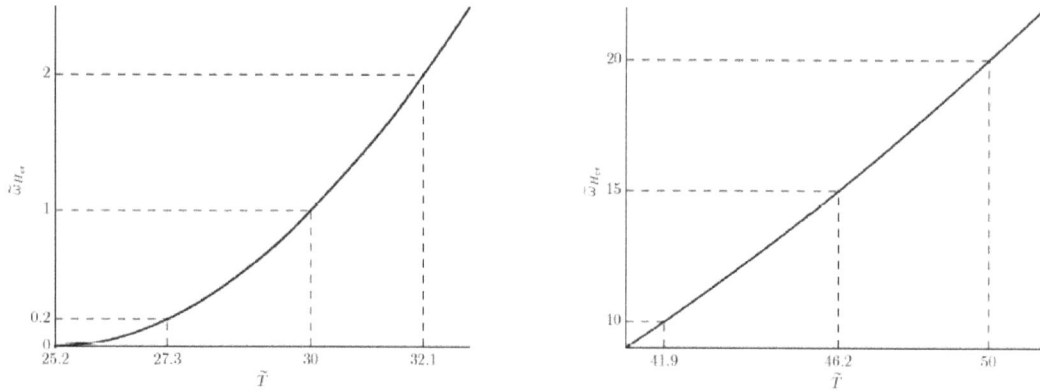

Figure 2. Solutions of Equation (25) with $C_{bp} = 331.35$ and $\tilde{\omega}_i = \{0.2; 1; 2; 10; 15; 20\}$, which correspond to $\tilde{T}_{c_i}$: $\tilde{T}_{c_1} = 27.3$; $\tilde{T}_{c_2} = 30$; $\tilde{T}_{c_3} = 32.1$; $\tilde{T}_{c_4} = 41.9$; $\tilde{T}_{c_5} = 46.2$; $\tilde{T}_{c_6} = 50$.

It is seen from Figure 2 that the critical temperature grows with increasing phonon frequency ω_0. It is also evident from Figure 2 that an increase in the concentration of TI-bipolarons n will lead to an increase in the critical temperature, while a gain in the electron mass m^* to its decrease. For $\tilde{\omega} = 0$ the results go over into the limit of ideal Bose-gas (IBG). In particular, (25) for $\tilde{\omega} = 0$, yields the expression for the critical temperature of IBG:

$$T_c = 3.31\hbar^2 n^{2/3}/M_e. \tag{26}$$

It should be stressed, however, that (26) involves $M_e = 2m^*$, rather than the bipolaron mass. This resolves the problem of the low temperature of condensation which arises both in the small radius polaron theory and in the large radius polaron theory in which expression (26) involves the bipolaron mass [50–57]. Another important result is that the critical temperature T_c for the parameter values considerably exceeds the gap energy ω_0. Such values as energy, free energy, heat capacity and the transition heat of TI-bipolaron gas were calculated for the case $H = 0$ in [39,40].

5. Current States of a TI-Bipolaron Gas

As is known, the absence of a magnetic field inside a superconductor is caused by the existence of surface currents compensating this field. Thus, from condition (8) it follows that:

$$\vec{\mathcal{P}}_R = -\frac{2e}{c}\vec{A}_R, \tag{27}$$

i.e., in a superconductor there is a persistent current $\vec{j}$:

$$\vec{j} = 2en_0\vec{\mathcal{P}}_R/M_e^* = -\frac{4e^2 n_0}{M_e^* c}\vec{A}_R, \tag{28}$$

(where M_e^* is bipolaron effective mass), providing Meissner effect, where n_0 is a concentration of superconducting charge carriers: $n_0 = N_0/V$. Comparing (28) with the well-known phenomenological expression for the surface current $\vec{j}_S$ [58]:

$$\vec{j}_S = -\frac{c}{4\pi\lambda^2}\vec{A}, \tag{29}$$

and putting $\vec{A} = \vec{A}_R$, with the use of (28), (29) and equality $\vec{j} = \vec{j}_S$ we will get a well-known expression for London penetration depth λ:

$$\lambda = \left(\frac{M_e^* c^2}{16\pi e^2 n_0} \right)^{1/2}, \tag{30}$$

The equality of 'microscopic' expression for current (28) to its 'macroscopic' value cannot be exact. Accordingly, the equality $\vec{A} = \vec{A}_R$ is also approximate since $\vec{A}_R$ represents a vector-potential at the point where the center of mass of two electrons occurs, while in London theory $\vec{A}$ is a vector-potential at the point where a particle resides. For this reason, these two quantities should better be considered proportional. In this case the expression for the penetration depth has the form:

$$\lambda = \text{const} \left(\frac{M_e^* c^2}{16\pi e^2 n_0} \right)^{1/2}, \tag{30'}$$

where the constant multiplier in (30') (of the order of unity) should be determined from a comparison with the experiment.

Expression (27) was obtained in the case of isotropic effective mass of charge carriers. However, actually, it has a more general character and does not change when anisotropy of effective masses is taken into account. Thus, for example in layered HTSC materials kinetic energy of charge carriers in Hamiltonian (1) should be replaced by the expression:

$$T_a = \frac{1}{2m_\parallel^*} \left(\hat{P}_{1\parallel} - \frac{e}{c} \vec{A}_1 \right)^2 + \frac{1}{2m_\parallel^*} \left(\hat{P}_{2\parallel} - \frac{e}{c} \vec{A}_2 \right)^2 +$$
$$\frac{1}{2m_\perp^*} \left(\hat{P}_{1\perp} - \frac{e}{c} \vec{A}_{1z} \right)^2 + \frac{1}{2m_\perp^*} \left(\hat{P}_{2\perp} - \frac{e}{c} \vec{A}_{1z} \right)^2,$$

where $\hat{P}_{1,2\parallel}$, $\vec{A}_{1,2\parallel}$ are operators of the momentum and vector-potential in the planes of layers (ab planes); $\hat{P}_{1,2\perp}$, $\vec{A}_{1,2\perp}$ are relevant quantities in the direction perpendicular to the planes (along c-axis); $m_\parallel^*$, $m_\perp^*$ are effective masses in the planes and in the perpendicular direction.

As a result of transformation:

$$\tilde{x} = x, \quad \tilde{y} = y, \quad \tilde{z} = \gamma z \tag{31}$$
$$\tilde{A}_{\tilde{x}} = A_x, \quad \tilde{A}_{\tilde{y}} = A_y, \quad \tilde{A}_{\tilde{z}} = \gamma^{-1} A_z,$$
$$\hat{\tilde{P}}_{\tilde{x}} = \hat{P}_x, \quad \hat{\tilde{P}}_{\tilde{y}} = \hat{P}_y, \quad \tilde{\mathcal{P}}_{\tilde{z}} = \gamma^{-1} \mathcal{P}_z,$$

where $\gamma^2 = m_\perp^* / m_\parallel^*$, γ is the anisotropy parameter kinetic energy $\tilde{T}_a$ appears to be isotropic. It follows that: $\hat{\tilde{P}}_R + (2e/c)\tilde{\vec{A}}_{\tilde{R}} = 0$. Then (31) suggests that relation (27) appears to be valid in the anisotropic case too. It follows that:

$$\vec{\mathcal{P}}_{R\parallel} = -\frac{2e}{c} \vec{A}_{R\parallel}, \quad \vec{\mathcal{P}}_{R\perp} = -\frac{2e}{c} \vec{A}_{R\perp},$$

$$\vec{j}_\parallel = 2e n_0 \mathcal{P}_{R\parallel} / M_{e\parallel}^*, \quad \vec{j}_\perp = 2e n_0 \vec{\mathcal{P}}_{R\perp} / M_{e\perp}^*.$$

The magnetic field directed perpendicular to the plane of layers will induce currents running in the plane of layers. Having penetrated into a sample, such a field will decrease along the plane of layers. Let us write $\lambda_\parallel$ for the London penetration depth of the magnetic field perpendicular to the plane of layers ($H_\perp$) and $\lambda_\perp$ for that of the magnetic field parallel to the plane of layers ($H_\parallel$).

This suggests expressions for London depths of the magnetic field penetration into a sample:

$$\lambda_\perp = \left(\frac{M_{e\perp}^* c^2}{16\pi e^2 n_0} \right)^{1/2}, \lambda_{||} = \left(\frac{M_{e||}^* c^2}{16\pi e^2 n_0} \right)^{1/2}. \tag{32}$$

For $\lambda_{||}$ and $\lambda_\perp$, designations λ_{ab} and λ_c are also used. From (32) it follows that:

$$\frac{\lambda_\perp}{\lambda_{||}} = \left(\frac{M_{e\perp}^*}{M_{e||}^*} \right)^{1/2} = \gamma^*. \tag{33}$$

From (32) it also follows that the London penetration depth depends on temperature:

$$\lambda^2(0)/\lambda^2(T) = n_0(T)/n_0(0). \tag{34}$$

In particular, for $\omega = 0$, with the use of (26) we get: $\lambda(T) = \lambda(0) \left(1 - (T/T_C)^{3/2}\right)^{-1/2}$. Comparison of the dependence obtained with those derived within other approaches is given in Section 7.

It is generally accepted that the Bose system became superconducting due to the inter-particle interaction. The existence of a gap in TI-bipolaron spectrum can drive their condensation and the Landau superfluidity condition:

$$v < \hbar\omega_0/\mathcal{P}. \tag{35}$$

(where $\mathcal{P}$ is the momentum of bipolaron condensate) can be fulfilled even for noninteracting particles. From condition (35) it follows the expression for maximum value of current density $j_{\max} = env_{\max}$:

$$j_{\max} = en_0 \sqrt{\frac{\hbar\omega_0}{M_e^*}}$$

It should be noted that all the aforesaid refers to local electrodynamics. Accordingly, expressions obtained for λ are valid only on condition that $\lambda >> \xi$, where ξ is a correlation length determining the characteristic size of the pair, i.e., the characteristic scale of changes of the wave function $\psi(r)$ in (9). This condition is usually fulfilled in HTSC materials. In ordinary superconductors the inverse inequality is valid. Nonlocal generalization of superconductor electrodynamics was made by Pippard [59]. Within this approach relation between $\vec{j}_S$ and $\vec{A}$ in expression (29) can be written as:

$$\vec{j}_S = \int \hat{Q}(z - r') \vec{A}(r') d\vec{r}',$$

where Q is a certain operator whose radius of action is usually believed to be equal to ξ. In the limit of $\xi >> \lambda$ this leads to an increase in the absolute value of the length of the magnetic field penetration into a superconductor which becomes equal to $(\lambda^2 \xi)^{1/3}$ [58].

6. Thermodynamic Properties of a TI-Bipolaron Gas in a Magnetic Field

To start with, let us notice that expression for $\tilde{\omega}_H$ (25) suggests that for $\omega_0 = 0$ Bose-condensation appears to be impossible if $H \neq 0$. For an ordinary ideal charged Bose-gas, this conclusion was first made in [60]. In view of the fact that in the spectrum of TI-bipolarons there is a gap between the ground state of a TI-bipolaron gas and the excited one (Section 3), this conclusion becomes invalid for $\omega_0 \neq 0$.

Expression $\tilde{\omega}_H$ (25) suggests that there is a maximum value of the magnetic field H_{max} equal to:

$$H_{max}^2 = \frac{2\omega_0 \hbar^2 M_e}{\eta^2}. \tag{36}$$

As follows from (14), the value η consists from two parts: $\eta = \eta' + \eta''$. The value η' is determined by the integral entering into the expression for $\tilde{\omega}_k$ (11). For this reason η' depends on the form of sample surface. The value η'' is determined by the sum entering into the expression for ω_k (14) and weakly depends on surface form. This leads to the conclusion that the value η can be changed by changing sample surface and thus changing H_{max}. For $H > H_{max}$ a homogeneous superconducting state is impossible. With the use of (36), $\tilde{\omega}_H$ (25) will be written as:

$$\tilde{\omega}_H = \tilde{\omega}\left(1 - H^2/H^2_{max}\right). \tag{37}$$

For a given temperature T, let us write $H_{cr}(T)$ for the value of the magnetic field at which the superconductivity disappears. According to (37), this value of the field corresponds to $\tilde{\omega}_{H_{cr}}$:

$$\tilde{\omega}_{Hcr}(T) = \tilde{\omega}\left(1 - H^2_{cr}(T)/H^2_{max}\right). \tag{38}$$

The temperature dependence of the quantity $\tilde{\omega}_{H_{cr}}(T)$ can be found from Equation (25):

$$C_{bp} = \tilde{T}^{3/2}F_{3/2}\left(\tilde{\omega}_{Hcr}(\tilde{T})/\tilde{T}\right).$$

It has the form given in Figure 2 if we replace $\tilde{\omega}$ by $\tilde{\omega}_{H_{cr}}$ and $\tilde{T}_c$ by $\tilde{T}$.

Using (38) and the temperature dependence given in Figure 2 we can find the temperature dependence of $H_{cr}(\tilde{T})$:

$$\frac{H^2_{cr}(\tilde{T})}{H^2_{max}} = 1 - \omega_{Hcr}(\tilde{T})/\tilde{\omega}. \tag{39}$$

For $\tilde{T} \leq \tilde{T}_{ci}$, these dependencies are given in Figure 3.

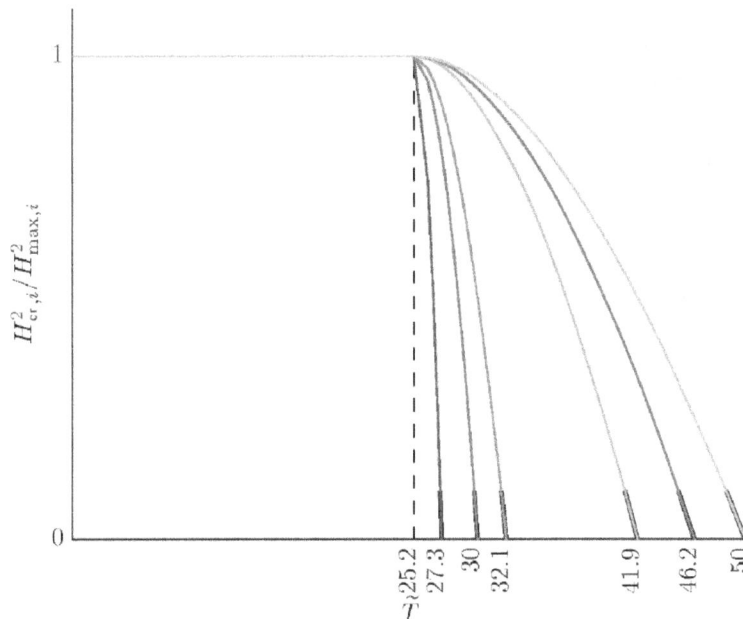

Figure 3. Temperature dependencies $H^2_{cr,i}/H^2_{max,I}$ on the intervals $[0; T_{c,i}]$ for the values of parameters $\tilde{\omega}_i$, given in Figure 2.

As is seen from Figure 3, $H_{cr}(\tilde{T})$ reaches its maximum at a finite temperature of $\tilde{T}_c(\tilde{\omega} = 0) \leq \tilde{T}_c(\omega_{0i})$. Figure 3 suggests that at temperature below $\tilde{T}_c(\tilde{\omega} = 0) = 25.2$ a further decrease of the temperature no longer changes the value of the critical field $H_{cr}(\tilde{T})$ irrespective of the gap value $\tilde{\omega}$.

Let us also introduce the notion of a transition temperature $T_c(H)$ in the magnetic field H. Figure 4 illustrates the dependencies $T_c(H)$ resulting from Figure 3 and determined by the relations:

$$C_{bp} = \tilde{T}^{3/2} F_{3/2} \left(\tilde{\omega}_{H,i} / \tilde{T}_{C,i}(H) \right), \quad \tilde{\omega}_{H,i} = \tilde{\omega}_{H=0,i} \left[1 - H^2 / H^2_{max,i} \right].$$

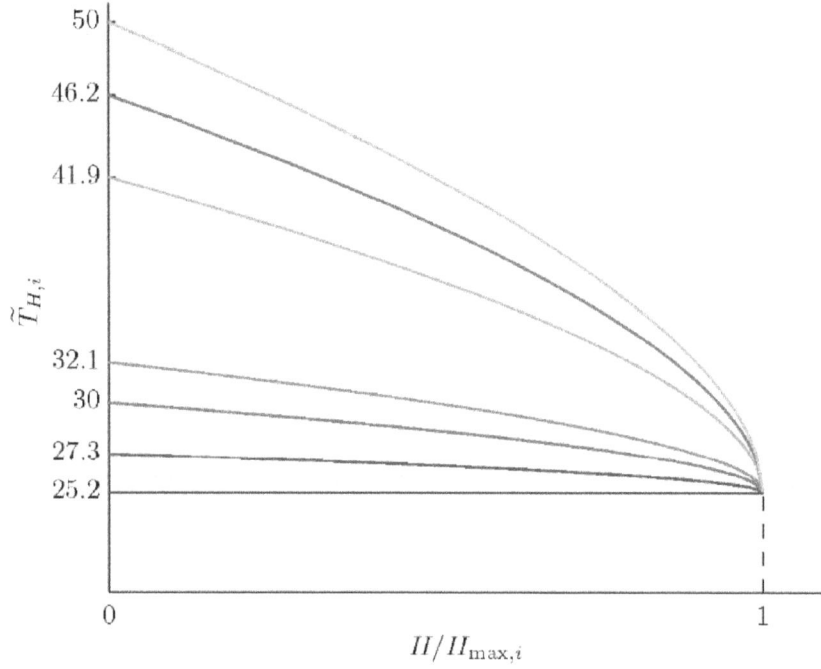

Figure 4. Dependencies of the critical transition temperature $\tilde{T}_{H,i}$ on the magnetic field H for the values of parameters $\tilde{\omega}_i$ given in Figure 2.

Figure 4 suggests that the critical transition temperature $\tilde{T}_c(H)$ changes stepwise as the magnetic field reaches the value $H_{max,i}$.

To solve the problem of the type of the phase transition in a magnetic field let us proceed from the well-known expression which relates free energies in superconducting and normal states:

$$F_S + \frac{H^2}{8\pi} = F_N, \tag{40}$$

where F_S and F_N are free energies of the unit volume of superconducting and normal states, respectively:

$$F_S = \frac{N}{V} E_{bp}(H = 0) - \frac{2}{3} \Delta E(\omega_{H=0}) \frac{N}{V},$$

$$F_N = \frac{N}{V} E_{bp}(H) - \frac{2}{3} \Delta E(\omega_H) \frac{N}{V},$$

where $\Delta E = E - E_{bp}$, $E = \omega^* \tilde{E}$, where $\tilde{E}$ is determined by formula (29) in [39]. Differentiating (40) with respect to temperature and taking into account that $S = -\partial F/\partial T$, we express the heat of transition q as:

$$q = T(S_N - S_S) = -T\partial(F_N - F_S)/\partial T = -T\frac{H_{cr}}{4\pi} \frac{\partial H_{cr}}{\partial T}, \tag{41}$$

Accordingly the difference of entropies $S_S - S_N$ will be written as:

$$S_S - S_N = \frac{H_{cr}}{4\pi}\left(\frac{\partial H_{cr}}{\partial T}\right) = \frac{H_{max}^2}{8\pi\omega^*}(\tilde{S}_S - \tilde{S}_N). \tag{42}$$

Figure 5 shows the temperature dependence of the difference of entropies (42) for various values of critical temperatures ($\tilde{\omega}_i$) given in Figure 2. These dependencies may seem strange in, at least, two respects:

1. In BCS and Ginzburg-Landau theory at the most critical point T_c the difference of entropies becomes zero in accordance with Rutgers formula. In Figure 5 entropy is a monotonous function $\tilde{T}$ which does not vanish for $T = T_c$.
2. Second, in absolute terms, the difference $|\tilde{S}_S - \tilde{S}_N|$, when approaching the limit point $\tilde{T}_c = 25.2$, which corresponds to the value $\tilde{\omega} = 0$, as it can be seemed should decrease rather than increase vanishing at $\tilde{\omega} = 0$.

As for the second point, this is really the case for $|S_S - S_N|$, since the value of the maximum field H_{max} and, accordingly, the multiplier $H_{max}^2/8\pi$ relating the quantities $S_S - S_N$ and $\tilde{S}_S - \tilde{S}_N$ becomes zero for $\tilde{\omega} = 0$.

As for the first point, as will be shown below, Rutgers formula appears inapplicable for Bose-condensate of TI-bipolarons.

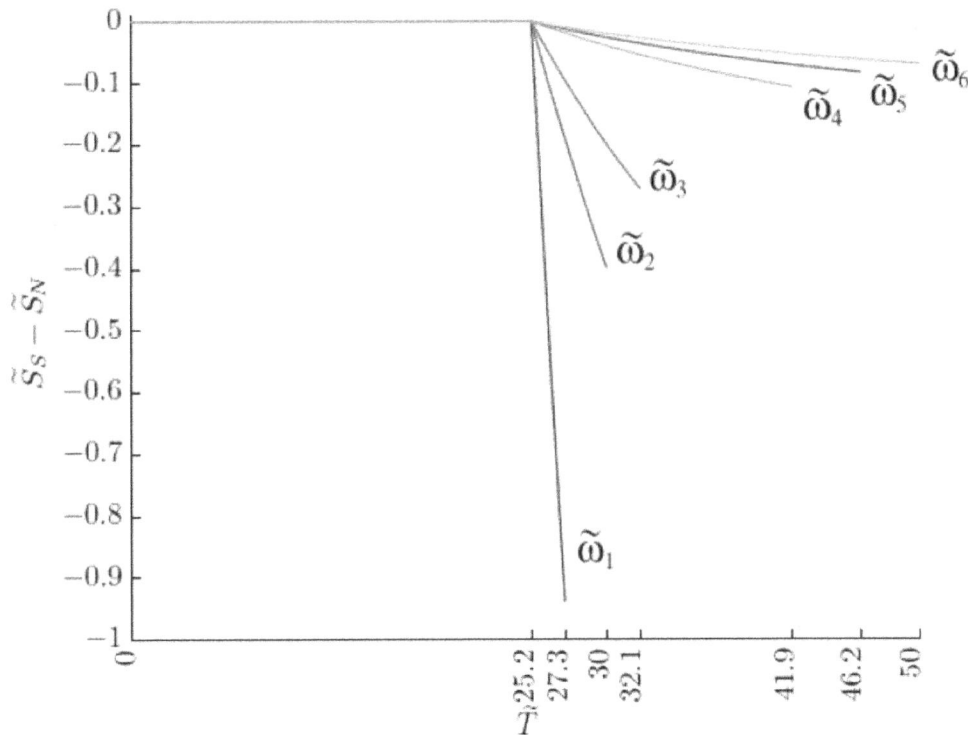

Figure 5. Temperature dependencies for the difference of entropies of superconducting and normal states for the values of parameters $\tilde{\omega}_i$ given in Figures 3 and 4.

Table 1 lists the values of the quantity $\tilde{S}_S - \tilde{S}_N$ for critical temperatures corresponding to different values of $\tilde{\omega}_{H_{cr,i}}$.

The results obtained suggest some fundamental conclusions:

1. The curve of the dependence $H_{cr}(T)$ (Figure 3) for $T = 0$ has a zero derivative, accordingly $dH_{cr}(T)/dT = 0$ for $T = 0$. This result is consistent with Nernst theorem which implies that entropy determined by (41) is equal to zero for $T = 0$.

2. According to Figure 3, $H_{cr}(T)$ is a curve monotonously drooping with increasing T for $T > T_c(\tilde{\omega} = 0)$, and a constant value for $T \leq T_c(\tilde{\omega} = 0)$. Hence $\partial H_{cr}(T)/\partial T < 0$ for $T > T_c(\tilde{\omega} = 0)$. Therefore on the temperature interval $[T_c(\tilde{\omega} = 0), T_c(\tilde{\omega})]$ $S_S < S_N$ and on the interval $[0, T_c(\tilde{\omega} = 0)]$ $S_S = S_N$.

This suggests some important conclusions:

1. Transition on the interval $[0, T_c(\tilde{\omega} = 0)]$ occurs without absorption or release of latent heat since in this case $S_S = S_N$. Experimentally it will be seen as a phase transition of the second kind. Actually, in the region $[0, T_c(\tilde{\omega} = 0)]$, a phase transition into a superconducting state is a phase transition of infinite kind, since in this region, according to (40) and Figure 3, any-order derivatives of the difference of free energies $F_S - F_N$, become zero.

2. Passing in a magnetic field from a superconducting state to a normal one on the interval $[T_c(\tilde{\omega} = 0), T_c(\tilde{\omega})]$, which corresponds to $S_S < S_N$, occurs with absorption of latent heat. On the contrary, passing from a normal state to a superconducting one takes place with release of latent heat. The phase transition on the interval $[0, T_c(\tilde{\omega} = 0)]$ is not attended by absorption or release of the latent heat being the phase transition of infinite kind.

With regard to the fact that the specific heat capacity of a substance is determined by the formula $C = T(\partial S/\partial T)$, the difference of specific heat capacities of superconducting and normal states, according to (42) will be written as:

$$C_S - C_N = \frac{T}{4\pi} \left[\left(\frac{\partial H_{cr}}{\partial T} \right)^2 + H_{cr} \frac{\partial^2 H_{cr}}{\partial T^2} \right]. \tag{43}$$

This relation is usually used to get the well-known Rutgers formula. To do this one assumes the critical field in (43) to be $H_{cr}(T_c) = 0$ for $T = T_c$ and leaves in the brackets on the right-hand side of (43) only the first term:

$$(C_S - C_N)_R = \frac{T_c}{4\pi} \left(\frac{\partial H_{cr}}{\partial T} \right)^2_{T_c}.$$

It is easily seen, however, that at the point $T = T_c$ the quantity $\omega_{H_{cr}}$ determined by Figure 2, for all the values of the temperature, has a finite derivative with respect to T and, therefore, according to (39), an infinite derivative $\partial H_{cr}/\partial T$ for $T = T_c$. Hence, the second term in the brackets (43) reduces to $-\infty$, leaving this bracket a finite value. As a result, a proper expression for the difference of heat capacities of the considered model of Bose-gas should be determined by the formula:

$$C_S - C_N = \frac{T}{8\pi} \frac{\partial^2}{\partial T^2} H_{cr}^2(T) = \frac{H_{max}^2}{8\pi\omega^*} (\tilde{C}_S - \tilde{C}_N), \tag{44}$$

$$\tilde{C}_S - \tilde{C}_N = \tilde{T} \frac{\partial^2}{\partial \tilde{T}^2} \left(H_{cr}^2(\tilde{T})/H_{max}^2 \right).$$

Table 1 lists the values of quantity $\tilde{C}_S - \tilde{C}_N$ for the values of critical temperatures corresponding to various values of $\tilde{\omega}_{H_{cr,i}}$. Notice that according to results obtained the capacity jump (44) has its maximum value at zero magnetic field and decreases as the magnetic field increases being equal zero at $H = H_{cr}$ in full accordance with the experimental data [39]. Comparison of the jumps in the heat capacity presented in [39] with expression (44) enables us co calculate the value of H_{max}. The values of H_{max} obtained by this means for various values of $\tilde{\omega}_i$ are given in Table 1. These values unambiguously determine the values of constants η in formulae (22′), (22″).

Table 1. The values of H_{max} entropy differences $\tilde{S}_S - \tilde{S}_N$ and heat capacities $\tilde{C}_S - \tilde{C}_V$ in superconducting and normal states determined by relations (42) and (44) are presented for transition temperatures $\tilde{T}_{C_i}$, for the same values of $\tilde{\omega}_{Hcr,i}$ as in Figure 2.

i	$\tilde{\omega}_{Hcr,i}$	$\tilde{T}_{C_i}$	$\tilde{S}_S - \tilde{S}_N$	$\tilde{C}_S - \tilde{C}_V$	$H_{max} \times 10^{-3}$, Oe
0	0	25.2	0	0	0
1	0.2	27.3	−0.94	−11.54	2.27
2	1	30	−0.4	−2.18	7.8
3	2	32.1	−0.27	−1.05	13.3
4	10	41.9	−0.1	−0.19	47.1
5	15	46.2	−0.08	−0.12	64.9
6	20	50	−0.07	−0.09	81.5

It follows from what has been said that Ginzburg-Landau temperature expansion for a critical field near the critical temperature T_c is not applicable for Bose-condensate of TI-bipolarons. Since the temperature dependence $H_{cr}(T)$ determines the temperature dependencies of all thermodynamic quantities, this conclusion is valid for all such values. As was pointed out in the Introduction, this conclusion follows from the fact that BCS theory, in view of its nonanalyticity on coupling constant, on no condition passes on to the theory of bipolaron condensate.

Above we dealt with an isotropic case. In the anisotropic case formulae (22′), (22″) yield:

$$H_{max}^2 = H_{max \perp}^2 = \frac{2\omega_0 M_\perp \hbar^2}{\eta^2}, \quad \vec{B} || \vec{c}, \tag{45}$$

i.e., in the case when the magnetic field is directed perpendicular to the plane of layers and:

$$H_{max}^2 = H_{max \, ||}^2 = \frac{2\omega_0 M_{||} \hbar^2}{\eta^2}, \quad \vec{B} \perp \vec{c}, \tag{46}$$

in the case when the magnetic field lies in the plane of layers. From (45) and (46) it follows that:

$$\frac{H_{max \perp}^2}{H_{max \, ||}^2} = \sqrt{\frac{M_\perp}{M_{||}}} = \gamma. \tag{47}$$

With the use of (39), (46), (47), the critical field $H_{cr}(\tilde{T})$ (in the directions perpendicular and parallel to the plane of layers) will be:

$$H_{cr \, ||, \perp}(\tilde{T}) = H_{max \, ||, \perp} \sqrt{1 - \tilde{\omega}_{H_{cr}}(\tilde{T})/\tilde{\omega}}. \tag{48}$$

From (48) it follows that the relations $H_{cr \, ||}(\tilde{T})/H_{cr \, \perp}(\tilde{T})$ are independent of temperature. The dependencies obtained are compared with experimental data in Section 7.

7. Comparison with the Experiment

By way of example let us consider HTSC $YBa_2Cu_3O_7$ with the temperature of transition $90 \div 93$ K, volume of the unit cell 0.1734×10^{-21} cm^3, concentration of holes $n \cong 10^{21}$ cm^{-3}. According to estimates [61], Fermi energy is equal to: $\varepsilon_F = 0.37$ eV. Concentration of TI-bipolarons in $YBa_2Cu_3O_7$ is found from equation (24):

$$\frac{n_{bp}}{n} C_{bp} = f_{\tilde{\omega}}(\tilde{T}_c), \tag{49}$$

with $\tilde{T}_c = 1.6$.

Among experiments with the use of an external magnetic field, of importance are experiments concerned with measurements of London penetration depth λ. In $YBa_2Cu_3O_7$ for λ for $T = 0$ the authors of [62] obtained $\lambda_{ab} = 150 \div 300$ nm, $\lambda_c = 800$ nm. The same order of magnitude of these quantities is given in a lot of papers [63–66]. The authors of [65] (see also references therein) demonstrate that anisotropy of lengths λ_a and λ_b in cuprate planes can be 30% depending on the type of the crystal structure. If we take the value $\lambda_a = 150$ nm and $\lambda_c = 800$ nm obtained on most papers, then, according to (33) the anisotropy parameter will be $\gamma \approx 30$, which is the value usually used for for $YBa_2Cu_3O_7$ crystals.

The temperature dependence $\lambda^2(0)/\lambda^2(T)$ was studied in many papers (see [66] and references therein).

Figure 6 shows a comparison of various curves for $\lambda^2(0)/\lambda^2(T)$. In paper [66] it is shown that in high quality crystals of $YBa_2Cu_3O_7$ the temperature dependence $\lambda^2(0)/\lambda^2(T)$ is well approximated by a simple dependence $1 - t^2$, $t = T/T_c$.

Figure 7 demonstrates a comparison of the experimental dependence $\lambda^2(0)/\lambda^2(T)$ [66] with the theoretical one:

$$\frac{\lambda^2(0)}{\lambda^2(T)} = 1 - \left(\frac{T}{T_c}\right)^{3/2} \frac{F_{3/2}(\omega/T)}{F_{3/2}(\omega/T_c)}, \tag{50}$$

which follows from (34), (25). Hence there is a good agreement between experimental and theoretical dependencies (50).

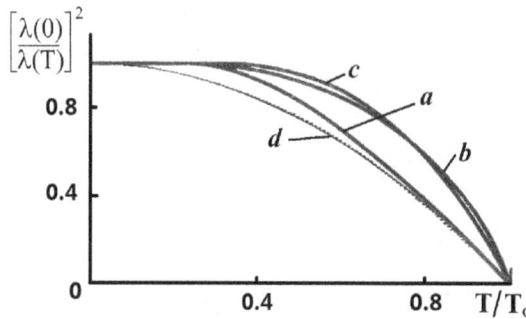

Figure 6. Penetration depth of the magnetic field found with the use of BCS theory (*a*-local approximation, *b*-nonlocal approximation); on empirical law λ^{-2} $1 - (T/T_c)^4$ (*c*) [67]; in $YBa_2Cu_3O_7$ (*d*) [66].

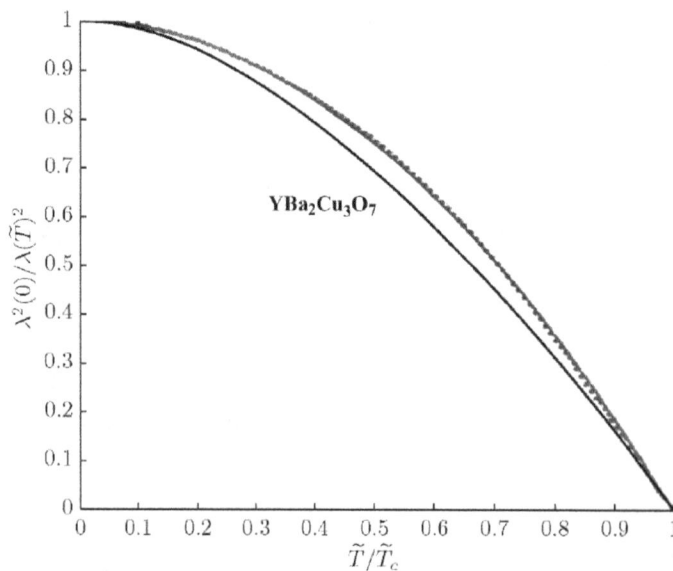

Figure 7. Comparison of the theoretical dependence $\lambda^2(0)/\lambda^2(\tilde{T})$ (solid line) obtained in the present article with the experimental one [66] (dotted line).

The theory developed enables us to compare the temperature dependence of the value of the critical magnetic field in $YBa_2Cu_3O_7$ with experimental data [68]. Since the theory constructed in Section 6 describes a homogeneous state of a TI-bipolaron gas, then the critical field under consideration corresponds to a homogeneous Meissner phase. In paper [68] this field is denoted by H_{c1} which is related to denotations of Section 6 as: $H_{c1} = H_{cr}$, $H_{c1||} = H_{cr\perp}$, $H_{c1\perp} = H_{cr||}$. To make a comparison with the experiment we use parameter values obtained earlier for $YBa_2Cu_3O_7$: $\tilde{\omega} = 1.5$, $\tilde{\omega}_c = 1.6$. Figure 8 shows a comparison of experimental dependencies $H_{c1\perp}(T)$ and $H_{c1||}(T)$ [68] with theoretical dependencies (48), where for $H_{max\,||,\perp}(T)$, we took the following experimental values: $H_{max\,||} = 240$, $H_{max\,\perp} = 816$. The results presented in Figure 8 confirm the conclusion (Section 6) that relations $H_{cr\perp}(T)/H_{cr||}(T)$ are independent of temperature.

Relations (33), (45), (46) yield:

$$(\gamma^*)^2 = \frac{M_\perp^*}{M_{||}^*} \propto \frac{\lambda_\perp^2}{\lambda_{||}^2}; \quad \frac{H_{max\,\perp}^2}{H_{max\,||}^2} = \gamma^2 = 11.6. \tag{51}$$

The assessment of anisotropy parameters $\gamma^2 = 11.6$ determined by relations (51) differs from the value $(\gamma^*)^2 = 30$ used above. This difference is probably caused by difference in anisotropy of polaron effective mass $M_{||,\perp}^*$ and electron band mass $m_{||,\perp}^*$.

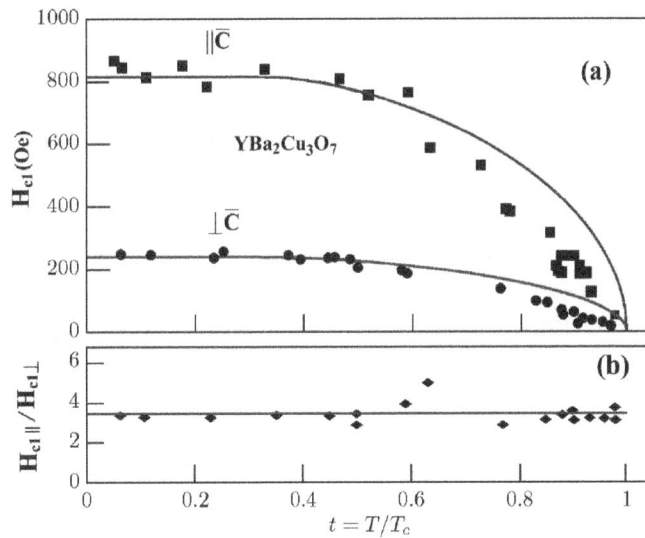

Figure 8. Comparison of calculated (solid line) and experimental values of H_{c1} (squares, circles, rhombs [68]) for the cases $||\vec{c}$ and $\perp\vec{c}$.

8. Scaling Relations

Scaling relations play an important role in the theory of superconductivity assisting the search for new high-temperature superconductors with record parameters. These relations can emerge as a result of numerous experiments lacking any reliable theoretical substantiation. Or else they can be derived from insufficiently reliable theoretical considerations, but subsequently be supported by a lot of experiments. By way of example we refer to Uemura law considered in the previous section.

The theory presented here enables us to give a natural explanation to some important scaling relations. In particular, in this section we will derive Alexandrov's formula [69,70] and Homes's scaling law.

- **Alexandrov's formula** As was mentioned in [39], in an anisotropic case formula (49) takes on the form:

$$\tilde{T}_c = F_{3/2}^{-2/3}(\tilde{\omega}/\tilde{T}_c)\left(\frac{n_{bp}}{M_\parallel}\right)\frac{2\pi\hbar^2}{M_\perp^{1/3}\omega^*}. \tag{52}$$

It is convenient to pass on in formula (52) from quantities n_{bp}, $M_\parallel$, $M_\perp$ which can hardly be determined in experiments to quantities which are easily measured experimentally:

$$\lambda_{ab} = \left[\frac{M_\parallel}{16\pi n_{bp}e^2}\right]^{1/2}, \quad \lambda_c = \left[\frac{M_\perp}{16\pi n_{bp}e^2}\right]^{1/2}, \quad R_H = \frac{1}{2en_{bp}}, \tag{53}$$

where $\lambda_{ab} = \lambda_\parallel$, $\lambda_c = \lambda_\perp$ are London lengths of penetration into the planes of layers and in perpendicular direction, respectively; R_H is Hall coefficient. In expressions (53) the light velocity is assumed to be equal to unity: $c = 1$. With the use of relations (53) and (52) we get:

$$k_B T_c = \frac{2^{1/3}}{8}F_{3/2}^{-2/3}(\tilde{\omega}/\tilde{T}_c)\frac{\hbar^2}{e^2}\left(\frac{eR_H}{\lambda_{ab}^4\lambda_c^2}\right)^{1/3}. \tag{54}$$

In formula (54) the quantity eR_H is measured in cm^3, λ_{ab}, λ_c in cm, T_c in Kelvin.

Taking into account that in most HTSC materials $\tilde{\omega} \approx \tilde{T}_c$ and the function $F_{3/2}(\tilde{\omega}/\tilde{T})$ changes near $\tilde{\omega} = \tilde{T}_c$ only slightly, with the use of the value $F_{3/2}(1) = 0.428$ and expression (53) we present T_c in the form:

$$T_c \cong 8.7\left(\frac{eR_H}{\lambda_{ab}^4\lambda_c^2}\right)^{1/3}. \tag{55}$$

Formula (55) differs from Alexandrov's formula [69,70] only in numerical coefficient which in [69,70] is equal to 1.64. As is shown in [69,70], formula (55) practically always properly describes relation between the parameters for all known HTSC materials. In [69,70] it is also shown that Uemura relation [71,72] is a particular case of formula (55).

- **Homes's law** Homes's law holds that scaling relation are valid for superconducting materials [73,74]:

$$\rho_S = C\sigma_{DC}(T_c)T_c, \tag{56}$$

where ρ_S is the density of a superfluid component for $T = 0$, $\sigma_{DC}(T_c)$ is the conductivity of direct current for $T = T_c$, C is a constant equal to $\approx 35cm^{-2}$ for ordinary superconductors and HTSC materials for a current running in the plane of layers.

The quantity ρ_S involved in (56) is related to plasma frequency $\omega_p = \sqrt{4\pi n_S e_S^2/m_S^*}$ (where n_S is a concentration of superconducting charge carriers; m_S^*, e_S are a mass and charge of superconducting charge carriers) by a well-known expression $\rho_S = \omega_p^2$ [75]. Using this expression, relation $\sigma_{DC} = e_n^2 n_n \tau/m_n^*$ (where n_n is the concentration of charge carriers for $T = T_c$), m_n^*, e_n are the mass and the charge of charge carriers, relation $\tau \sim \hbar/T_c$ (where τ is the minimum Planck time for scattering of electrons at the critical point [75]) and also expression (56), on the assumption $e_S = e_n$, $m_S = m_n$, we get:

$$n_S(0) \cong n_n(T_c). \tag{57}$$

This result is confirmed by experimental data [76].

In our scenario of Bose-condensation of TI-bipolarons, Homes's law in the form of (57) becomes almost obvious. Indeed, for $T = T_c$ TI-bipolarons are stable formations (they decay at temperature equal to the pseudogap energy which considerably exceeds T_c). Their concentration for $T = T_c$ is equal to n_n and therefore these bipolarons for $T = T_c$ start forming condensate whose concentration $n_S(T)$ reaches its maximum $n_S(0) = n_n(T_c)$ for $T = 0$ (i.e., when bipolarons fully pass on to condensed state) which corresponds to relation (57). Notice that in the framework of BCS theory Homes's law can hardly be explained.

9. Summary

It is generally accepted that super-flow, superfluidity and superconductivity are collective phenomena that are driven by inter-particle interactions. Here we state an opposite suggestion that the above phenomena are mainly determined by the specific properties of separate boson particles.

In this paper we have presented conclusions emerging from consistent translation-invariant consideration of EPI. It implies that pairing of electrons, for any coupling constant, leads to a concept of TI-polarons and TI-bipolarons. Being bosons, TI-bipolarons can experience Bose condensation when $H \neq 0$ leading to superconductivity. Let us list the main results following from this approach.

First and foremost the theory resolves the problem of the great value of the bipolaron effective mass (Section 4). As a consequence, formal limitations on the value of the critical temperature of the transition are eliminated too. The theory quantitatively explains such thermodynamic properties of HTSC-conductors as availability and value of the jump in the heat capacity (details in [39]) lacking in the theory of Bose condensation of an ideal gas. The theory also gives an insight into the occurrence of a great ratio between the width of the pseudogap and T_c ([39,40]). It accounts for the small value of the correlation length [36] and explains the availability of a gap and a pseudogap ([39,40]) in HTSC materials. Accordingly, isotopic effect automatically follows from expression (25) where the phonon frequency ω_0 acts as a gap. The conclusion of the dependence of the temperature of the transition T_c on the relation $n_{bp}/M_{||}$ (see [39]) correlates with Alexandrov-Uemura law (Section 8) universal for all HTSC materials. It is shown that Homes's scaling law is a natural consequence of the theory presented (Section 8). The theory explains a wide variety of phenomena observed in a magnetic field (Section 6). In particular:

1. It is shown that the occurrence of a gap in the spectrum of TI-bipolarons makes possible their condensation in a magnetic field.
2. It is demonstrated that there exists a critical value of the magnetic field above which homogeneous Bose-condensation becomes impossible.
3. The temperature dependence of the critical magnetic field and London penetration depth obtained in the paper are in good agreement with the experiment.

At the same time the theory presented shows that:

1. Rutgers formula cannot be applied to describe Bose-condensation of TI-bipolarons.
2. Ginzburg-Landau expansions do not suit to describe Bose-condensation of TI-bipolarons.

The theory predict some phenomena such as:

1. Isotopic effect for a jump of heat capacity in passing from the normal phase to superconducting one.
2. A possibility of the existence of a phase transition of infinite kind in a magnetic field at low temperatures.
3. Identity of the energy gap with phonon frequency.
4. Existence of superconducting TI-bipolarons whose concentration is much less than the total concentration of charge carriers.

Application of the theory to 1D and 2D systems leads to qualitatively new results since the occurrence of a gap in the TI-bipolaron spectrum automatically removes divergences at small momenta, inherent in the theory of ideal Bose gas. An important consequence of this fact is the existence of a superconducting phase in homogeneous 1D and 2D systems [77].

In conclusion it should be said that in the theory developed the TI-bipolarons are equivalent of Cooper pairing in BCS. In distinction from the latter, this theory does not impose the upper limit of critical temperature for SC transition. Of course there are a lot of other mechanisms of pairing. Each theory of superconductivity needs to explain three basic effects: the zero resistance at $T < T_c$, Meissner effect and the isotop effect. As was shown in the paper the considered EPI is sufficient for such an explanation. Recent experiments on H_3S and LaH_{10} (with $T_c = 203$ K for H_3S [78] and $T_c = 260$ K for LaH_{10} [79,80]) possessing a record temperature of SC transition (under high pressure) are in accordance with the idea of the important role of EPI mechanism and TI-bipolarons considered in the paper.

Funding: This research received no external funding.

Conflicts of Interest: The author declares no conflict of interest.

Appendix A

Hamiltonian H_1 involved in (12) has the form:

$$H_1 = \sum_k (V_k + f_k \hbar \omega_k)(a_k + a_k^+) + \sum_{k,k'} \frac{\vec{k}\vec{k'}}{m} f_{k'}(a_k^+ a_k a_{k'} + a_k^+ a_{k'}^+ a_k) + \frac{1}{2m} \sum_{k,k'} \vec{k}\vec{k'} a_k^+ a_{k'}^+ a_k a_{k'} ,$$

Let us apply the operator H_1 to functional $\hat{R}|0\rangle$, where $\hat{R}$—operator that generates Bogolyubov-Tyablikov canonical transformation (15). We will show that $\langle 0|\hat{R}^+ H_1 \hat{R}|0\rangle = 0$. Indeed, the action of $\hat{R}$ on H_1 terms containing an odd number of operators in H_1 (i.e., the first and second terms in H_1) will always contain an odd number of terms and mathematical expectation for these terms will tend to zero.

Let us consider mathematical expectation for the last term in H_1:

$$\langle 0|\hat{R}^+ \sum_{k,k'} \vec{k}\vec{k'} a_k^+ a_{k'}^+ a_k a_{k'} \hat{R}|0\rangle . \tag{A1}$$

The function $\langle 0|\hat{R}^+ a_k^+ a_{k'}^+ a_k a_{k'} \hat{R}|0\rangle$ represents the norm of vector $a_k a_{k'} \hat{R}|0\rangle$ and will be positively defined for all k and k'. If we replace $\vec{k} \to -\vec{k}$ in (A1) than the whole expression will change the sign and, therefore, (A1) is also equal to zero. Hence $\langle 0|\hat{R}^+ H_1 \hat{R}|0\rangle = 0$.

As was shown in [42] the explicit form for operator $\hat{R}$ is:

$$\hat{R} = C \exp \left\{ \frac{1}{2} \sum_{k,k'} a_k^+ A_{kk'} a_{k'}^+ \right\} ,$$

where C is the normalizing constant and matrix A satisfies the conditions:

$$A = M_2^* (M_1^*)^{-1}, \quad A = A^T ,$$

where M_1 and M_2 are matrices involved in (15).

References

1. Bardeen, J.; Cooper, L.N.; Schrieffer, J.R. Theory of Superconductivity. *Phys. Rev.* **1957**, *108*, 1175. [CrossRef]
2. Schrieffer, J.R. *Theory of Superconductivity*; Westview Press: Oxford, UK, 1999.
3. Moriya, T.; Ueda, K. Spin fluctuations and high temperature superconductivity. *Adv. Phys.* **2000**, *49*, 555–606. [CrossRef]
4. Sinha, K.P.; Kakani, S.L. Fermion local charged boson model and cuprate superconductors. *Proc. Natl. Acad. Sci. India Sect. A Phys. Sci.* **2002**, *72*, 153–214.
5. Alexandrov, A.S. *Theory of Superconductivity from Weak to Strong Coupling*; IoP Publishing: Bristol, UK, 2003.
6. Manske, D. *Theory of Unconventional Superconductors*; Springer: Heidelberg, Germany, 2004.

7. Benneman, K.H.; Ketterson, J.B. (Eds.) *Superconductivity: Conventional and Unconventional Superconductors 1–2*; Springer: New York, NY, USA; Berlin, Germany, 2008.

8. Gunnarsson, O.; Rösch, O. Interplay between electron-phonon and coulomb interactions in cuprates. *J. Phys.* **2008**, *20*, 043201. [CrossRef]

9. Kakani, S.L.; Kakani, S. *Superconductivity*; Anshan: Kent, UK, 2009.

10. Plakida, N.M. *High Temperature Cuprate Superconductors: Experiment, Theory and Applications*; Springer: Heidelberg, Germany, 2010.

11. Cooper, L.N.; Feldman, D. (Eds.) *BCS: 50 Years*; World Sci. Publ. Co.: Singapore, 2011.

12. Tohyama, T. Recent progress in physics of high-temperature superconductors. *Jpn. J. Appl. Phys.* **2012**, *51*, 010004. [CrossRef]

13. Askerzade, I. *Unconventional Superconductors: Anisotropy and Multiband Effects*; Springer: Berlin, Germany, 2012.

14. Bardeen, J. Developments of concepts in superconductivity. *Phys. Today* **1963**, *16*, 19. [CrossRef]

15. Keldysh, L.V.; Kozlov, A.N. Collective Properties of Excitons in Semiconductors. *Sov. Phys. JETP* **1968**, *27*, 521.

16. Eagles, D.M. Possible Pairing without Superconductivity at Low Carrier Concentrations in Bulk and Thin-Film Superconducting Semiconductors. *Phys. Rev.* **1969**, *186*, 456. [CrossRef]

17. Nozières, P.; Schmitt-Rink, S. Propagation of Second sound in a superfluid Fermi gas in the unitary limit. *J. Low Temp. Phys.* **1985**, *59*, 195. [CrossRef]

18. Loktev, V.M. Mechanisms of high-temperature superconductivity of Copper oxides. *Fizika Nizkih Temperatur* **1996**, *22*, 3.

19. Randeria, M. Precursor Pairing Correlations and Pseudogaps. *arXiv* **1997**, arXiv:cond-mat/9710223.

20. Uemura, Y.J. Bose-Einstein to BCS crossover picture for high-T_c cuprates. *Phys. C Supercond.* **1997**, *282*, 194–197. [CrossRef]

21. Drechsler, M.; Zwerger, W. Crossover from BCS-superconductivity to Bose-condensation. *Ann. Phys.* **1992**, *1*, 15. [CrossRef]

22. Griffin, A.; Snoke, D.W.; Stringari, S. (Eds.) *Bose-Einstein Condensation*; Cambridge University Press: New York, NY, USA, 1996.

23. Eliashberg, G.M. Interactions between Electrons and Lattice Vibrations in a Superconductor. *Sov. Phys. JETP* **1960**, *11*, 696.

24. Marsiglio, F.; Carbotte, J.P. Gap function and density of states in the strong-coupling limit for an electron-boson system. *Phys. Rev. B* **1991**, *43*, 5355. [CrossRef]

25. Micnas, R.; Ranninger, J.; Robaszkiewicz, S. Superconductivity in narrow-band systems with local nonretarded attractive interactions. *Rev. Mod. Phys.* **1990**, *62*, 113. [CrossRef]

26. Zwerger, W. (Ed.) The BCS-BEC Crossover and the Unitary Fermi Gas. In *Lecture Notes in Physics*; Springer: Berlin, Heidelberg, 2012.

27. Bloch, I.; Dalibard, J.; Zwerger, W. Many-body physics with ultracold gases. *Rev. Mod. Phys.* **2008**, *80*, 885. [CrossRef]

28. Giorgini, S.; Pitaevskii, L.P.; Stringari, S. Theory of ultracold atomic Fermi gases. *Rev. Mod. Phys.* **2008**, *80*, 1215. [CrossRef]

29. Chen, Q.; Stajic, J.; Tan, S.; Levin, K. BCS-BEC crossover: From high temperature superconductors to ultracold superfluids. *Phys. Rep.* **2005**, *412*, 1–88. [CrossRef]

30. Ketterle, W.; Zwierlein, M.W. Making, probing and understanding ultracold Fermi gases. In *Ultra-cold Fermi Gases*; Inguscio, M., Ketterle, W., Salomon, C., Eds.; IOS Press: Amsterdam, The Netherlands, 2007; p. 95.

31. Pieri, P.; Strinati, G.C. Strong-coupling limit in the evolution from BCS superconductivity to Bose-Einstein condensation. *Phys. Rev. B* **2000**, *61*, 15370. [CrossRef]

32. Gerlach, B.; Löwen, H. Analytical properties of polaron systems or: Do polaronic phase transitions exist or not? *Rev. Mod. Phys.* **1991**, *63*, 63. [CrossRef]

33. Lakhno, V.D. Translation invariant theory of polaron (bipolaron) and the problem of quantizing near the classical solution. *JETP* **2013**, *116*, 892–896. [CrossRef]

34. Gor'kov, L.P. Microscopic derivation of the Ginzburg-Landau equations in the theory of superconductivity. *Sov. Phys. JETP* **1959**, *9*, 1364–1367.

35. Lakhno, V.D. Energy and Critical Ionic-Bond Parameter of a 3D Large-Radius Bipolaron. *J. Exp. Theor. Phys.* **2010**, *110*, 811–815. [CrossRef]

36. Lakhno, V.D. Translation-invariant bipolarons and the problem of high-temperature superconductivity. *Sol. State Commun.* **2012**, *152*, 621–623. [CrossRef]

37. Kashirina, N.I.; Lakhno, V.D.; Tulub, A.V. The Virial Theorem and the Ground State Problem in Polaron Theory. *J. Exp. Theor. Phys.* **2012**, *114*, 867–869. [CrossRef]

38. Lakhno, V.D. Pekar's ansatz and the strong coupling problem in polaron theory. *Phys. Usp.* **2015**, *58*, 295. [CrossRef]

39. Lakhno, V.D. Superconducting Properties of 3D Low-Density Translation-Invariant Bipolaron Gas. *Adv. Condens. Matter Phys.* **2018**, *2018*, 1380986. [CrossRef]

40. Lakhno, V.D. Superconducting properties of a nonideal bipolaron gas. *Phys. C Supercond. Its Appl.* **2019**, *561*, 1–8. [CrossRef]

41. Lakhno, V.D. Spin wave amplification in magnetically ordered crystals. *Phys. Usp.* **1996**, *39*, 669. [CrossRef]

42. Tulub, A.V. Slow Electrons in Polar Crystals. *Sov. Phys. JETP* **1962**, *14*, 1301.

43. Heisenberg, W. Die selbstenergie des elektrons. *Z. Phys.* **1930**, *65*, 4–13. [CrossRef]

44. Rosenfeld, L. Über eine mögliche Fassung des Diracschen Programms zur Quantenelektrodynamik und deren formalen Zusammenhang mit der Heisenberg-Paulischen Theorie. *Z. Phys.* **1932**, *76*, 729–734. [CrossRef]

45. Lee, T.D.; Low, F.; Pines, D. The motion of electrons in a polar crystal. *Phys. Rev.* **1953**, *90*, 297. [CrossRef]

46. Tyablikov, S.V. *Methods in the Quantum Theory of Magnetism*; Plenum Press: New York, NY, USA, 1967.

47. Miyake, S.J. Bound Polaron in the Strong-coupling Regime. In *Polarons and Applications*; Lakhno, V.D., Ed.; Wiley: Leeds, UK, 1994; p. 219.

48. Levinson, I.B.; Rashba, E.I. Threshold phenomena and bound states in the polaron problem. *Sov. Phys. Usp.* **1974**, *16*, 892–912. [CrossRef]

49. Porsch, M.; Röseler, J. Recoil Effects in the Polaron Problem. *Phys. Status Solidi B* **1967**, *23*, 365–376. [CrossRef]

50. Alexandrov, A.S.; Mott, N. *Polarons & Bipolarons*; World Sci. Pub. Co.: Singapore, 1996.

51. Alexandrov, A.S.; Krebs, A.B. Polarons in high-temperature superconductors. *Sov. Phys. Usp.* **1992**, *35*, 345, 383. [CrossRef]

52. Ogg, R.A., Jr. Superconductivity in solid metal-ammonia solutions. *Phys. Rev.* **1946**, *70*, 93. [CrossRef]

53. Vinetskii, V.L.; Pashitskii, E.A. Superfluidity of charged Bose-gas and bipolaron mechanism of superconductivity. *Ukr. J. Phys.* **1975**, *20*, 338.

54. Pashitskii, E.A.; Vinetskii, V.L. Plasmon and bipolaron mechanisms of high-temperature superconductivity. *JETP Lett.* **1987**, *46*, 124–127.

55. Emin, D. Formation, motion, and high-temperature superconductivity of large bipolarons. *Phys. Rev. Lett.* **1989**, *62*, 1544. [CrossRef] [PubMed]

56. Vinetskii, V.L.; Kashirina, N.I.; Pashitskii, E.A. Bipolaron states in ion crystals and the problem of high temperature superconductivity. *Ukr. J. Phys.* **1992**, *37*, 76.

57. Emin, D. In-plane conductivity of a layered large-bipolaron liquid. *Philos. Mag.* **2015**, *95*, 918–934. [CrossRef]

58. Schmidt, V.V. *The Physics of Superconductors*; Muller, P., Ustinov, A.V., Eds.; Springer: Berlin/Heidelberg, Germnay, 1997.

59. Pippard, A.B. Field variation of the superconducting penetration depth. *Proc. Roy. Soc. (Lond.)* **1950**, *A203*, 210–223.

60. Schafroth, M.R. Superconductivity of a Charged Ideal Bose Gas. *Phys. Rev.* **1955**, *100*, 463. [CrossRef]

61. Gor'kov, L.P.; Kopnin, N.B. High-T_c superconductors from the experimental point of view. *Sov. Phys. Usp.* **1988**, *31*, 850. [CrossRef]

62. Buckel, W.; Kleiner, R. *Superconductivity: Fundamentals and Applications*, 2nd ed.; Wiley-VCH: Weinheim, Germany, 2004.

63. Edstam, J.; Olsson, H.K. London penetration depth of YBCO in the frequency range 80-700 GHz. *Phys. B* **1994**, *194–196 Pt 2*, 1589–1590. [CrossRef]

64. Panagopoulos, C.; Cooper, J.R.; Xiang, T. Systematic behavior of the in-plane penetration depth in d-wave cuprates. *Phys. Rev. B* **1998**, *57*, 13422. [CrossRef]

65. Pereg-Barnea, T.; Turner, P.J.; Harris, R.; Mullins, G.K.; Bobowski, J.S.; Raudsepp, M.; Liang, R.; Bonn, D.A.; Hardy, W.N. Absolute values of the London penetration depth in $YBa_2Cu_3O_{6+y}$ measured by zero field ESR spectroscopy on Gd doped single crystals. *Phys. Rev. B* **2004**, *69*, 184513. [CrossRef]

66. Bonn, D.A.; Liang, R.; Riseman, T.M.; Baar, D.J.; Morgan, D.C.; Zhang, K.; Dosanjh, P.; Duty, T.L.; MacFarlane, A.; Morris, G.D.; et al. Microwave determination of the quasiparticle scattering time in $YBa_2Cu_3O_{6.95}$. *Phys. Rev. B* **1993**, *47*, 11314. [CrossRef]

67. Madelung, O. *Festkörpertheorie I, II*; Springer: Berlin/Heidelberg, Germany; New York, NY, USA, 1972.

68. Wu, D.H.; Sridhar, S. Pinning forces and lower critical fields in $YBa_2Cu_3O_y$ crystals: Temperature dependence and anisotropy. *Phys. Rev. Lett.* **1990**, *65*, 2074. [CrossRef] [PubMed]

69. Alexandrov, A.S. Comment on Experimental and Theoretical Constraints of Bipolaronic Superconductivity in High T_c Materials: An Impossibility. *Phys. Rev. Lett.* **1999**, *82*, 2620. [CrossRef]

70. Alexandrov, A.S.; Kabanov, V.V. Parameter-free expression for superconducting T_c in cuprates. *Phys. Rev. B* **1999**, *59*, 13628. [CrossRef]

71. Uemura, Y.J.; Luke, G.M.; Sternlieb, B.J.; Brewer, J.H.; Carolan, J.F.; Hardy, W.N.; Kadono, R.; Kempton, J.R.; Kiefl, R.F.; Kreitzman, S.R.; et al. Universal correlations between T_c and ns/m (carrier density over effective mass) in high-T_c cuprate superconductors. *Phys. Rev. Lett.* **1989**, *62*, 2317. [CrossRef] [PubMed]

72. Uemura, Y.J.; Le, L.P.; Luke, G.M.; Sternlieb, B.J.; Wu, W.D.; Brewer, J.H.; Riseman, T.M.; Seaman, C.L.; Maple, M.B.; Ishikawa, M.; et al. Basic similarities among cuprate, bismuthate, organic, Chevrel-phase, and heavy-fermion superconductors shown by penetration-depth measurements. *Phys. Rev. Lett.* **1991**, *66*, 2665. [CrossRef]

73. Homes, C.C.; Dordevic, S.V.; Strongin, M.; Bonn, D.A.; Liang, R.; Hardy, W.N.; Komiya, S.; Ando, Y.; Yu, G.; Kaneko, N.; et al. A universal scaling relation in high-temperature superconductors. *Nature* **2004**, *430*, 539. [CrossRef]

74. Zaanen, J. Superconductivity: Why the temperature is high. *Nature* **2004**, *430*, 512. [CrossRef]

75. Erdmenger, J.; Kerner, P.; Müller, S. Towards a holographic realization of Homes law. *J. High Energy Phys.* **2012**, *10*, 21. [CrossRef]

76. Balakirev, F.F.; Betts, J.B.; Migliori, A.; Ono, S.; Ando, Y.; Boebinger, G.S. Signature of optimal doping in Hall-effect measurements on a high-temperature superconductor. *Nature* **2003**, *424*, 912–915. [CrossRef]

77. Lakhno, V.D. A Translation invariant bipolaron in the Holstein model and superconductivity. *SpringerPlus* **2016**, *5*, 1277. [CrossRef]

78. Drozdov, A.P.; Eremets, M.I.; Troyan, I.A.; Ksenofontov, V.; Shylin, S.I. Conventional superconductivity at 203 kelvin at high pressures in the sulfur hydride system. *Nature* **2015**, *525*, 73–76. [CrossRef] [PubMed]

79. Somayazulu, M.; Ahart, M.; Mishra, A.K.; Geballe, Z.M.; Baldini, M.; Meng, Y.; Struzhkin, V.V.; Hemley, R.J. Evidence for Superconductivity above 260 K in Lanthanum Superhydride at Megabar Pressures. *Phys. Rev. Lett.* **2019**, *122*, 027001. [CrossRef] [PubMed]

80. Drozdov, A.P.; Kong, P.P.; Minkov, V.S.; Besedin, S.P.; Kuzovnikov, M.A.; Mozaffari, S.; Balicas, L.; Balakirev, F.; Graf, D.; Prakapenka, V.B.; et al. Superconductivity at 250 K in lanthanum hydride under high pressures. *arXiv* **2018**, arXiv:1812.01561.

Permissions

List of Contributors

Andrea Tononi
Dipartimento di Fisica e Astronomia "Galileo Galilei" and CNISM, Università di Padova, Via Marzolo 8, I-35131 Padova, Italy

S. V. G. Menon
Shiv Enclave, 304, 31-B-Wing, Tilak Nagar, Mumbai 400089, India

Bishnupriya Nayak
High Pressure and Synchroton Radiation Physics Division, Bhabha Atomic Research Centre, Mumbai 400085, India

Francesco Petiziol and Sandro Wimberger
Dipartimento di Scienze Matematiche, Fisiche e Informatiche, Università di Parma, Parco Area delle Scienze 7/A, 43124 Parma, Italy
Italian Institute for Nuclear Physics (INFN), Sezione di Milano Bicocca, Gruppo Collegato di Parma, Parco Area delle Scienze 7/A, 43124 Parma, Italy

Milind N. Kunchur
Department of Physics and Astronomy, University of South Carolina, Columbia, SC 29208, USA

Takashi Yanagisawa
National Institute of Advanced Industrial Science and Technology 1-1-1 Umezono, Tsukuba, Ibaraki 305-8568, Japan

Maren Hellwig and Martin Köppen
Independent researchers, Heinersdorfer Str. 52, 13086 Berlin, Germany

Albert Hiller, Hans Rudolf Koslowski and Andrey Litnovsky
Forschungszentrum Jülich GmbH, Institut für Energie—und Klimaforschung—Plasmaphysik, 52428 Jülich, Germany

Klaus Schmid
Max-Planck-Institut für Plasmaphysik, Boltzmann-straße 2, 85748 Garching, Germany

Christian Schwab and Roger A. De Souza
Institut für Physikalische Chemie, RWTH Aachen University, Landoltweg 2, 52074 Aachen, Germany

Klaus M. Frahm and Dima L. Shepelyansky
Laboratoire de Physique Théorique, IRSAMC, Université de Toulouse, CNRS, UPS, 31062 Toulouse, France

Leonardo Ermann
Departamento de Física Teórica, GIyA, Comisión Nacional de Energía Atómica, CP1650 Buenos Aires, Argentina

Fábio Luís de Oliveira Paula
Complex Fluid Group, Institute of Physics, University of Brasilia—UnB, Campus Darcy Ribeiro, Brasilia (DF) 70919-970, Brazil

Gaetano Campi
Institute of Crystallography, CNR, via Salaria Km 29.300, 00015 Monterotondo, Roma, Italy

Antonio Bianconi
Institute of Crystallography, CNR, via Salaria Km 29.300, 00015 Monterotondo, Roma, Italy
Rome International Center for Materials Science Superstripes RICMASS, Via dei Sabelli 119A, 00185 Roma, Italy

Nicola Poccia
Institute for Metallic Materials, Leibniz IFW Dresden, 01069 Dresden, Germany

Boby Joseph
Elettra Sincrotrone Trieste. Strada Statale 14 - km 163.5, AREA Science Park, I-34149 Basovizza, Trieste, Italy

Alessandro Ricci
Rome International Center for Materials Science Superstripes RICMASS, Via dei Sabelli 119A, 00185 Roma, Italy
Deutsches Elektronen-Synchrotron DESY, Notkestraße 85, D-22607 Hamburg, Germany

James Lee and Sujoy Roy
Advanced Light Source, Lawrence Berkeley National Laboratory, Berkeley, CA94720, USA

Shrawan Mishra
Advanced Light Source, Lawrence Berkeley National Laboratory, Berkeley, CA94720, USA
School of Materials Science and Technology, Indian Institute of Technology, Banaras Hindu University, Varanasi 221005, India

Agustinus Agung Nugroho
Faculty of Mathematics and Natural Sciences Intitut Teknologi Bandung, Jl. Ganesha 10 Bandung, Jawa Barat 40132, Indonesia

Marcel Buchholz and Markus Braden
II. Physikalisches Institut, Universität zu Köln, Zülpicher Str. 77, 50937 Köln, Germay

Christoph Trabant and Christian Schüßler-Langeheine
Helmholtz-Zentrum Berlin für Materialien und Energie GmbH, Institute Methods and Instrumentation in Synchrotron Radiation Research, Albert-Einstein-Str. 15, 12489 Berlin, Germany

Leonard Müller, Jens Viefhaus and Michael Sprung
Deutsches Elektronen-Synchrotron DESY, Notkestraße 85, D-22607 Hamburg, Germany

Alexey Zozulya
Deutsches Elektronen-Synchrotron DESY, Notkestraße 85, D-22607 Hamburg, Germany
European X-ray Free-Electron Laser Facility GmbH Holzkoppel 4, 22869 Schenefeld, Germany

Robert Pilemalm, Sergei Simak and Per Eklund
Department of Physics, Chemistry and Biology (IFM), Linköping University, SE-581 83 Linköping, Sweden

Leonid Pourovskii
Centre de Physique Théorique, Ecole Polytechnique, CNRS, Université Paris-Saclay, Route de Saclay, FR-91128 Palaiseau, France
Collège de France, 11 place Marcelin Berthelot, FR-75005 Paris, France

Igor Mosyagin
Materials Modeling and Development Laboratory, NUST "MISIS", RU-119991 Moscow, Russia

Levan Chkhartishvili
Engineering Physics Department, Georgian Technical University, 77 Kostava Ave., Tbilisi 0175, Georgia
Boron and Composite Materials Laboratory, Ferdinand Tavadze Institute of Metallurgy and Materials Science, 10 Mindeli Str., Tbilisi 0186, Georgia
Boron Metamaterials, Cluster Sciences Research Institute, 39 Topsfield Rd., Ipswich, MA 01938, USA

Rick Becker
Boron Metamaterials, Cluster Sciences Research Institute, 39 Topsfield Rd., Ipswich, MA 01938, USA

Ivane Murusidze
Institute of Applied Physics, Ilia State University, 3/5 Cholokashvili Ave., Tbilisi 0162, Georgia

Evgueni F. Talantsev
M. N. Miheev Institute of Metal Physics, Ural Branch, Russian Academy of Sciences, 18, S. Kovalevskoy St., Ekaterinburg 620108, Russia
NANOTECH Centre, Ural Federal University, 19 Mira St., Ekaterinburg 620002, Russia

Victor D. Lakhno
Keldysh Institute of Applied Mathematics of Russian Academy of Sciences, 125047 Moscow, Russia

Index